Robust Power Supply Design in a Supply-Chain-Challenged World

IEEE Press
445 Hoes Lane
Piscataway, NJ 08854

Robust Power Supply Design in a Supply-Chain-Challenged World

Ron Lenk

IEEE Press
WILEY

Published by John Wiley & Sons, Inc., Hoboken, New Jersey.

Library of Congress Cataloging-in-Publication Data

Names: Lenk, Ron, 1958- author.
Title: Robust power supply design in a supply-chain-challenged world / Ron Lenk.
Description: Hoboken, NJ : John Wiley & Sons, Inc., 2026. | Includes bibliographical references and index.
Identifiers: LCCN 2026006450 | ISBN 9781394343027 (hardback) | ISBN 9781394343041 (adobe pdf) | ISBN 9781394343034 (epub)
Subjects: LCSH: Electric power supplies to apparatus—Design and construction. | Power electronics. | Electronic apparatus and appliances Power supply—Design and construction. | Switching power supplies—Design and construction. | Voltage regulators—Design and construction.
Classification: LCC TK7868.P6 L+
LC record available at https://lccn.loc.gov/2026006450

Cover Design: Wiley
Cover Images: © Ron Lenk, © CoderRu/Shutterstock

To my Mother

Contents

List of Figures

List of Tables

About the Author

Ron Lenk, FRSS, was the co-founder and CEO of SwitchLight, and is now a power supply and power systems consultant. He is a Life Member of the IEEE, a Fellow of the Royal Statistical Society, a co-author of *Practical Lighting Design with LEDs, 2nd Edition*, and the author of *Practical Design of Power Supplies*. He currently has 36 US patents issued in his name.

Preface

"An oscillator is something you designed to be a power supply, and vice versa."

—Rolf Drucker, Vice President of Electronic Graphics at ABC, personal communication to the author c. 1972

How often has this happened to you? You're working on a power supply design. You've spent a couple of months collecting the specifications and designing the power supply. You even went the extra mile and have talked with the distributor that the main control IC you're using has a five-year lifetime production guarantee from the manufacturer. You continue with laying out the board, prototyping it, making measurements, and then revising the design and re-spinning the PCB. It's finally ready to go to production. Then the buyer sends you a message: due to high demand, the part you've designed in is on allocation, and your production isn't big enough to get parts. Can you design in a different part? No matter whether you decide to wait for the parts to be produced in higher volume, or go for a redesign, your power supply is going to be four months late.

Or suppose you're a production engineer for a power supply design. The product has been in production for a year, but suddenly, due to a new disease, the ports into the country are closed. Because of this, the control IC on your design is no longer available, and you have only one month's supply on hand. There aren't any pin-compatible replacement parts, so you want to get the engineer to re-layout the board and retest the supply. Unfortunately, the engineer who did the original design is no longer with the company. One month from now your production is going to go on hold for three, four, or even six months.

One more: suppose that for some reason you need or want to build a power supply while avoiding parts from one particular region. But almost all the parts you need for this particular type of design come from there. How are you going to find replacement parts? Of course, if you are designing or in production with a small production run, you could simply have purchased enough components for the entire run. But if you're planning on volume production for years, you can't financially afford to stock more than a month or two's worth of parts. And even if you did, there's no guarantee that you'll receive enough advance notice about issues to be able to redesign your supply before you run out of parts.

Such problems are going to become increasingly common in the future. Fortunately, there's a way to avoid them. There are some ICs, usually the older ones, that have been so popular that there are multiple manufacturers. However, the circuits that have been historically designed with them have historically had less than optimal performance: low switching frequency, poor regulation, incompatible inputs and outputs, and so on. After all, this is why IC companies keep designing new ICs!

In this book, we show you how to build modern, high-performance power supplies that are robust to supply-chain challenges. We use a variety of older, multisourced ICs, as well as discrete parts, in modernized circuits that provide high performance while avoiding the need to redesign your power supply if there's a supply-chain problem. If one manufacturer can't supply the parts you designed with, you can simply get drop-in replacements from another manufacturer! Topics in this book include linear regulator replacement circuits; bang–bang converters; buck and boost converters, including some with synchronous rectification for high efficiency; power factor correction (PFC) circuits; and isolated converters. This book provides detailed information on discrete component selection and how to robustly design a variety of support and control circuits, including gate drives, isolated and non-isolated feedback, and control loops. A major emphasis of the book is on building SPICE models for the circuits to verify their performance. Specific circuit examples of all these converters and circuits are designed and explained in detail, and SPICE models are constructed showing how to measure and tune the circuits' performance.

Throughout this book, the word "robust" will refer to components that are available from two or more significant manufacturers, with the parts being typically available through distributors. "Robust" will also refer to power supplies that are designed using exclusively robust components.

A word on terminology. A switch-mode power supply (SMPS) refers to a power supply where a transistor is in only one of two states[1] all the time, either fully on or fully off, not in its linear operation region. While these typically operate at a constant-frequency (with PWM) or (for PFC) a constant-on time, a special type of SMPS is the bang–bang controller, which does neither. We have a chapter about these also. Even though they are relatively uncommon these days, primarily for historical reasons, they can work very well.

---

1 Alright, sometimes three.

But the concept of a power supply in general is not limited to SMPS. There are also linear low dropout regulators (LDOs), which we robustly design in this book, and even simple shunt regulators. The word "converter" is also used in this book, fundamentally just as a synonym for SMPS.

There is some quite detailed analog design in this book: gain of bipolar transistors as a function of current, gate charge of MOSFETs, control loop compensation, etc. If you don't want to deal with this at such a detailed, complex level, you can still produce adequate designs. You can just use the designs presented as examples and modify them, rather than starting from scratch. Either way, you are designing power supplies that can be built even in the face of major supply-chain challenges!

A final note: This book is not intended to be an introduction to modeling in SPICE or an introduction to LTspice. It is assumed you already have at least some experience with modeling, though we have a chapter giving advice on how to avoid common problems. How to make specific less usual measurements is explained in the appropriate places. This book uses modeling to measure the performance of the robust power supplies we design.

As for the specific contents of this book after this brief preface, we have a chapter on Safety, Chapter 1. Many of the power sources we deal with in this book are potentially lethal, both to you and to others around you. Make sure to read this!

Chapter 2 is an introduction to power supplies, their fundamental concepts, and their function in the world. This is followed by Chapter 3 (Components). Though they seem familiar, there are a lot of intricacies that aren't well-known except to specialists. The author has given a three-hour webinar just on capacitors! This chapter looks at both passive and active components and then briefly at some other items useful for power supply design, such as EMI and MTTF.

In Chapter 4 (Modeling), we consider how to build models of the various components we just discussed. Even something as simple as an inductor has some intricacies, and we characterize the performance of a number of important active components and compare them with datasheet[2] specifications. We look at how to find suitable replacement models for components that aren't available in LTspice. In one case (the 9910), we build a new model from scratch. We also look at the concept of state-space averaging.

We start actually designing circuits with Chapter 5 (LDOs). While standard voltages and currents are, of course, available robustly, we look at more unusual requirements such as extremely low dropout voltage, high-voltage regulation, and negative voltage LDOs.

---

2 Refer to the datasheets on the companion site.

In Chapter 6, we look at bang–bang-controlled SMPS. These are not even taught anymore as they are considered as antique as vacuum tubes. But the reality is that with modern parts, they can be high-speed, high-precision SMPS with really minimal size and cost. You may find that a bang–bang converter, despite its age, may suit your needs very well.

In Chapter 7, we look at the traditional fixed-frequency PWM converter. We show how to design a number of significant components of such a design, such as a ramp and a clock, and then look at a number of designs that use multisourced parts to achieve high performance, including a 22 kW four-slice buck for EV charging.

Chapter 8 is about PFC. First we show a little-known passive PFC, which can work over a surprising range of power levels. Next, a novel method is revealed using a 555 timer to do constant-on time PFC. There is also a constant-frequency design, invented by the author, using the 9910. We further look at how to design an EMI filter for these latter two, as they have very different noise characteristics.

Chapter 9 is about isolated power supplies. The control systems are similar to those already developed in previous chapters, but the design must now take into account the addition of an extra variable, the turns ratio of a suitable transformer.

Chapter 10 deals with some special topics in robust power supply design, such as high-speed power, adding in current limit to designs and system optimization, and generation of a negative voltage to name just a few.

Each chapter also has information targeted at the hobbyist, who wants to build one or a few power supplies for a project. We consider which designs are simple enough and cheap enough to meet most such needs, with a special focus on safety.

So, is robust design necessary for your project? Perhaps it might have seemed somewhat esoteric to worry about whether you could get parts ... until COVID came along and shut down the ports through which parts are delivered. Nowadays, it is something that all engineers and their managers should be very actively thinking about. Are you perhaps unworried because your IC vendor is one of the big players? That won't help you if parts go on allocation and you aren't a major customer. And it need not be that the company goes out of business, though that would certainly be a big problem. There could just be issues with all the other companies involved with making your part: the wafer manufacturer, the packaging house, equipment vendors, and so on. This book is intended to make your design robust against all of these types of issues.[3]

Throughout this book there are numerous examples covering a wide variety of designs, which are summarized in Table 1. Power levels range from 5 mW to 22 kW

---

3 Are you perhaps thinking that you're immune because you're in a special field where parts are guaranteed to be produced? That piece of paper is great ... until the company can't get what it needs to produce them.

(that's a range of four million), input voltages as low as 900 mV to as high as three-phase 480 VAC and designs including an output of 1 V at 100 A, a four-slice interleaved buck for EV charging, and a 5 kW PFC for LED stadium lighting. There are a variety of topologies and control schemes, all designed with robustly available parts and verified by simulation. Enjoy!

Table 1 Table of examples in this book.

Power	V_{in}	V_{out}	Chapter	Control	Topology
0.005 W	3.3 VDC	5 V	6	B–b	Boost
0.06 W	700 VDC	12 V	7	B–b	Buck
0.06 W	12 VDC	−12 V	10	None	Capacitor
0.24 W	12 V	±12 V	10	None	Capacitor
0.3 W	−12 VDC	−5 V	5	LDO	LDO
1.5 W	0.9 VDC	0.5 V	5	LDO	LDO
1.8 W	84 VDC	76 V	5	LDO	LDO
3.6 W	5 VDC	1.8 V	6	B–b	Buck
5 W	5 VDC	5 V	5	LDO	LDO
6 W	5 VDC	12 V	6	B–b	Boost
10 W	12 VDC	5 V	7	CVm	Buck
24 W	120 VAC		8	None	Passive
100 W	48 VDC	1 V	6	B–b	Buck
100 W	200 VDC	48 V	9	3845	Boost
100 W	120 VAC	200 V	8	555	Boost
100 W	230 VAC	375 V	8	555	Boost
100 W	120 VAC	1 V	9	Various	System
200 W	48 VDC	1 V	6	B–b	Buck
300 W	375 VDC	1000 V	9	555	Boost
900 W	120 VAC	36 V	9	9910	Boost
5 kW	277 VAC	450 V	8	9910	Boost
22 kW	480 VDC	400 V	7	CVm	Buck

B–b = bang-bang; CVm = clocked voltage mode.

Notes for the Hobbyist

While this book is primarily intended for power supply engineers who need to design converters that are robust against supply-chain interruptions, its use of common components means that it could also be readily used to design power supplies with modern performance by hobbyists.

Throughout this book, we have written sections on hobbyist use of the example designs. While it would be helpful for the hobbyist to read and understand the other sections in this book, quite a few designs can be "home-brewed" by simply modifying the designs given, and we give some specific suggestions on how to do this in the sections throughout the book entitled "Notes for the Hobbyist."

Acronyms

APL	Approved Parts List
BCM	Boundary-conduction Mode
BJT	Bipolar Junction Transistor
BOM	Bill of Materials
BW	Bandwidth
CCM	Continuous Conduction Mode
CMOS	Complementary Metal Oxide Semiconductor
DC	Duty Cycle
DCM	Discontinuous Conduction Mode
DCR	DC Resistance (of an inductor)
DMM	Digital Multimeter
DVM	Digital Voltmeter
EMI	Electromagnetic Interference
EMV	Electromagnetic Vulnerability
ESR	Equivalent Series Resistance (of a capacitor)
FFT	Fast Fourier Transform
FIT	Failure in Time
FPGA	Field Programmable Gate Array
Fsw	Switching Frequency
GBW	Gain-bandwidth Product (of an op-amp)
GDT	Gas Discharge Tube
GUI	Graphical User Interface
Ib	Input Bias Current (of an op-amp)
IC	Integrated Circuit
Ic	Collector Current (of a BJT)
ICC	Power supply current (of an IC)
Id	Drain Current (of a MOSFET)
IGB	T Insulated Gate Bipolar Transistor
Ilim	Current Limit

Irms	Root-mean-square Current
Isat	Saturation Current (of an inductor)
LED	Light Emitting Diode
LISN	Line Impedance Stabilization Network
MOSFET	Metal Oxide Semiconductor Field-effect Transistor
MOV	Metal Oxide Varistor
MTBF	Mean Time Between Failures
MTTF	Mean Time to Failure
NTC	Negative Temperature Coefficient (of a thermistor)
PCB	Printed Circuit Board
PF	Power Factor
PFC	Power Factor Correction
PRBC	Pseudo-random Binary Code
PSRR	Power Supply Rejection Ratio
PTC	Positive Temperature Coefficient (of a thermistor)
PWM	Pulse Width Modulation
Qg	Gate Charge (of a MOSFET)
RHPZ	Right-half Plane Zero
RMS	Root Mean Square
SMPS	Switch Mode Power Supply
SMT	Surface-mount Technology
SoC	State of Charge (of a cell or battery)
SSS	Spread-spectrum Switching
THD	Total Harmonic Distortion
Tj	Junction Temperature (of an IC)
Trr	Reverse Recovery Time (of a diode)
Vbe	Base-emitter Voltage (of a BJT)
VCC	Power supply voltage (of an IC)
Vce	Collector-emitter Voltage (of a BJT)
VDD	Power supply voltage (of an IC)
VDS	Drain-source Voltage (of a MOSFET)
Veb	Emitter-base Voltage (of a BJT)
Vf	Forward Voltage (of a diode)
VGS	Gate-source Voltage (of a MOSFET)
Vios	Input Offset Voltage (of an op-amp)
Vpp	Peak-to-peak Voltage
Vrms	Root-mean-square Voltage
VSS	Negative power supply voltage (of an IC)

About the Companion Website

Robust Power Supply Design in a Supply-Chain-Challenged World is accompanied by a companion website

www.wiley.com/go/lenk

The website includes:

- LTspice Model files
- Data sheets

1

Safety

We begin this book with a short chapter on Safety. While for low-voltage power converters running from lab bench supplies the danger is low, some of the designs in this book run from 120 VAC, 230 VAC, or even 480 VAC. One of the designs goes up to 22 kW. The author considers it his duty to warn the reader about the dangers of these sorts of voltages and to offer advice on how to remain safe when working with them. This chapter is a bit long because the author considers it to be so important. Obviously, none of this applies when you're designing on paper or simulating.

1.1 AC Safety

We were all taught as children not to stick a paperclip into the wall outlet. There's a really good reason for this. **The AC line can kill you** if you don't deal with it appropriately. This includes things that are attached to the AC line, such as power supplies that you are debugging.

To repeat a story, the author was once in a lab working on lighting ballasts for fluorescent tubes, which in this case ran off 277 VAC. One of the engineers in the room seems to have accidentally come into contact with the AC line while probing the ballast with an oscilloscope. He was knocked over backwards in his chair and remained for perhaps 10 s on the floor, convulsively twitching while everyone else stood around staring without knowing what to do.

The engineer survived, but it wouldn't have been that surprising if he had gone into cardiac arrest and died. And he wasn't doing anything much different than what engineers do all the time.

To avoid events like this, the author over the years has assembled the following list of things to do and not do when dealing with AC voltage or other high-voltage sources, including ones you generate.

1. Doing it safely doesn't really slow you down much. And it's much faster than time spent in the hospital.
2. Every lab dealing with high voltage should have installed in it a "Big Red Button" that, when depressed, turns off all the power in the lab except for the lights. Ask your lab manager to have one installed; it's not that expensive. Have each new hire told what it's for.
3. The most likely method of being killed by AC is for the current to go up your arm, through your heart, and back through your other arm. Reading this isn't enough; the danger is that in the excitement of the work you're doing, you'll forget. Always work on AC and circuits attached to any high voltage *with only one hand*. Put the other hand in your pocket or behind you, holding your belt. If you don't have a pocket or belt, hold something in your nondominant hand, something nonmetallic and nonliquid (of course), either your cell phone or maybe a copy of this book!
4. The second most likely method of being killed by AC is for the current to go up your arm, down through your heart, and then down your legs to the ground. The author recommends always wearing shoes that are insulated with rubber or plastic soles, not leather, when working on AC.
5. You need to attend to equipment when it's energized. If you go off to the bathroom, make sure you turn the power off. If someone comes on a tour through the lab, caution them to stand back and not touch anything. And it's best not to leave a test running overnight. If you just have to leave it on, make sure everything is covered, and put up warning signs (in multiple languages!) so the cleaning staff doesn't have an accident.

Here are some additional recommendations for operating AC and high-voltage equipment without hurting yourself or damaging the unit:

- Disconnect the power before doing *anything*. Even if you just want to put an oscilloscope and its ground on the circuit, or move it from where it presently is, or even just tweak a pot, *turn the power off first!* Then move the probe, and then turn the power back on.
- From this, it should be obvious: Don't have big wrenches, or anything metallic, remotely near the circuitry. It could be bumped into and fall on something live.
- Oscilloscopes always have a three-wire plug. One of these is for earth ground. Don't try to bypass it with what is called a "cheater," converting the three prongs to two. If you do this, the ground of the oscilloscope's probe will be connected to the ground in your converter, and if this is connected to anything on the AC side, *your whole oscilloscope will be connected to AC*. Touching the oscilloscope chassis will then (at the least) blow up your probe.

- The main thing is caution, not to become complacent after the first few days of working with the power. Familiarity breeds contempt and carelessness.
- Although we've already said to turn everything off before doing anything to the circuit, things may not be safe even after the power is gone. Components in the power supply can get very hot without failing and may take a while to cool. Don't immediately stick your hand in so that you don't get burned.

1.2 Battery Safety

From a safety standpoint, batteries can be even *worse* than AC lines. Of course, I'm not talking about a 1.5 V AA cell, which just gives an unpleasant tingle if you hold on to it the right way. But in this book, we'll be dealing with some quite high-voltage batteries, for example, the 400 V battery in an electric car. So let's start this section like the previous one. **A battery can kill you** if you don't deal with it appropriately. So can a power supply that you're debugging, that is, or is going to be, attached to a battery.

The safety recommendations for the AC line also apply to batteries. If you haven't read that yet, the author urges you to do so. There are some additional safety considerations for batteries.

1. Of course, every system ought to have a fuse. But you should be aware that fuses rated for AC have a very different (and lower) rating for DC. This is because of the way fuses work; they interrupt the current when the AC line crosses zero, which current from a battery doesn't do.
2. The fuse for a battery should be located as close as possible to one of the terminals. If, as is common, the battery consists of many cells, each cell really ought to have its own fuse, attached as close as possible to its terminal.
3. As a rule of thumb, make your fuse be twice the maximum current you expect. This leaves some margin. Fast-blow is best for electronics.
4. A fuse is your first line of defense, but there's no guarantee that it will protect you. It only takes 20 mA to kill you. You need two layers of protection. The author was once testing a relay using a marine battery. When the relay failed, the fuse didn't blow—the current was maliciously slightly lower than the fuse current![1] A manual disconnect such as a switch should also be in series with the battery.
5. During this same test, the currents were so high that the Lorentz force became important. The zero-gauge cables, which are big and heavy, started out draped on the floor, straightened out in the air, and were rigid! Had anyone or anything been in their way, they would certainly have been knocked over.

1 The author knows this because of a current probe attached to an oscilloscope set to one-shot.

6. In the case of the wrench, you may have to use it to put on or off connections to the battery terminals. The author strongly recommends using (electrically) insulated gloves when doing this. The wrench needs to have an insulated handle as well, as it could otherwise even *weld* across the terminals!

A few more safety recommendations in general are as follows:

- There can be danger inside a power supply even after you've turned off or disconnected the power source. Large capacitors can become charged up, and may retain their high voltage for seconds or even minutes. The right thing to do is to design any large or high-voltage capacitors with a bleed resistor. The power dissipation doesn't have to be excessive, just enough to discharge the capacitor to a safe level in a second or two.
- If the bleed power is too much for normal operation, you can build a circuit that only turns on the resistor (through a MOSFET) when the input power is removed.
- Batteries, especially lithium, really don't like over-voltage or under-voltage. In either condition, there could be a fire or even an explosion. You need to have overcharging and overdischarging circuitry attached to the battery, preferably some sort that will turn off the offending power source or power load that is causing the problem.
- In a battery without cell balancing, it's possible for the battery voltage to be almost normal, and yet one of the cells will be close to critical danger. Especially for development work, it's really very much better to have monitoring on each cell.
- It's conventional these days to have cells in a battery separated by a flame-retardant material. If one cell overheats and starts to burn, everything is much worse if the other cells also catch on fire. Separators are highly recommended, at least for lab work.

1.3 Notes for the Hobbyist

The author urges you to carefully read and understand everything in this chapter.

In case you're unaware, typical oscilloscope probes are only good up to 300 V or even only 150 V. For higher voltages, such as some of the inputs and outputs that we deal with in this book, there are probes available at 1000 V or more. Decent quality DMMs will read up to 1000 VAC and 1000 VDC, but read the specifications before trying it.

In general, the author isn't telling the reader to be afraid but merely to exercise appropriate caution when working with high voltage or high power and to make sure that others in the vicinity are aware that what you're doing is potentially more hazardous than working on, say, software. It's not "just engineering."

If you have any doubts about safety, do the research or contact a professional before you hook up anything!

2

Background Material

This brief introductory chapter is intended to provide a quick overview of the most important aspects of power supplies: what they do, why they're needed, and the challenges facing a designer in a world where parts can no longer be assumed to be readily available. It's assumed you already know something about electricity, and if you already have experience with power, you can just skip over this chapter.

2.1 Why a Power Supply?

The modern world runs on electricity. Your phone, your computer, your car, your lights ... pretty much everything uses electricity. However, you can't just "put some electricity in." It follows some very specific physical rules, and has to be the right quantity measured in some specific ways, or else it won't work.

As an example, think of charging your phone. There's a little box that converts the electricity in the wall outlet to the type of electricity that the phone needs. If you were able to connect your phone directly to the wall outlet (no, don't try this!), your phone would explode and probably catch on fire. The quantities determining the *type* of electricity in the outlet are incompatible with the quantities needed by your phone.

2.2 Characterizing Power

What characterizes these differences? Electricity comes in both AC and DC. AC means that the voltage is sinusoidal: it goes to zero 50 or 60 times a second (depending on where in the world you are). Common AC voltages are 120 VAC, 230 VAC, 277 VAC, 400 VAC, and 480 VAC. DC means that the voltage is constant. Common DC voltages are, for example, 5 and 12 V, with examples in this book ranging from 500 mV to 1000 V. Corresponding to these two types, there are four possible types of conversion. A power supply can convert one DC voltage to a different one, either higher or lower, and we will have many examples of how to do this. A power supply can convert AC to DC voltage, and this is what your phone charger does. There can also be conversion from DC to AC, and AC–AC, but they are less common, and we won't be showing examples of this. A power supply can also have multiple differing output voltages at the same time, and some have multiple inputs as well.

Another distinction between power supplies is whether they are isolated or not. An isolated power supply is one where the output isn't connected by a component or a piece of wire to the input, and thus forms a special category of power supplies. The main example of where this is required is for power supplies that come from the wall outlet. Since contact with this can be lethal, power supplies coming from an outlet are almost always isolated. That's not to say that the output of an isolated is necessarily safe, but it at least means that it isn't like sticking a paperclip in the wall outlet.

In addition to voltage differing between various power supplies, so does the power level. To give an idea of the range of power levels with which we're concerned, the examples in this book range from a 5 mW DC/DC to a 22 kW AC/DC. There are obviously very many power supplies in that range. In this book, we will not be looking at the extremely high voltage and power on, for example, the power grid.

Yet another quantity that varies between power supplies is their switching frequency, which refers to how often per second a transistor or other switch turns on and off. In the old days, a power supply with a switching frequency of 10 kHz was considered exciting; nowadays, 500 kHz or even higher is considered standard. In this book, we will have power supplies running over that whole spectrum of switching frequencies, with one example going up to an effective 2 MHz.

From this overview, it is apparent that there is a gigantic range of possible power supplies. The result is that essentially every power supply is unique, and each one needs engineering. Of course, there are many common components between differing supplies. They all use resistors and capacitors ... even though these have variations beyond merely their values, and not all the variations are multi-source available, as we will discuss in Chapter 3 on Components. MOSFETs and diodes also have a wide range of choices, and pretty much anything you want is available robustly. Magnetics, however, are almost always custom for power supplies over about 100 W.

The worst thing about the components used for power supplies is their control ICs. Due to the way the market developed historically, there are literally thousands of SMPS control ICs, and almost none of them have second sources. This is one area where the robust design methodology presented in this book particularly shines. The only control ICs we use are those with multiple sources, guaranteeing their availability despite supply-chain disturbances.

2.3 Topologies and Controls

Having detailed the basic quantities and components of power supplies, we now turn very briefly to consider the various topologies and controls used in power supplies' design. What is a topology in this context? A topology is an arrangement of power components (MOSFETs, diodes, inductors, and transformers) that can be controlled to make the power-handling part of a power supply.

While we defer detailed explanations to the appropriate chapters, we briefly mention those in which we are primarily interested. An LDO is a transistor that is controlled in its linear region to regulate an output, basically using an op-amp to measure the difference between what we have and what we want. We have a chapter on how to design these robustly to avoid relying on single-source parts.

LDOs are controlled linearly; the rest of our designs are switch-mode, again meaning that the transistor(s) is either on or off, not in its linear region. For SMPS, we start with bang–bang control, in which power is transferred to the output when the output voltage is low, and power stops being transferred when the output voltage is high. Topologies used in the examples include buck and flyback (sometimes referred to as a boost), whose operation is discussed in the chapter on bang–bangs.

Next, we have a chapter on PWM-controlled SMPS, which are typically operated at a constant frequency, while modulating what percentage of the time the switch is on, known as the duty cycle. The same topologies are used here.

Our next chapter is on PFC controllers. We show two different methods of controlling power factor correction of the input power from an AC line: one using constant-on time, the other operating at constant frequency. All the design examples are flybacks.

We have one more type of SMPS, those that are isolated, usually because they are the output of a PFC converter, which itself is connected to the AC line. These are mostly isolated flyback topologies, though there is also an example of a forward. There is a chapter on isolated SMPS.

You might have noticed that we really are only using a few basic topologies and control schemes. Why? After all, there are new ones in every issue of the IEEE Transactions. The reason is that most of these are only academic. In real-world power supply design, the ones we show in this book (with the notable exception of bang–bang control) are the most common, and, as the examples show, cover most of the low- to medium-power SMPS you could ever want.

Of course, there are a few variations on these basic topologies. One is to replace the diode rectifier by a MOSFET that is controlled to act like a diode, known as a synchronous rectifier.[1] It is turned on when the diode would be conducting, and turned off when the diode would be blocking. We show examples of how to do this robustly.

---

1 The idea itself dates back to vacuum tube days [1], when thyratrons were used as the rectifiers.

One more variation is the multi-slice converter. In this, several identical sub-supplies are built and then operated in parallel. The reason this is preferred over combining all the power components into a single big slice is that the switching can be out of phase from slice to slice. This reduces the output ripple voltage. We have an example of how to do this robustly.

As for control of these topologies, in this book, most are controlled by voltage-mode, which uses negative feedback from the output voltage to control the switch to get the desired output. We do show an example of the other type of control, current-mode. This has an inner control loop controlling the output current, with this inner loop in turn being controlled by a voltage-mode control loop.

Current-mode control was quite popular in its day because the inner control loop effectively removes the inductor from the outer voltage-mode control, enabling higher bandwidths. An example of this is shown in the PWM chapter. However, with switching frequencies now available in the multi-megahertz range, the advantage gained by having current-mode control is, in the author's opinion, no longer worth the extra components. Very high bandwidth can be achieved with voltage-mode control, and we have an example of this.

2.4 Measuring Power Supplies

With so many variations of power supplies, we also need a way to quantitatively measure them. DC inputs are straightforward to measure. They have a voltage and a current, and the product of the two is their power.

AC inputs are more complex because they are continuously changing. Here, you need to know the frequency with which they repeat, their frequency, which is usually (but not always) 50 or 60 Hz. Once you know the frequency, their period, the amount of time until they repeat is one over this frequency (either 20 msec or approximately 16.666 msec[2]). To measure the voltage or current of an AC source, you have to find the RMS over one or more of these periods. For power, it's the average over the period. Once you have these three numbers, you can also find the power factor, which is the input power divided by the product of the RMS input current and the RMS input voltage.[3] Power factor is a measure of how closely the power supply resembles a resistor, which would be ideal; power factor varies from zero to one, with a resistor being exactly one. Power factor is regulated by governments.

---

2 Don't be confused, 2π isn't relevant here.

3 RMS current times RMS voltage does not necessarily equal power because current and voltage need not be in phase.

The same measurements can be done both for the internal components of the power supply as well as the power supply's output(s). The output power divided by the input power is the efficiency, which obviously must be strictly less than 100%.

There are additional time scales associated with the power supply itself. The power supply undergoes changes at different time scales. For example, when you turn it on or plug it in, it starts operation. And while it is important that the operation of the power supply be designed to be appropriate during this period of time, that is not what is of central interest. Rather, the measurements that are of interest are (for the most part) when the power supply has reached a steady-state. This doesn't mean that it isn't doing things inside that are at high speed, but that the overall operation is constant on a human timescale, perhaps like a second or so. This is the timescale of thermal effects.

2.5 Modeling

To test a power supply's design by building it, you would have to select every component, not only every value, but also a lot of the other things that characterize the components, their voltage and power rating, the type of material, the package they come in, and so on. And, of course, you'd have to lay out and fabricate a PCB, and these two can take weeks. So building a model on a computer can save quite a bit of time for an initial investigation.

However, these same things that make simulation handy are also potentially problems. Models don't exactly reproduce the behavior of real parts, as we detail in the chapter on Modeling. And there are many things that can go wrong in a simulation that you don't notice, things such as excessive power or voltage in or on a component.

Simulations are useful; you just have to pay attention and ultimately build hardware to test. We have a lengthy chapter dedicated to the ins and outs of modeling power supplies.

2.6 Notes for the Hobbyist

This is obviously not enough information to qualify you as a power supply engineer! You can, however, start with the numerous examples in this book and find something similar to what you want. The examples are fairly straightforward to modify.

When you come to actually designing and building something, you need the extra knowledge about the selection of components, so you need to read that chapter. And of key importance is your personal safety. **Don't skip the chapter on safety!**

3

Components

This chapter gives an introduction to the various types of components that we will be using in making robust designs of power supplies and how to go about selecting them. We consider in detail the parameters that are important in the selection for each type. The components used include passive components, active components (including integrated circuits [ICs]), magnetic components, and a selection of others, including PCBs and connectors. In addition to the parametric selections of importance for each type, we also look at questions (when appropriate) about their Mean Time to Failure (MTTF). We defer discussion of making models of the various components until Chapter 4.

3.1 Passive Components

3.1.1 Resistors

A resistor is a resistor is a ... Well, no, not so fast. There are many different kinds of resistors: thin film and thick film, carbon comp and wire-wound, not to mention potentiometers and rheostats, etc. In this book, we will be exclusively using the surface-mount film type. And we will be indifferent to the differences between the two types of film resistor.

But there are still a number of other considerations in selecting a resistor of which you need to be aware. Generically, it is best to use resistors that are $\pm 1\%$ tolerance (hereinafter referred to as 1%), because they cost the same as 5% parts, so you get enhanced performance at no cost. However, for high-precision outputs, we will use 0.1% resistors, though only in places where the tight tolerance is useful. As an example, suppose we have a dozen 10 kΩ, 1% resistors in a design, and a single 10 kΩ, 0.1%. The extra cost of the tight tolerance part probably doesn't make it worthwhile to consolidate the Bill of Materials (BOM), and so we just have two types of 10 kΩ resistors. When we draw a schematic, 1% is assumed for resistors, but if it's some other value, then that is explicitly shown on the schematic, not just the BOM.

As for the resistance values of the resistors, of course, they vary from 1 mΩ to gigaohms. But very high-value resistors are not only uncommon, presenting an issue for multi-sourcing, but they also tend to have performance issues. For example, there can be leakage current around high-value resistors from dirt on the Printed Circuit Board (PCB), and there can be increased electrical noise from them as well. In general, we will not use any resistor higher than 1 MΩ of resistance. So, for example, when we need 3 MΩ of resistance for sensing a 400 VDC output of a power factor correction (PFC), we will put three 1 MΩ in series. This avoids both these electrical issues as well as the supply chain issues.

Then there is the power rating of the resistor, which for film resistors can range from 10 mW to 250 W! The common resistors, though, are in 0603 or 0805 sizes, which generically support power ranges of roughly 60–125 mW. (We don't consider smaller part sizes because they can be difficult to deal with by hand in a prototype. Naturally, you can use them in production.) Suppose you need a half-watt resistor? One-watt resistors (we of course need some derating), while not uncommon, are less common than smaller-value types, and so it is more robust for your design to parallel (or series) (say) four 0805 resistors.

And finally, what of the voltage rating? While it may seem surprising that resistors have a voltage rating, it makes sense if you think about it: the separation of the pads determines whether an arc can occur across the resistor. An 0603 resistor typically has a voltage rating of 75 V, while an 0805 is 100–150 V. While you could go (again) to larger-size resistors, which have a higher voltage rating, a more robust plan is to put a number of 0805 resistors in series. This often also satisfies the recommendation above to limit the maximum value of the individual resistors by putting multiple resistors in series, since the high voltage usually requires high resistance values to minimize power dissipation. We show a table of ratings (Table 3.1) for typical surface-mount film resistors.

Table 3.1 Resistor size, typical rated power, and voltage rating.

Size	Power (mW)	Voltage (V)
0402	62	50
0603	100	75
0805	125	150
1206	250	200

3.1.2 Potentiometers

Potentiometers are usable in a development environment. They should not be used in production, and we won't be using them in any of our designs in this book. The reason is that they can shift in value after production is complete, for example, due to vibration, potentially causing issues in your design.

While you could in principle glue the potentiometer once it has been set, this has yet more potential issues. The author recommends that if you need to set a resistance value during production test, the test department should have a selection of fixed resistors that can be soldered down at test time.

3.1.3 Capacitors

For the purpose of robust design, we will, for the most part, only use two types of capacitors: surface-mount ceramic capacitors and aluminum electrolytic capacitors. We consider tantalum capacitors to be almost obsolete, since ceramic capacitors are now available in the large values that used to be the province of tantalums, and ceramic capacitors don't have the poor MTTF of tantalums.

There are really three primary applications of capacitors in power supplies. They are used for small-signal filtering, for timing, and for bulk filtering. Starting with the first, small-signal filtering involves things such as the 100 nF ceramic capacitor you will see as a bypass capacitor on the power pin of every IC in our designs. The exact value doesn't matter, and so we will be using a generic capacitor, say 0603, or larger for higher voltages, in an X7R temperature coefficient and a +80/−20% tolerance. Similarly, for other standard values, there are many manufacturers.

For timing capacitors, we need better tolerance and temperature coefficient, so we turn to C0G/NPO ceramic capacitors. There are many manufacturers of such parts, and we can get 5% or even 1% parts if needed. Of course, there is an additional cost to get tight tolerance, but it's not huge; you use it if necessary, as we will be in setting the switching frequency of a pulse width modulation (PWM). Low-voltage timing capacitors (say ≤63 V) are more common than higher voltage and are all that will be required for this book's applications.

Finally, for bulk filtering, we will use, where possible, ceramic capacitors. They are readily available from multiple sources up through 22 μF, and a few of these in parallel will cover most low-power power supplies.

However, at higher power and voltage, ceramic capacitors become impractical. So, we turn to aluminum electrolytic capacitors. Just as for ceramic capacitors, there is a very wide selection of aluminum electrolytics from multiple manufacturers.

These capacitors have four key specifications. Of course, you have to have the right capacitance and voltage ratings. But you also have to have (or at least know) the equivalent series resistance (ESR) of the capacitor. The ESR determines how much ripple voltage the capacitor will produce given its ripple current. And further, the ESR pretty much really determines the power dissipation of the electrolytic capacitor, which affects how hot it gets.[1] So, the ripple current rating (which reflects both the ESR and the size of the package) is also important.[2]

Less obvious, though, is that you have to select the correct lifetime at the correct temperature to ensure that the aluminum electrolytic capacitor's ESR doesn't go up so high over the unit's lifetime that it makes the power supply ripple voltage fail specification. (ESR increase is the typical lifetime-limiting parameter for aluminum electrolytic capacitors.) You normally look at the table of values of lifetime at temperature and select the highest lifetime using the "factor of 2" criterion: for every 10 degrees of additional temperature rating specified, the lifetime increases by a factor of two.

Here's an example. If you have two aluminum electrolytic capacitors, one of which has a lifetime of 10,000 hours at 85 °C and the other 5,000 hours at 105 °C, the first one has lower reliability than the second. We perform the analysis by calculating on the second capacitor, though, of course, the result would be the same using the first. We go down from 105 to 85 °C, which is 20 degrees. Twenty degrees is two factors of 10 degrees, and each factor of 10 degrees multiplies the MTTF by a factor of 2. Since there are two factors of 10 degrees, we have two times two equals four; so, the 5,000 hours at 105 °C is equal to 5,000 times four equals 20,000 hours at 85 °C, which is higher than the first capacitor's 10,000 hours at 85 °C. Higher lifetime is, of course, better (though it perhaps costs somewhat more); you just have to verify that there are multiple vendors manufacturing all the same (minimum required) parameters.

It should be noted that aluminum electrolytic capacitors are available in three package styles. The most common is through-hole. For smaller values, surface-mount aluminum electrolytics are now fairly common and can be a good choice for pick and place. Finally, for large values, there is a chassis-mount type, used with screws. We try hard to avoid the use of these.

Two types of capacitors that we will use infrequently are aluminum polymer and film capacitors. Aluminum polymer capacitors, just like aluminum capacitors, are multi-source available, have a lifetime rating, and are available in both through-hole and surface mount. They are desirable in certain cases, as they can have rather less ESR than regular aluminum electrolytics.

---

1 Yes, aluminum electrolytic capacitors dissipate power and even get hot.

2 The ripple current rating of aluminum electrolytic capacitors intended for power supply bulk will sometimes have a frequency multiplier. It's best to use the type that has this multiplier, as that way you know that the capacitor is suited for high-frequency SMPS bulk decoupling. The rating means that if the switching frequency is high enough, say above 10 kHz, the capacitor can take a somewhat greater amount of ripple current than its nominal 60 Hz value.

Film capacitors are multi-source available and have *much* better MTTF than aluminum electrolytic capacitors. They also have much higher voltage ratings available than aluminum electrolytic. Their downsides are

1. They are really big, much bigger than aluminum electrolytics for a similar value.
2. They are mostly through-hole.
3. Their cost is considerably higher than aluminum electrolytics.

Their higher cost and size will generally limit their use. However, they can certainly be considered at voltages of over 630 V, where you would need to put multiple aluminum electrolytics in series. In this book, the only type of film capacitor we will use will be metallized polypropylene film capacitors.

3.1.4 Metal-oxide Varistors

Metal-oxide varistors (MOVs) are used to protect AC power supplies from surges on the AC lines. They're multi-source available (although be aware that some of the manufacturers simply rebrand other companies' products; in this case, a shortage might mean that all of those suppliers get drawn down at the same time). It's best to pick one of the disc-type MOVs since they're more readily available than the surface-mount type; but you probably will anyway since the disc-types handle a lot bigger surges than the surface-mount types and are thus more suitable for AC line protection.

3.1.5 Gas Discharge Tubes

Gas Discharge Tubes (GDTs) also are used to protect AC power supplies from surges on the line. They're multi-source available (although again be aware that some of the manufacturers simply rebrand other companies' products). The only thing to be careful of is that the high-power GDTs (along with those in unusual packages) may be single-source.

3.1.6 Connectors

Connectors come in a huge variety from many vendors. They have different numbers and arrangements of pins, different types of interlock, and so on. The real issue is ensuring that you don't, for some reason, select a connector that is single-sourced. What we typically do is to select a header pin arrangement. Since this is the most basic type, and since the header can be easily trimmed to almost any number of pins, this gives a good multi-sourced option. You can also key the connector by putting a pin where there isn't a mating hole, so that the pin won't fit in, ensuring correct orientation of the connector.

3.1.7 Switches

Switches as part of a SMPS were a bad design idea even before considering supply chain issues. (Naturally, if they're something like an ON/OFF switch for use by the user, there's less of an issue, though even then an electronic interface signal would be better.) Switches on PCBs have a nasty tendency to flip when you don't expect it (though you could glue them in the proper position once they're set).

As with potentiometers, switches are usable in a development environment. They should not be used in production except possibly for user interface. If you need to turn something on or off during the test, have a 0 Ω resistor that can be soldered down or removed at that time.

3.1.8 Thermistors

Thermistors are of two types: Positive and Negative Temperature Coefficients (PTC and NTC), respectively. NTCs are typically used for inrush-current limiting: when they're cold at turn-on, they have high resistance, which limits the current, and then, as current passes through them, they heat up, reducing their resistance, and thus power dissipation, to small values.

PTCs are also used for current limiting, but in the other direction. When they are cold at turn-on, their resistance is small. So if, for example, they are in series with the gate of a metal-oxide semiconductor field-effect transistor (MOSFET) pass transistor in a low dropout regulator (LDO), there's not much effect. But when they become hot, for example, if the MOSFET becomes hot and they are physically located right next to the MOSFET, their resistance rises to very high values. This can shut off the MOSFET or signal a PWM to turn off until the PTC (and thus the MOSFET) cools down again.

There are many vendors for both NTCs and PTCs. In power supply boards, the most common type of thermistor is surface mount. Just ensure that the one you choose has second sources; the ones with linear response are somewhat less available than those that are just basically on or off, and there is a pretty limited selection of 25 °C resistance values available.

3.1.9 Fuses

Every AC power supply is going to require a fuse, and they can be a good idea even on DC power supplies.

A fuse really has only two ratings. There's the voltage it's designed to run on (typically either 125 VAC or 250 VAC, or 32 VDC),[3] and then its rated current. The rule of thumb is to select the rated current to be about twice the current of normal operation, to prevent short-term overloads from blowing the fuse.

We try to always use surface-mount fuses, though there may be circumstances in which you need to repair the blown fuse in the field, in which case a fuse that fits into a holder might be used. In any case, there are multiple vendors providing footprint-compatible fuses, so you just pick one with the correct ratings, add another one to the Approved Parts List (APL), and you're done.

3.1.10 Batteries

Many power supplies use batteries, either as a power source or as a load, or to back up operation when some other power source becomes temporarily unavailable. So selecting batteries that are robustly available is an important task.

Before we can turn to this, though, we need to take a terminological diversion. What is commonly referred to as a "battery" is actually a "cell." If it's something you can buy in a grocery store, it's almost certainly a cell, not a battery.

A battery consists of two or more cells, almost always in series.[4] We will always refer to cells and batteries with this appropriate terminology.

When you build a system with a cell or a battery, you need to ensure that there are at least two vendors in order to have a robust design. For cells, this is not so hard. Common multi-source sizes include AAA, AA, D, and maybe C and 9 V for alkaline cells, and, of course, the 18650 for lithium cells, and probably a few others.

Batteries, since they consist of more than one cell, are more difficult to find multi-sourced. The problem isn't just the number of cells, but also the physical shape of the battery. If you have to redesign the busbars because the only replacement battery has terminals in a different place, your design isn't robust.

Hopefully, it goes without saying that different cell chemistries don't count as possible replacements. Alkaline cells are 1.55 V when fully charged; lithium cells are 3.6 V or so. One can't be substituted for the other. And there are different chemistries of cells that are all called "lithium," and they have differing full and empty voltages, different cycle lives, different temperature responses, and so on. Fundamentally, you need to test the cell or battery from your second-source vendor to verify that it really does work as a replacement in your power system.

3 Ensure you get the right kind. The DC voltage rating is much less than the AC voltage rating. This is because the AC goes through zero every line cycle, which interrupts the current flow when the fuse opens. With DC the voltage is always there, so the fuse has to "work harder" to interrupt it.

4 Don't try to put cells in parallel; that takes special equipment only their manufacturer has.

3.2 Magnetic Components

Although magnetic components are passive, we've put this section on magnetic components separately from the section on the other passive components. The reason is the availability issues surrounding magnetic components.

Magnetic components, including inductors, power transformers, current-sense transformers, and common-mode chokes (along with various other names that we won't be dealing with in this book), are to a large extent custom: You get them built specially for your design. Now there are exceptions, as discussed later, particularly for lower-power inductors. There are a number of these that are multi-source available, and more if you consider dual footprints. But viewed from the highest level of generality, there are just too many choices for how to construct a given magnetic component. And since there's no standardization, there are also very few second sources.

To understand this issue, consider some of the design possibilities for magnetics. There can be one or multiple windings on a given magnetic component. Each winding has a unique number of turns of wire. And the wire itself can be any of a considerable number of gauges, or it can be litz, again with many different choices of size.

The windings are coupled to each other, usually (in this book, always) through a magnetic core. Cores are available in a variety of shapes, and each shape has a significant number of sizes. Some cores can also have small gaps inside them. And then, the cores themselves come in a considerable variety of materials, each with different magnetic characteristics; the core materials tend to be similar between manufacturers but are *not* identical.

The windings can have different numbers of turns per layer, and, even for non-isolated windings, may have differing amounts of insulation between layers.

Finally, some magnetics are built on bobbins, especially split; other magnetics are wound directly on the core.

Each of these numerous options impacts the performance of the magnetic component. The slightest difference in the requirements changes the optimal magnetic structure, and there are actually a number of different structures that will give close to the same performance. At this point, it should be no surprise that almost all magnetic components are unique to a given design. How to deal with this is one of the topics we deal with in this section.

3.2.1 Inductors

In this book, single-winding inductors are going to be used in all non-isolated switch-mode power supply (SMPS). The word "single-winding" is used here because there are devices used in power supplies that have the word "inductor" in their name, but aren't—for example, the "coupled inductor," which sort of is a multi-winding inductor, but also acts as a transformer. We won't be using coupled inductors in this book.[5]

Now for the purposes of this book, we can separate out two varieties of inductor: roughly speaking, there are those that are used in SMPS that are less than ~100 W; and then there are those that are used in higher-power SMPS. What's the difference? Although distributors list hundreds of thousands of inductors, only something like 10% of those are actually stocked—the rest typically have 12–16 week lead times. This is because they are not used commonly enough for the manufacturer or the distributor to stock them, and probably the manufacturer doesn't even have the core material to make them. So, if something happens to the supply chain, those inductors are not going to be available.

So as for the break at ~100 W, this roughly corresponds to the transition from stocked to unstocked inductors. Since the energy stored in an inductor is related to its size, the higher-power ones would require physically bigger inductors—and these, naturally, are the ones that are less stocked.

In short, up to something like a 100 W SMPS, you can frequently use inductors from distribution, and there will be multiple sources. The most common ones are those with the same value sequence that capacitors come in 1, 1.5, 2.2, etc., microhenries, millihenries, etc. And if you parallel or series two inductors, you can get some other values and perhaps somewhat more power. The 100 W is naturally intended as a guide, not a firm rule.

For higher-power inductors, there are very few multisourced parts. What everyone does for these designs, of course, is to go to a manufacturer and tell them what you want (inductance, current rating, saturation current,[6] physical size, etc.), and they'll make them for you. Since this is now a custom part, what you need to do to ensure availability is to go and talk with a *second* manufacturer, get him to make you a part you can test, and then get him on the APL as well. Now you at least have two sources.

Earlier we mentioned current rating and saturation current rating for inductors. What's the difference? Current rating typically refers to how much current can be put through the inductor to generate a 40 °C temperature rise.[7] You can put more current through an inductor than its current rating, provided you're willing to live with a higher temperature.

5 Also in this category of things with inappropriate names is the "Flyback Transformer," which during a portion of its operating cycle acts as an inductor, while in another portion, it acts as a transformer. The author once suggested that a better name would be a "Transductor." We *will* be using flyback transformers in designs in this book.

6 See below for the difference in meaning between these two.

7 Although 40 °C is typical, some vendors define it differently. You have to look at the datasheets for the parts to tell.

Saturation current refers to how much current you can put through the inductor before its inductance falls by typically 10%.[8] You can put more current through an inductor than its saturation current, provided you're willing to live with a lower inductance while it's operating at a high current.

As an afternote, you might wonder about the components that go into building an inductor itself. The core material has standard shapes, but the core material from every manufacturer is somewhat different (most significantly in core loss). So, it might be a good idea to get your vendor and your second vendor to both also submit samples to you using a different core material source than your first set of samples, just in case ....

3.2.2 Power Transformers

Transformers are even worse from a multi-source point of view than high-power inductors: there are *no* standard parts whatsoever![9] Perhaps this should make the reader glad, as it is what makes power supply engineers irreplaceable.

In any event, just as for high-power inductors, for transformers, you have to go to a vendor with your design and tell them what you want (primary inductance, turns ratio, peak primary current, secondary current, tolerable power losses, etc.) and have them make you a sample. This is true for *all* transformers, regardless of their power level. And since this is custom, you need to ensure a second vendor also produces a transformer for you. Don't forget that you have to actually *test* all the designs![10]

There are quite a number of parameters that need to be specified in a transformer.[11] The primary inductance of a transformer has to be specified, as this affects the volt-seconds that can be applied to it, or for the power level that can be used in a flyback. You also have to specify the turns ratio (primary to secondary), which sets the duty cycle.[12]

---

8 Although 10% is typical, some vendors define it differently. You have to look at the datasheet for the parts to tell.

9 Alright, there are a few specialty transformers designed for specific ICs that can be multisourced. Since the ICs themselves are single source, they are of no consequence for robust design.

10 Due to the intricacies of transformer construction, it's possible to have two vendors produce transformers from the same specification and yet work differently, even with well-written specifications. The author once had transformers from a *single* vendor that worked differently: it turned out that those produced on the night shift worked differently than those produced on the day shift!

11 Before starting on the list, the author wishes to observe that the transformers in this book will only have two windings as a matter of principle. More windings than this, it becomes very difficult to get reliable multiple sources.

12 You don't have to specify the secondary inductance, as this is given by the primary inductance and the turns ratio.

For the primary, you need to specify the maximum allowable leakage inductance (which, in turn, is related to the coupling coefficient, which is what you usually specify, either in nanohenries or as a percentage of the winding inductance), as this can have a significant effect on the power transistor losses. Losses are, of course, also dependent on the current, so you need to specify both the primary and secondary currents. Core losses should also be specified, or at least you have to tell the vendor what your switching frequency is, since core losses are nonlinear in frequency. However, rather than individually specifying the losses of each portion of the transformer's design, your vendor will be happier if you just specify total losses. In any case, neither you nor the vendor will really know what the losses are until you measure the transformer in-circuit; neither models nor datasheet calculations do more than get you sort-of close.

A final thing to note in a transformer design is the placement of the "dots." These refer to the winding direction, either clockwise or counter-clockwise. Putting the dots one way makes a forward converter; putting them the other way makes a flyback.

3.2.3 Current-sense Transformers

A special type of transformer used in SMPS is the current-sense transformer. As the name states, you use it to sense current, usually in the main switching path. (Not in the output current; since it's a transformer, it can only sense AC current, not DC.) Current-sense transformers are typically used when the power dissipation of a resistor would be too high due to the high current.

There are several vendors of these devices and even a few multi-vendor compatible parts. In many cases, though, you're going to be doing the same as for the power transformer, getting a custom-built to your specifications from two vendors. (Typically, all they need is the peak current and the turns ratio, and, of course, the footprint.)

Later in this book, we will be talking about the obsolescence of current-mode control, which was one of the major reasons for using current-sense transformers. So, to the extent this is true, current-sense transformers are also obsolete.

3.2.4 Common-mode Chokes

Yes, a choke is another name for an inductor. And in some ways, a common-mode choke does act like two inductors, one in each of two (or more) AC lines. However, when you consider how they actually work, they can be seen to really be a form of a transformer. They are used as part of an Electro-Magnetic Interference (EMI) filter for an AC line, filtering out the current that appears simultaneously on the lines, that is, the common-mode current.

Unfortunately, there is no second source readily available for common-mode chokes. So, yet again, you need to design one (or pick one) and then find a second manufacturer who's willing to build it for you.

3.3 Mag-amps

For completeness, let's mention mag-amps, which used to be used fairly extensively. Their idea was to present a high-impedance path (an inductance) to a voltage source (typically the secondary of an isolation transformer) during a portion of a cycle, and then a low impedance during the remaining portion. The low impedance was created and controlled by putting current through a secondary winding that had many turns relative to the main current path, so that the current required was low.

This design would surely be considered robust, as it involves a transformer and probably something like an error amp, a reference voltage, and a 555 timer to implement.

However, since there are many good transistors now, there seems to be no reason to add in yet another magnetic component. Mag-amps may be considered obsolete.

3.4 Active Components

We can now turn to active components. The reader will at once notice that our list is short: diodes, Bipolar Junction Transistors (BJTs), and MOSFETs (no Insulated Gate Bipolar Transistors [IGBTs], since these are almost obsolete); a couple common op-amps and comparators; the 555 timer, the 431 voltage reference, a couple quite old PWM ICs, a half-bridge gate drive, a few old opto-couplers, and a couple very common logic gates.

The reason for such an abbreviated list is due to the way the semiconductor industry, and more particularly the power controller IC market, has historically developed. There are many different power supplies required in both consumer and industrial markets. There were many different companies making products for these markets, and each developed its own power products and ICs for each of its many market segments. And since each company wanted to win out over the others, there were almost no second sources, particularly for control ICs.

Today, the problem of second sources for discrete active components is really not significant due to common packages. All companies, for example, make MOSFETs in DPAK. Only the very most popular have true second sources, but the rest are more-or-less comparable and in the same package.

As for ICs, The 431 shunt reference, the 358 op-amp, the 339 comparator, and the 555 timer immediately come to mind. And that is why they appear on our short list of active components to be used in robust power supply design.

You will notice that there are no microcontrollers in this list—there simply aren't second sources for almost any of them. And without a microcontroller, you also don't need analog-to-digital converters (A/Ds) and digital-to-analog converters (D/As). We really don't think a micro is necessary in most power supplies. But if you just have to have something (perhaps for very detailed status communication), you might consider using the venerable 8051 family. The CMOS version, the 80C51, is multi-source available, especially in a 44-pin PLCC. You can probably find multi-source A/Ds and D/As. And there are lots of crystals, or better, you can use an RC timer.

Generally, you'll see that essentially all of your power supply requirements can be met with this short list. If you simply *have to* have a component that isn't multi-sourced, be sure to read the information at the end of this chapter about using dual footprints to allow for the use of similar parts in different packages.

3.4.1 Diodes

There is a huge assortment of different types of diodes. This is because of the significant number of parameters that characterize a diode, the most important for our designs being the maximum current, forward voltage, reverse voltage, reverse recovery time, and package and cost. Which specific parameters we need to ensure exist in our second source for a given diode depends on its application.

Fortunately, there are a substantial number of manufacturers of diodes. What this means for robust design is that, to a reasonable degree, you can just pick the diode that suits your application, and chances are that there will be multiple sources (though, of course, you always have to check). In example designs in this book, we will be using a 4148 type diode for small-signal applications, a 1N4005 diode for AC-line bridge rectifiers, and an 860 type Schottky (60 V 8 A with a V_f of something like 500–600 mV at rated current) for medium-current output rectifiers. For higher-current rectification applications, we will use MOSFETs as synchronous rectifiers instead of using diodes, in order to keep the efficiency relatively high.

3.4.2 Zeners

Zener diodes have several possible uses in power supplies: voltage references, voltage-dropping components, and voltage clamps. The good news is that there are many vendors, and you can get tolerances down to 2%; however, 1% zeners are not multi-sourced and should not be used.

The tolerance specified assumes that you run the zener at its rated current (which can be anywhere from 50 μA to 20 mA) and doesn't include temperature variation. You can minimize the latter by selecting a 5.1 V or 5.6 V zener, if that works in your design. If you need a reference source better than zeners offer, you can go to a reference voltage: see Section 3.4.8.

3.4.3 Light-emitting Diodes

There are basically two categories of light-emitting diodes (LEDs): indicator lights and those intended for illumination. The author's previous book [2] dealt with LEDs used for illumination; in this book, we will only be concerned with indicator lights such as those used to signal Power Good.

There are, of course, a variety of indicator LEDs. They come in a wide variety of different colors, so you should try the specific one you pick before settling on it. Not all reds are the same! They also have a variety of current requirements, 5, 10, and 20 mA being the most common. However, their luminous efficacy varies all over the place; you may be able to run one at a milliamp and still be perfectly visible in office lighting. Again, you just have to try it. Finally, indicator LEDs also come in a wide variety of sizes. The author prefers surface-mount and probably not bigger than 2.5×2.5 mm so as not to use up a lot of board space.

3.4.4 Metal-oxide Semiconductor Field-effect Transistors

MOSFETs are, of course, the workhorse of modern power supplies. They have (almost) completely replaced BJTs in power switching applications and are far along the road of replacing IGBTs as well. In our designs, they will be used as pass transistors in LDOs; switches in SMPS and synchronous rectifiers. Just as with diodes, there are a huge number of MOSFET types available, since there are many characterizing parameters: p- vs. n-channel; $R_{\mathrm{DS,on}}$; drain-source voltage (V_{DS}); gate capacitance; and package and cost, just to name the most common.

Just as with diodes, there is a very substantial number of manufacturers of MOSFETs. And again, this means that to a reasonable degree, you can just pick the MOSFET that suits your application, and chances are that there will be multiple sources (though again, you, of course, have to check). As of this writing, silicon MOSFETs remain more common than MOSFETs made of other semiconductor materials; there's no particular reason at present to use the other types, as silicon has also been improving at the same time that the other types have been.

3.4.5 Bipolar Junction Transistors

BJTs[13] may be readily classified into power devices and small-signal devices. BJTs really aren't used anymore for power devices. The one exception you'll find in this book is in LDOs with very low input voltages. The reason is that only about

---

13 The "J" stands for Junction.

700 mV is necessary to activate the emitter-base junction of a Positive–Negative–Positive (PNP) BJT. Of course, there are p-channel MOSFETs available with a gate threshold voltage this low, but they are a lot less common than the standard type, which requires at least 2 V drain-to-gate. A low-voltage input may not have adequate drive to run a p-channel MOSFET, and an n-channel requires even higher voltage, which would have to be specially generated.

For small-signal applications, a wide variety of BJTs is still available from many sources. For simple inversion, we will use the very standard 2222 in a SOT-23. For a totem-pole pair for gate drives, we will use the 3904 and 3906 complementary pair.

For power devices in some LDOs, we will use the multi-sourced BJT MJD32, which is a PNP with 100 V V_{CE} and with a gain of about 8 at 3 A (and a typical V_{CE} of 100 mV and V_{BE} of 900 mV, both at 1 A of collector current).

3.4.6 Op-amps

Operational amplifiers are the workhorse of analog design. Whole books have been written about exotic circuitry that can be designed with them [3–6]. For our purposes, their most common use will be as error amplifiers, though they also show up as gain stages for low-level signals and as differential amplifiers for the same.

An error amplifier is basically a subtractor. You subtract the actual voltage (say) from the desired voltage and multiply the result by some frequency-dependent amount. The result, the output of the op-amp, is used to appropriately increase or decrease the activity of the power circuitry it controls.

From this short description, the sorts of parameters that are necessary for our use of op-amps become clear. First off, the op-amp has to be able to run (V_{CC}) from the voltage that's available to it. Then, depending on what it's controlling, the output needs to swing high enough and low enough for the controlled device to respond appropriately; for LDOs controlling a p-channel transistor, the op-amp should be able to swing to ground, though not necessarily up to the supply voltage. If there are steps in the load, how fast the op-amp responds may be important, and this will involve not only the frequency-dependent gain network, but also both the GBW (Gain Bandwidth) of the op-amp for the small-signal response and the slew rate for the large-signal response. For many of our circuits, parameters such as the input offset voltage and input bias current won't matter much; we can just select the resistors at the inputs to the op-amp appropriately to minimize their effect.

For the circuitry in this book, we will mostly be using the multi-sourced 358 type op-amp, which can run from a 3 to 36 V single supply and can output down to 100 mV above ground and 1.5 V below positive rail. It has a nonspectacular GBW of 1.2 MHz and a slew rate of 500 mV/μsec, but this will be good enough for many purposes.

For a high-performance error amplifier, we'll use the TLV9062. It has a GBW of 10 MHz and a slew rate of 6.5 V/μsec, runs from 1.8 to 5.5 V and has rail-to-rail outputs. While there isn't a second source with the same part number, there are many parts that have the same or better specifications from a variety of manufacturers and with the same pin-out. Of course, it's more expensive than the 358, so it's only used when the extra performance is necessary.

We should mention that the one thing the old op-amp books were never really able to do adequately with an op-amp was multiplication of two signals. This is important in PFCs, as in the most common present-day method of doing PFC, we want to multiply the reference by the input voltage, to force the PFC to draw current proportional to the input voltage. We will be doing something similar to this with novel circuitry using multi-source ICs controlled by op-amps, rather than op-amps by themselves. See Chapter 8 for details.

3.4.7 Comparators

You should never use an op-amp as a comparator. They act differently; that's why manufacturers make them both.

One of the two primary uses of comparators in this book will be to control the on/off state of a switch transistor in a bang–bang converter. For low-power converters, we may possibly use the output of a comparator directly to run the switch; but in most applications, the base or gate drive requirement will be substantial enough that we will have a separate driver to do that, which, in turn, will be run from the comparator. We discuss driver design in Chapter 5.

The other primary use of comparators in this book will be for over-voltage protection in SMPS. If the bandwidth of the converter is too low (which is particularly the case for PFC circuits), the output voltage may go so high as to endanger the bulk capacitor. To prevent this, an over-voltage protection comparator detects this circumstance and immediately shuts down the switching transistor; it only turns back on when the output voltage reaches a safe level, which is set by the hysteresis around the comparator.

These uses suggest what the important parameters of the comparator are going to be. Of course, it has to work from an available voltage. And as with the op-amp, the input bias current and offset voltage aren't going to be very important. The main things are going to be the response time of the comparator and its ability to switch between output states rapidly. This last is particularly important if the output of the comparator is open-collector.

For the designs in this book, we'll be using the multi-source 339 quad comparator, or else the dual version of it, the 393. The generic version runs from 2 to 36 V and has a response time somewhat less than 1 μsec, depending on the signal level it receives. While not spectacular, this is good enough for many of our applications.

If much faster is needed (perhaps in order to run the bang–bang at higher frequencies), we could choose the open-drain 9022 type comparator in an SO-8 package, which runs from 1.65 to 5.5 V and has a response time on the order of 100 nsec; while the part itself is not multi-sourced, there are a number of manufacturers making comparable parts in the same package and pin-out.

3.4.8 References

There have been voltage references available since before transistors [1]. In the very old days, they used a neon bulb as a voltage reference. This had an almost constant voltage across several orders of magnitude of current (from perhaps 100 μA to 10 mA); similar, actually, in this respect, to the operation of a fluorescent tube. You could even get a radioactive add-in to help it work properly in the dark [7]!

As we've noted earlier, about the best zener you can get from multiple sources is going to have a tolerance rather worse than 2%, when you additionally consider temperature. If you need to do better than that, you can use an IC voltage reference.

The most common voltage reference is the 431 type. They are all adjustable, though you can choose whether the minimum voltage is 2.5 V or 1.24 V; we usually choose the latter. The 431s come with a minimum cathode current of as little as 100 μA, which can be very convenient when you're deriving their cathode current from a high voltage. And 431s are available multi-source down to 0.2%, which is good enough for almost all applications.[14]

One of our designs in this book will have even tighter tolerance requirements. For this sort of situation, you can use the 4041 family, which is multi-source available down to 0.1%.

Let's just note here what to do if you want a voltage different than the normal 1.24 V produced by the 431 shorted cathode to reference pin. For higher voltages, it's straightforward to add in the two resistors in the divider that produces the higher voltage as shown in the datasheet. Naturally, the tolerance of the resistors should be better than the tolerance of the part, so you don't get into trouble.

For voltages *lower* than the minimum voltage of the 431, you have to be slightly more careful. You can stick a resistor divider on the cathode of the reference, and, of course, it too needs to be of suitably low tolerance, but don't forget that it will be pulling current from the cathode: you should increase the cathode current to account for this.

If, in order to do this and have a low-impedance reference, you need to use an op-amp in unity gain, be sure that the input offset voltage and input bias

14 Watch out! There are some ICs out there whose schematic symbol is shown as being the same as a 431 but are not adjustable!

current of the op-amp are accounted for, as otherwise this may degrade the tolerance of the reference. And, of course, ensure that the op-amp is rated to be stable at unity gain!

3.4.9 ICs Used for Control

The underlying problem with making a robust Power Supply design is that most control ICs are unique to their manufacturers; an extraordinarily small percentage of such ICs are sourced by more than one company. Although this has been accepted as normal in the power supply industry, if one of the manufacturers has a factory problem, or there is a supply-chain issue with the wafers or the packaging, or if the manufacturer were to go out of business, all the designs using their parts would have to be redesigned. There aren't enough power-supply engineers left in the world to do that! The goal of this book, then, is to show how to design power supplies of all types that use only multi-sourced components; and in particular, multi-sourced ICs that can be used for SMPS control.

The primary IC that we consider is also the oldest. The 555 timer was first brought out by Signetics before many readers of this book were born: in 1971! Since then, countless designs have used it in a huge variety of ways, with entire books being written about possible application circuits.[15] It is available from many sources and also as two circuits in one package, the 556.

The basic functional mode of the 555 of use to us in this book will be as an astable vibrator, that is to say, an oscillator. Its timing is set by a capacitor and two resistors, one for the on-time and one for the off-time, as shown in Figure 3.1. The output is high for a time that is approximately given by

$$T_{\text{high}} \approx 0.693(R_{\text{A}} + R_{\text{B}})C, \tag{3.1}$$

and the output is low for a time that is approximately given by

$$T_{\text{low}} \approx 0.693R_{\text{B}}C. \tag{3.2}$$

Since the high time is greater than the low time, the duty cycle will always be greater than 50%. For power supplies, we will want the duty cycle to be less than 50%, so that we will be inverting the output of the 555.

We will also use the 555 as a one-shot, known as a monostable. It produces a single pulse when there it receives a trigger signal. Its pulse duration is set by a capacitor and a single resistor, as shown in Figure 3.2. The output is low for a time approximately given by

$$T_{\text{low}} \approx 1.1R_{\text{A}}C. \tag{3.3}$$

15 [10–12], see also [13–16].

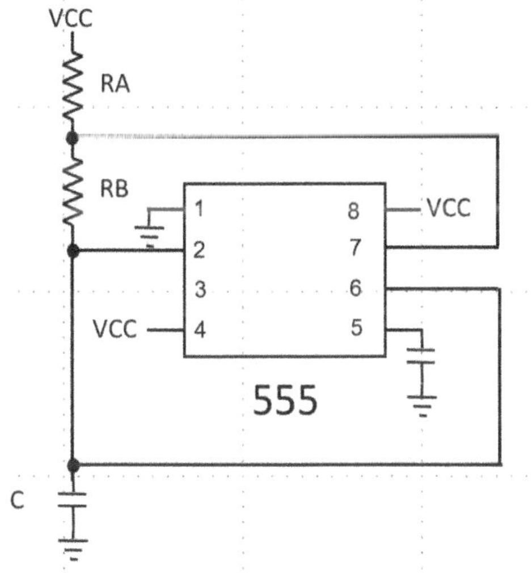

Figure 3.1 555 astable oscillator.

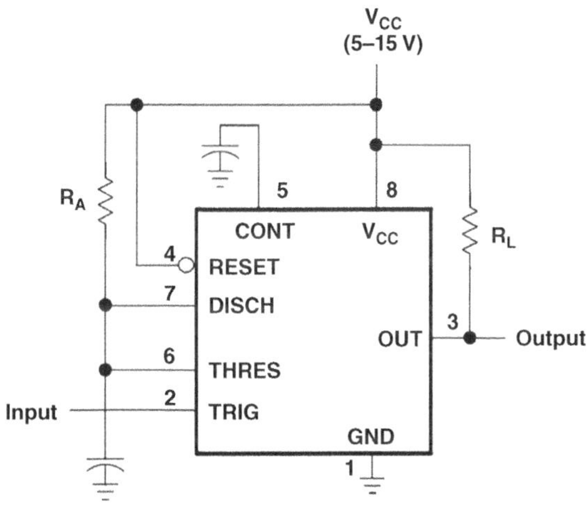

Figure 3.2 555 monostable oscillator.

Again, the output is normally high, and the pulse goes low, so we will be inverting the output of the 555.

In this book, we will be showing some novel circuits in which the duty cycle of the 555 output can be modulated, thus enabling it to function as a PWM controller. While not particularly accurate, the timing is good enough to accomplish switch-mode control. We'll also be demonstrating a method new to the literature to operate the 555 as a constant on-time controller for a PFC front-end converter.

For the PWM application of the 555, we will consider operating at several hundred kilohertz to reduce the size of the SMPS. For this reason, we will be using one of the many high-frequency versions of it; in particular, any of those that will do at least 1 MHz switching.

Only slightly more modern than the 555, the 3842 family of PWM ICs (including the 3842, 3843, 3844, and 3845) was introduced in the mid-1980s by Unitrode. Despite its age, it's still used in offline power supplies not requiring PFC and has a half dozen vendors. It has a few old-fashioned problems, and we'll show how to address these in our designs.

A much more modern IC (relatively speaking) is the 9910, which was introduced by Supertex in 2006. Due to its popularity, it has multiple manufacturers, and we'll use it to construct a PFC LED driver.

Finally, although not a controller, the 431 shunt voltage regulator, mentioned earlier as a voltage reference, was introduced way back in 1977. It can serve as a low-power adjustable voltage supply, requiring much less current than a zener. Today, aside from its use as a reference, it is also widely used as an error amplifier on the secondary side of an isolated power supply. It pulls current to try to force its reference voltage to be 1.24 V (or 2.5 V for other types), and this current is pulled through the opto-diode of an opto-coupler to cross the isolation barrier and provide feedback to the primary-side controller. We will give an explicit example of this in Chapter 9.

3.4.10 Gate Drivers

As suggested earlier, for most of our converters, it will be sufficient to have the controller (or timer, or comparator) drive the MOSFET directly. The one time this isn't true is when, due to high currents, we have a converter with a synchronous rectifier.[16] Then we need to drive both MOSFETs out of phase, and in particular, we need to ensure that they are never on at the same time—not even for tens of nanoseconds. If they were on simultaneously, that would present a direct path from the input voltage to ground, allowing extraordinarily high currents and probably destroying the power supply.

To avoid this, we have to ensure that on both edges of the gate drive, there is a delay so that the one MOSFET is off before the other one turns on. This could be done using the 555 timer as a delay circuit (or rather with two 555s, one for each edge of their respective MOSFETs) and then buffering the delayed signals with a totem pole pair of transistors to act as the gate drives.

Fortunately, it turns out that there are multi-sourced synchronous gate drivers that do this. If we search for suitable devices (look under half-bridge) capable of sinking and sourcing at least 1 A, bootstrapping at least 100 V, and in an SO-8

16 Synchronous rectification was already in use in the vacuum tube age; see [1] .

package, there turn out to be at least 15 such devices offered by 5 companies, so this is robust. Even if we want a 600 V boost, there are still five devices from three companies, and even 1,200 V parts are multisourced, so this is a reasonable choice to use for a robust design. As an example of such a part, we choose the IR2213, which is rated to 1,200 V.

3.4.11 Opto-couplers

The most important use of opto-couplers (or opto-isolators) for our purposes will be to provide feedback across an isolation barrier in an isolated power supply. This feedback can be either analog or digital. The analog feedback is the standard transfer of the analog control signal, usually from a 431, across the isolation barrier. Since the feedback loop ought to be designed in such a way that the initial tolerance on the transfer gain of the opto-coupler isn't an issue, we will not be overly concerned with its gain ... though naturally, you'll want to be sure that the gain range of second sources is similar.

Digital feedback is used in bang–bang controllers and just tells the primary side whether to be on or off. As such, it is even easier to design than the analog feedback.

There are two parameters of interest to us in using an opto-coupler. For the analog feedback, we care about the gain range, which is usually 2:1. The control loop using the opto-coupler must be designed so that it works over the whole range of gains.

For digital feedback, we care about the speed: the propagation delay (the time till something comes out after something goes in) and the rise and fall times of the output. Multi-source opto-couplers are available where both of these are in the tens of nanoseconds; faster seems unnecessary for our designs.

One other thing that could be of importance to us for an opto-isolator is the degradation of the gain with temperature and age. To avoid having to deal with this, we will use an opto-coupler with a 125 °C rating. We choose the multi-source open-collector 6N137. It has a BJT (base and collector) output but is very fast. This should work for all feedback designs; just ensure there are multiple sources for the package you choose.

3.4.12 Logic Gates

We will be using several types of logic gates for various applications, in particular SR flip-flops. A wide variety of types of gates are multi-source available. We will also be using an inverter. Because we may have to generate the power supply to run these gates internally to our design, we select the types that run from 5 V, and whose ICC is less than 100 μA.

3.5 Printed Circuit Boards

PCBs are used universally in power supplies. And there are many vendors around, so you don't have to worry about PCBs when making a robust design. Right?

Unfortunately, this is not necessarily true. Power supplies very frequently use heavier-weight copper, say 2-ounces, than, say, digital boards, which are often 1/2 ounce. Now there are plenty of vendors who make 2-ounce copper PCBs. But as the current goes up, so does the amount of copper required to carry it, and there's only so much board area. The other choice is to increase the copper weight, and as this goes up, the number of vendors who offer it goes down. On a recent project, it turned out there were only two vendors *in the whole world* who offered to make a PCB with 6-ounce copper!

The right solution here is that if you need to go much above 2-ounce copper weight on a PCB, you should use a wire or a busbar instead of copper traces. Now that involves considerable work from a Mechanical Engineer, what with bolts and nuts, and even to properly design the busbar takes some work. But the good news is that there are no supply limitations on those mechanical types of components.[17]

The other question, though perhaps not an issue, for PCBs is the number of layers. Sometimes backplanes for very complex designs involving multiple cards go up to 12 layers. But power supplies don't have 64-bit digital buses. In general, there is rarely a reason to go above four layers on a power supply PCB. Indeed, two layers are preferred, though it may take some extra work for the person doing the layout. Sometimes you just can't get from here to there on a PCB because of (mathematical) topology; in that case, there are 0 Ω resistors that can be used to jump from one spot to another.

3.6 Other Topics in Component Selection

In this final section of the chapter, we want to touch on some other aspects of component selection in making a robust design.

3.6.1 Derating

Of course, it's important to use parts that have multiple sources, so the next time a supply chain issue pops up, your design is immune. But that doesn't mean that just selecting multi-sourced components is enough: of course, they have to be the *right* components.

---

17 Though specialty busbars, such as laminated ones, have some limitations on number of vendors.

Table 3.2 Suggested component deratings.

Component	Parameter	Derating (%)
Resistor	Power	60
Ceramic capacitor	Voltage	75
Aluminum electrolytic	Voltage	75
Aluminum electrolytic	Ripple current	100
Connector	Current	80
Regular diode	Reverse voltage	75
Schottky diode	Reverse voltage	50
All diodes	Current	75
MOSFET	V_{DS}	90
Avalanche-rated MOSFET	V_{DS}	100
BJT	V_{CE}	85
IC	V_{DD}	90
Fuse	Current	50

What we're getting at is that components have to be derated; you mustn't dissipate 250 mW into a 250 mW resistor just because that's the part's rating! What will actually happen if you try this is that the resistor will heat up so much that its lifetime (as calculated by the MTTF, refer to Section 3.6.2) will dramatically shorten ... if it doesn't simply melt the solder and fall off the PCB entirely![18] Here, we show a table of the most critical parameters for good lifetime for the various components in this chapter, along with our suggested derating (Table 3.2). As a prototype of how to use this table, consider the resistor we just mentioned. We recommend a derating to 60% for surface-mount resistors. For a 250 mW resistor, the derating means that you should (in average circumstances) not dissipate more than 60% × 250 mW = 150 mW in this resistor. If no derating is listed, it isn't necessary; you can use the part at its rated parameters in the design, as long as you don't exceed them.

The power dissipation allowed in transistors and diodes isn't listed here and isn't based on their datasheet rating, because it depends very heavily on the package they're in and how that package is mounted and to what ... it's usually practically impossible to get anything like the rated power into these components; they'll self-destruct. Here in Table 3.3 are a few suggestions of maximal power dissipation for the most common surface-mount packages in their most common PCB mounting situations:

18 This is one of the things that UL looks for during its tests of your power supply. It can be a bit embarrassing.

Table 3.3 Suggested package power limits.

Package	Power limit
SOT-23	100 mW
SOT-223	150 mW
SO8	150 mW
DPAK	1 W
D2PAK	1.5 W

These powers are for a normal, crowded board with 2-ounce copper. If you have more copper area around a component or heavier copper, you can put some additional power into it. In any case, at the end of the day you need to measure the temperature of the component, especially for components such as TO-220s, which are wholly dependent on their heatsink and on what the heatsink in turn is connected to, for power dissipation limits.

3.6.2 Mean Time to Failure

We've referred several times in this chapter to Mean Time to Failure, MTTF. Why not MTBF? The reason is in the third letter of the acronym. "B" stands for "between," whereas "T" stands for "till" (or "to"). The difference is whether the unit you're looking at is going to be repaired and then reused. All the parts in your design age; when something fails, you can repair it. But repairing the part that failed doesn't reset all the other parts' age; they continue aging. Then you want to know the time until the *next* failure, hence the "between." For almost all designs these days, when something breaks, you throw it away. So, what matters is the time until the *first* failure. That's given by the MTTF.

Now before we get into how to calculate the MTTF, a comment is in order. Notice that the first two letters, "MT," stand for "mean time." That is to say, what we're calculating is how long it takes *on average* for the unit to fail.

Now while that's a useful piece of information, it's not all it could be: it means that after that many hours of operation, half of all units will have failed. It would really be better if we knew how long until, say, 10% of units would likely fail, which is to say, we'd like to know the standard deviation of the failure time.[19] This information isn't calculable with the tools we have. The best we can offer is that studies of the actual failure times at least suggest that the standard deviation is relatively

19 Note that the author, and everyone else, tacitly assumes that the distribution of the failures is Gaussian distributed. Since it will turn out that only a few components are the usual cause of failure, the distribution may in fact be multimodal.

small, and so most units will start failing not much before or after the calculated MTTF.

Yet another observation worth making: the MTTF calculates how long until the unit "fails." Failure could, of course, mean that it explodes or produces no output. But it turns out that for many power supplies, the items that first fail are the electrolytic capacitors. And they typically don't fail by exploding—they fail because their ESR goes out of specification.

Now if you're using an electrolytic capacitor as the bulk capacitance in your power supply, having higher ESR means that you will have higher ripple voltage. This may indeed mean that the supply is now out of spec ... or it may well be that there is a lot of margin on the ripple voltage, and the unit is just fine. The point here is that even if a unit reaches its MTTF, it may still work well for the application.

Let's turn to how to actually calculate the MTTF. The standard calculation remains based on MIL-HDBK-217F Notice 1 (it's available online [8]). The start of a sample calculation is shown in Figure 3.3.

In Column C, we list the part description, usually directly from the BOM. (We've hidden many of the nonessential columns to make the figure legible.) In Column G, we put the reference designator. The manufacturer's part number is shown in Column I. These are all important so that later on it can be checked if any components were changed since the last time the MTTF was calculated.

Column L specifies which Section of the MIL-HDBK we used for the calculation. Column M shows the value of (in this case) the capacitance (in pF), which will be used in the formula for MTTF. Column N shows the power dissipation in a resistor, which will again be used in a calculation. Column O shows stress, which for a capacitor is its typical voltage divided by its rated voltage.

Column Q is the formula for calculating λ_b, which is a formula given in the MIL-HDBK to determine the effects of operating temperature on the lifetime. The formulae are given in the MIL-HDBK and are typically hairy; ensure to enter them carefully![20,21] Of course, they're mostly copies of each other, so when you've done it once, you're in good shape for the next time.

	C	G	I	J	K	L	M	N	O	Q	R	AI
1					Ambient Temperature			25			MTTF	32,244
2	Description	Ref Di	Manufacturer #	Time	Temp	Section	Value	Power	Stress	λb	πCV	λp
3	CAP,CERAMIC,0.01UF,50V,10%,X7R,0603	C100	C0603C103K5RACTU			10.10	10000		0.3	0.001269	1.129234	0.0143
4	CAP,CERAMIC,0.01UF,50V,10%,X7R,0603	C101	C0603C103K5RACTU			10.10	10000		0.3	0.001269	1.129234	0.0143
35	CAP,10UF, POLY TANT, 35V 10UF 1210 PK	C12	TCJB106M035R0200	2000	105							1.953125

Figure 3.3 A portion of an MTTF calculation.

20 Don't try to interpolate from the examples given in MIL-HDBK; eyeballing numbers doesn't work that well. Since you have a spreadsheet, why not use it?

21 There are some programs around that already have the formulae entered into them. The author doesn't like using these programs; how do you know if the person typing in the formula did *his* typing correctly? At least when *you* do the typing, you can double-check it.

Column R shows π_{CV}, which is specific to capacitors. There are many more such component-specific parameters; you just enter a new column for each. Adding in all these parameters is how we end up at column AI, where λ_p is calculated, which is the expected lifetime of the component.

And finally, we end up with the MTTF of the entire design in Cell AI1, which is the inverse of the sum of the inverses of the λ_ps. The MTTF is given in hours and is readily converted to years.

One more entry should be noted. Cell N1 has the average expected operating temperature. In reality, if the unit is going to operate any substantial amount of time at a higher temperature, you should probably use that instead, as the lifetime is exponential in the temperature.

We skipped over Columns J and K. There are a number of components where the MIL-HDBK formulae may be considered to be superseded. We're showing an aluminum electrolytic capacitor, which in this particular case is rated 2,000 hours at 105 °C. We use this and the rule-of-two to directly calculate the λ_p. The same holds true for many ICs, MOSFETs, and diodes, where the vendor has done measurements on parts, obviating the need for the MIL-HDBK calculation.[22]

That's the basics. If you put the operating temperature in (say) Cell A1 and then use that in all the formulae, then it will be very easy to answer questions about what happens to the reliability if the temperature is decreased, for instance, using a fan. The table is organized the way it is so that it's easy to change temperature or component types once the components that have the dominant effect on the MTTF have been identified.

Finally, this whole process of calculation of MTTF can take an entire day, especially if you're entering all of the formulae by hand, rather than copying and pasting from a previous spreadsheet. Need a fast guess? MIL-HDBK suggests just counting components, but the author's experience suggests a better method for power supplies. Most power supplies' MTTFs are dominated by their electrolytic capacitors, if they have any. You'll probably get within a factor of two for the MTTF if you just do the calculation for the electrolytic capacitors alone.

3.6.3 Schematic and BOM

There are a few things you can do with the schematic and BOM to make your design more robust. The most important is not to wait with adding in multiple vendors. By the time the manufacturing floor runs out of parts and asks you for a substitute, it's too late; you've already wasted 24 hours or more.

Add alternate parts to your BOM right when it first goes through layout. It may be convenient to list the part you usually use as the first one, or you could somehow mark the alternates with a special symbol. As long as you state what it is you're

22 Many IC vendors give their failure rate in billions of hours, rather than millions. Be sure to divide these numbers by 1,000 to be compatible with the rest of the spreadsheet.

doing at the top, it will be fine. Or better yet, establish a company standard method for designating on a BOM, which are the preferred parts.

For the schematic, adding in (at least one of) the alternate parts in addition to the preferred parts makes things much easier than having to go the BOM every time you want to see what the other choices were. The goal on both the BOM and the schematic is to make both manufacturing and your debugging go smoothly with component swapping.

3.6.4 Dual Footprints

Another possibility for using multiple devices is the dual footprint. Even if the exact same package isn't second source, there may be a vendor who has a package that's similar enough that it can be substituted. This is especially useful for things like 0x0x packages, where you can simply put a big footprint on top of or in parallel with a smaller one and stuff whichever is available. Of course, you should have a preferred item on the BOM and schematic.

This idea can also work for slight differences in, for example, MOSFET packages. Where there are significant differences (for example, the original part is in a SOT-23, and the backup part is in SOT-223) if you have some space, you can parallel them by putting one right next to the other. Put the smaller one down first, and then the bigger one can sometimes have one of the traces go underneath its package.

3.6.5 Stockpiling

Finally, the often overlooked but sometimes best choice for a robust design is simply to stockpile the one or two parts that have few good replacements. While the goal is, of course, to avoid such parts, sometimes you just don't know in advance. Or you may know in advance that both the preferred part and the substitute may have problems.

Now CFOs truly hate inventory. They account for it as though it were money stuck in the bank, not earning anything. But if your production is relatively small, you may be able to convince them anyway. In a case that recently happened to the author, a certain special MOSFET was new, and while it was anticipated that other manufacturers would soon come out with second-source parts, there was no guarantee. Each PCB needed two of the devices, each of which cost US$0.35, and production was 1,000 pieces a month. For US$0.35 per transistor × 2 transistors per board × 1,000 boards per month × 3 months = US$2,100, we stockpiled enough parts for three months. This amount of money is small enough that, once you explain to the CFO the risk of not being able to produce the board at all, chances are they will agree to the creation of a stockpile.

3.7 Notes for the Hobbyist

A lot of hobbyists still use point-to-point hand soldering of wires or wire wrap. While these *can* be made to work, they have some serious issues.

One common issue with soldering wires and wire wrap is that connections are often not done well. You can jiggle the wire a bit and the "connection" falls off. This is obviously bad for trying to determine what the circuit is doing.

Another issue, more common when the circuit is more complex, is that the connections become so jumbled ("spaghetti" is the technical name) that you can't tell what is connected to what or what doesn't have any connections at all.

The solution for both is to just bite the bullet and do a PCB layout. There are free tools out there to do the layout, and it's well worth your time to get one, learn it, and do it right. PCBs can be home fabricated, but often PCB companies offer special introductory rates. If you take the longest lead time (usually two weeks), the cost can be totally affordable.

Since you're paying for a PCB, you might as well take advantage of that to make your measurements easier. Professional layout people always have numerous test points on the PCB. You should too. No, almost never will the test point interfere with circuit operation (with the exception noted above), and probably not when you probe it with an oscilloscope probe either.

Another issue with components is your selection of where to get them. Yes, 555 timers are all over old circuit boards, and you can find them even in flea markets. But honestly, is US$0.12 for a part really so bad? Get a brand-new part from a distributor, and save the worry about whether the PCB at the flea market was there because the part was blown up.[23]

Transformers are a problem, because, as we discussed above, there really aren't any standard parts. If you stick with the recommendations in this chapter, however, and only have two windings on a transformer, there are a certain number of SMPS transformers available from distribution. They tend to be for specific ICs, but you may be able to find one that works for your application. If not, you can try calling around to small manufacturers and see if they'd be willing to produce (and possibly design) one for you. Beyond that, some fairly detailed information is available in the author's book on Power Supplies [9].

Do you need a meter to test your components? If you're buying them new, probably not. That said, you probably have a digital multi-meter (DMM), and modern ones include not only resistance but also capacitance and (sometimes) inductance measurements. And no, you don't need a curve tracer for transistors, and there's no way to test an IC anyway. You put it in the circuit, and either it works or it doesn't.

---

23 Also, the ones on old PCBs tend to be the old low-frequency type.

4

Modeling

4.1 Precautionary Introduction to Modeling

Everyone learns how to model circuits while in university. And, of course, modeling is a necessary tool for design; the old days when you had to build something to see whether it worked are long gone. Or are they?

There are a number of potential issues with modeling that need to be kept firmly in mind while you're building and testing your circuit model, and which ultimately necessitate building actual hardware. The first issue you have probably encountered in university. You throw a bunch of components into the simulator's GUI. You run the model ... and it doesn't work. What do you do? Why, of course, you change that 1 kΩ resistor to a 2 kΩ. That should fix it! And when it doesn't, you change it to a 5 kΩ. And then a 10 kΩ.

We all do it. But the first problem on our list is that modeling encourages hacking[1]—let's try this, let's try that. What you *should* be doing instead is to figure out *why* the original 1 kΩ didn't work. Maybe it was supposed to be 1 Ω and you entered it incorrectly. Maybe it's attached to the wrong node. Maybe the circuit (gasp!) *doesn't work.* Modeling can be a crutch that you use to avoid thinking, but thinking is what a design needs in order to be successful. Do the design work before you make the model—the model is there to verify the design, not to create it.

This anecdote hints at another problem with modeling. You're busily entering all those resistors into the schematic. You go quickly. It's very easy to type in a 10 kΩ value for one of them rather than 1 kΩ. It's even easier to copy a 10 kΩ from somewhere and paste it in and forget to change the value. Or worse yet, tell yourself that you're *going* to change it later, and then never get around to it and forget.

1 This word is used in its older, pejorative sense. A hacker is someone who tries different things to see what happens rather than thinking through what the actual problem is.

Here's another problem that you may not encounter very often, but that the author has on more than one occasion. *A pre-built model component doesn't work properly*. The author once placed an open-collector comparator in a simulation circuit. The circuit didn't seem to work as expected. Some poking around in the simulation revealed that the comparator didn't transition from high to low when the inverting input voltage went above the voltage of the non-inverting input. Supposing the issue to be the collector current, the author increased the value of the pull-up resistor, but the comparator still didn't transition. *Removing* the pull-up resistor left the output of the comparator *still* high—even though it was supposed to be an open-collector comparator! It eventually turned out that the vendor-provided model of this comparator hadn't been adequately tested by the vendor. The model was, of course, fixed when the author brought it to the vendor's attention. But the moral of this anecdote is that just because your model *says* that something works this way doesn't mean that that's the way the circuit will *actually* work when you build it.

One more: just because it simulates correctly doesn't mean that it will work correctly when you build it. The most common problem here is that the simulation didn't really reach steady state, you just ran out of memory or got bored. Running actual hardware for five minutes tells you things your simulation never can.

Also, electrical simulation isn't enough; you need to also simulate, or better measure, temperature, which affects $R_{DS,on}$, V_f etc. SABER could do this, but of course the results will not be any better than the thermal models you entered.

There's one more to add to this long list of issues with simulations. For a given generic part, different manufacturers have different specifications for what are claimed to be identical parts. As an example, in the comparison tables in this chapter, one particular datasheet has been used. It doesn't claim to be representative; it's merely a single-point sample. If that parameter turns out to be critical for your design, what you *ought* to do is to redesign the circuit so that it isn't. Failing that, you need to make an APL to select only those parts that are going to work.

4.2 Overview of Simulating

There have historically been a number of simulation tools, some better, some worse. But the most popular has always been the first, Simulation Program with Integrated Circuit Emphasis (SPICE). It was created in 1969 and has never outgrown the problems of its origin in that primitive era, despite repeated upgrades to its interface.

SPICE's current most popular incarnation is offered for free by Analog Devices and is known as LTspice.[2] It has a reasonably nice GUI compared to its competitors, and a number of companies now offer models of their products for it.

Less nice is its list of issues. The most fundamental is directly traceable to its ancient origin: it is slow and consumes vast amounts of computer memory. When SPICE was created, it was natural (and still is) to represent a circuit by a matrix, whose values represent the interactions between nodes and which evolve with simulation time. To solve Kirchhoff's laws at each time step, they had to find the inverse of the matrix. Thus, as the number of circuit elements increased, the number of matrix elements increased as the *square*.

This has a variety of bad effects that are still present in SPICE-based simulation today. For larger circuits, the amount of dynamic memory required to store the matrix grows quadratically. For power supply circuits with hundreds of parts, this can be a problem even for modern computers with large memories.

The speed is also dependent on the matrix size, since the matrix inversion time depends on the matrix size, so it too grows quadratically. Worst of all, the size of the data file generated by SPICE grows linearly with the amount of time being simulated. So for stiff simulations, those where the ratio of total simulation time to minimum time step is large, the file size can grow so large as to overwhelm the memory storage even of modern computers. Even some of the relatively simple circuits shown in this book can easily generate gigabyte files. IC designers frequently have to run simulations for weeks on large workstations.

Naturally, workarounds have been devised over the years. Reducing the required accuracy helps—unless, of course, you need accurate results. Reducing the time being simulated helps—but not in intrinsically stiff simulations such as PFC, which has to run at least one line cycle with switching edges in the nanoseconds. There is also state-space averaging, which can give certain very useful types of information and which we describe in the final section of this chapter.

The real solution was embodied in a simulator known as SABER. This simulator was built from the ground up, not derived from SPICE. Their fundamental insight was that all circuits[3] are block-diagonal: they are built in blocks that are designed to do a certain function (such as provide a sawtooth or generate a reference) and have only one or a few connections with other blocks. This means the simulation matrix is sparse: there are blocks along the diagonal where the sub-circuits interact internally, and then predominantly zeroes elsewhere, representing the predominant lack of interaction between the sub-circuits. And dealing with sparse matrices is much faster and less memory intensive than with arbitrary matrices.

2 This is the official spelling.

3 All *human* circuits. NASA had what would now be called an AI do an evolutionary design of a circuit which was just a jumble of things. It worked, but no one could intuitively understand why.

Both speed and memory requirements were greatly improved. Regrettably, SABER is not widely used, presumably because of cost. If only there were a free version ....

4.3 Robust Simulation

We'll be using LTspice for simulations in this book, due to its availability, familiarity, and convenient GUI. Now, given the issues with SPICE described in the previous section, we're going to need to be a little careful with what we model and how we model it.

The key to successful modeling is to build a model suitable to get the information you want—you probably will be building more than one model of your circuit to get different pieces of information. How you model things determines how fast the model is going to run, which in turn has an impact on how huge the data file is going to be. For example, if you wish to determine switching losses in a switch mode power supply (SMPS), you will build a model with detailed metal-oxide semiconductor field-effect transistor (MOSFET) and diode models and run it for one switching cycle; while if you want to know the bandwidth and phase margin of the same SMPS, this approach wouldn't work at all, because sweeping the frequency response could take millions of switching cycles. What you would do instead would be to build a second model, a state-space average model. To repeat, which models you build is determined by what information you want to get from the model.

From the author's perspective, in SPICE, there are basically three types of component models available: primitives, ideal models, and device models. Primitives are models such as **V**, which is a voltage source, or **I**, which is a current source, both including a variety of options including a DC value, a sinusoid, piecewise linear etc., as well as sweeping sinusoidally through a range of frequencies. The most general of these is the **B**, which can have a function written to describe its behavior, including delays, integer-only, if-then-else statements, and just about everything you could want.[4] Primitives also include **E**, **F**, **G**, and **H**, which are voltage- or current-sources controlled by other voltages or currents in the schematic, though the **B** model can replace them all.

Ideal models are closer to real components. They include such models as **R**, **C**, and **L**, which are resistors, capacitors, and inductors, respectively, and are used in almost every design. These models also include optional arguments such as equivalent series resistance (ESR) for capacitors; the author strongly recommends that these extra parameters not be used. If you want to include ESR in a capacitor,

4 The author observes that if the expression in a **B** model becomes sufficiently complex, the simulator may fail.

make it a separate discrete component on the schematic so that it's visible. Otherwise, you're going to forget it's there and wonder why your model is acting weird.

Also included in ideal models are ideal diodes and ideal MOSFETs. You can put these down on your schematic and they more or less work. The problem with them is that you don't know what "ideal" means—how much V_f at 1 A? How much $R_{DS,on}$? There are additional, more complex ideal models available also, but these turn out to be totally uninformative.

SPICE was originally built for device engineers designing with bipolar transistors. For this reason, ideal models can be hopelessly complex. For example, there are three levels of model complexity for the ideal MOSFET models, for a total of *seventy*(!) parameters that need specifying. The author recommends not using any of these devices without selecting a device model for them.

Finally, in addition to primitives and ideal models, there are device models, such as the diodes and MOSFETs just mentioned with actual part numbers attached to them. There are similar models of op-amps, comparators, pulse width modulations (PWMs), etc.

How good is any particular device model? You have no way of knowing, you just have to test it yourself to see whether it has the parameters you're interested in, and whether they adequately approximate what the datasheet says. (Of course, sometimes the datasheet isn't right! The model may have been extracted by a machine from an actual device.) When we get to active component models, we'll be checking the model parameters we care about.

The variety of problems introduced in this and the previous sections lead us to propose some general guidelines for building robust simulation models.

1. Don't try to throw a schematic directly onto the simulator's GUI. Draw it out by hand on a piece of paper first, complete with values, just the way you would if it were going to a schematic capture person. And actually *design* it—don't just throw in a value with the thought that you'll figure out what it should be as a result of what your simulation shows.
2. Don't put off entering the values—do it for each component as you place it, rather than all at the end.
3. Place the components in a reasonable fashion: avoid lots of line crossings. Sometimes, the GUI will connect lines together that aren't supposed to be connected. And in any case, it's confusing.
4. Try to avoid using models you find on some non-vendor website. Who knows how much they've been tested (if any)?

5. While it probably sounds antediluvian, use primitive models instead of vendor models when possible. Vendor models are typically encrypted, so you can't see what their limitations are. Spice primitives certainly work, so if all you want to do is make sure that the comparator switches states, use an ideal comparator rather than the vendor-provided high-speed, low-offset comparator that has dual outputs and bias pins. In fact, why not use a **B** source? Using lower complexity parts will also simulate way faster when you have 200 of them.

In general, then, a simulation model is *not* going to be the exact circuit you design. Not only are the exact components you're designing not going to be in the component library, simulating with those components would generate huge amounts of data and would take forever, especially in power electronics, where the equations of state are very stiff.[5] Modeling isn't intended to replace prototyping; its usefulness is in verifying specific areas of performance, such as bandwidth and phase margin.

This leads us to the biggest single drawback of LTspice. It has all the models from Linear Technology, Maxim, and Analog Devices. But you want a device from TI? Not supported. Onsemi? Microlinear? Sorry. Some vendors provide models that will run on LTspice. For the rest, you will have to determine the key parameters you care about and look for something similar. We'll be doing that throughout the rest of this chapter. Or if it's unique, you'll have to build a model from scratch. We'll be doing that below for the 9910.

4.4 Passive Components

4.4.1 Resistors

As noted in the previous chapter, robust designs will be almost exclusively using film resistors, and the use of your simulator's resistor primitive is entirely fine.

That said, however, simulators tend not to do a good job in one particular area: power rating of the resistor. While LTspice allows you to enter the power rating of a resistor, it doesn't show up on the schematic. And if you exceed the power rating of the resistor, it doesn't issue a complaint. So you can make some pretty seriously wrong conclusions about a circuit if you don't check the power dissipation eventually.

---

5 Stiffness is the ratio of the highest relevant frequency in the model to the lowest. For example, if you need to model a 4 MHz SMPS for 10 msec, the stiffness is 4 MHz * 10 msec = 40,000 = 92 dB, which is very high. Running such a model takes a lot of time, possibly gigabytes of memory, and is very slow even to display.

A similar caveat applies to the voltage rating of the resistor. Simulators don't even provide a place to specify the resistor's voltage rating. Nonetheless, putting 400 V on an 0603 resistor will not work very well in practice.

4.4.2 Capacitors

As noted in the previous chapter, for the most part, designs use only two types of capacitors: ceramics and aluminum electrolytics. For a ceramic capacitor, you can just use a capacitor primitive in your simulator, because the ceramic is nearly ideal until you get to rather high frequencies.

Just as with resistors, simulators don't let you know if you've exceeded the voltage rating of the capacitor. You just have to check after the simulation has been run. (Obviously, if, for example, the capacitor is 16 V and is a bypass on a 5 V line, checking is unnecessary.)

Aluminum electrolytic capacitors sometimes need a little attention when you're modeling them. They typically have an ESR, which is not something that's fixed for a certain value of capacitance but varies by the construction used. It thus has to be looked up for each such device you used. Modeling ESR matters if you're interested in, for example, the ripple voltage at the switching frequency of a SMPS. It also matters if you're doing a state-space average model of a SMPS, as it forms a zero in series with the pole. (And it is for this reason that you have to model ceramic capacitors in parallel with the electrolytics separately. You can't just "add them together.")

Now, while LTspice offers the option to put an ESR directly into the capacitor model, the author considers this a bad idea. The trouble is that the ESR doesn't show up in the schematic. Did you put an ESR value in last week? What value was it, and did you remember to change it when you changed the physical device? The safe way to make the model is to put a discrete resistor model in series with the capacitor model and leave the internal ESR as zero.

4.4.3 Inductors

Inductors are ideal models in LTspice. They have optional series resistance, parallel capacitance, and so on. But just as for capacitors, the author recommends that, if you want any of these, you should put them as discrete components around the inductor, rather than burying them where you won't remember what you set. NOTE: Unlike capacitors, inductors in LTspice default to a series resistance value of 1 mΩ. If this is something that may affect your simulation, you should set it to zero, remembering to put it externally.

Transformers are covered in Section 4.7 on isolation.

4.4.4 MOV Model

Making a model of a metal-oxide varistor (MOV) is a little bit sophisticated, and we have moved the calculation and an example to the Appendix. The end result is that an MOV obeys an equation of the form

$$I = ae^{be^{cV}} \tag{4.1}$$

and the three parameters a, b, and c can be determined using Excel. LTspice supports this type of equation, and you can use a BI model to produce an MOV model.

A section on building a model of lightning is in Chapter 10 on Special Topics.

4.4.5 Batteries

Unsurprisingly, LTspice doesn't include any cell or battery models. While we'll be giving some specific detailed models in the following chapters, a few general comments about the simulation of batteries can be made here.

Perhaps the most important is that batteries are really not ideal voltage sources. They have something equivalent to ESR and ESL, and so their terminal voltage changes depending on how much current you're putting in or taking out. You can't simulate a battery with a voltage source except for the applications where you don't care about what the battery is actually doing.

Beyond this, it is well-known that cells and batteries change their voltage as a function of their state-of-charge (SoC). Since going from fully discharged to fully charged typically takes at least an hour, it is impossible to directly simulate the charging of a battery. What you have to do is to slowly sweep the voltage of your battery model to emulate the charging. (Here, slow means changing the voltage of the battery slowly enough that the power supply has time to reach steady-state at every point of the sweep.)

If you want to get yet more complex, cells and batteries also change their voltage as a function of their temperature and as a function of how many charge/discharge cycles they've been through. This level of fidelity can be achieved in a simulation, but takes a lot of detailed information from the lab about how the particular cells or batteries respond in a period of hours and months. This can be done, but you have to want the information pretty badly. And despite all your hard work, in the end, you'll find that seemingly identical cells and batteries *aren't* ... not even those from a single manufacturer. There's a limit to the extent of usefulness of even the best simulation.

4.5 Active Components

In this section, we turn to active components: diodes and transistors; op-amps, comparators, and voltage references; and models of the ICs we use in SMPS. The one exception is the 9910, for which LTspice doesn't have a comparable model at all, and for which we need to create our own model. This is accomplished in the following section.

As we work through this list, we'll see that all the active components except for the 9910 fall into two categories: those for which LTspice has a model, and those for which LTspice has similar models.

For those for which LTspice already has a model, this doesn't mean that we can just naively use it. While it might possibly meet all of our needs, perhaps the model doesn't encompass all the parameters we care about in our simulation. Or perhaps the model parameters differ from the datasheet. Does the slew rate of the op-amp model match the datasheet? And does it match the typical value or the minimum value or the maximum value? We need to compare the model's performance with its datasheet for every model and for every parameter of that model we are interested in.

For those components for which LTspice *doesn't* have a model, for many parts we can find close-enough models. For example, if we have selected a multisourced MOSFET and are interested in power losses, the parameters we most care about are $R_{DS,on}$ and Q_g. Chances are LTspice doesn't have a model of precisely the MOSFET we've chosen, but we can also probably find one that has close to the same parameters, and that will be good enough for obtaining loss information from the simulation.

As we go through the active component models for the multisourced components we selected in the previous chapter, we will systematically test the parameters of importance for our designs and verify them (or not) against the datasheet specifications.

4.5.1 Diodes

As we discussed above, how you model a diode depends on what information you want out of the model. The diodes we selected in the last chapter were the 4148 for small-signal applications, the 1N4005 family of diodes for AC-line bridge rectifiers, and an 860 Schottky for medium-power SMPS rectification.

Before diving into measuring the models of these, let's consider one more option. There's an ideal diode in the simulator; maybe you should just use it? If we do a quick sweep of the ideal diode with a current source (Figure 4.1), we see that the diode is just an ideal exponential. It has a forward voltage that starts at about 480 mV at a very low current of 1 μA, rising to about 830 mV at 1 A. So it's a pretty

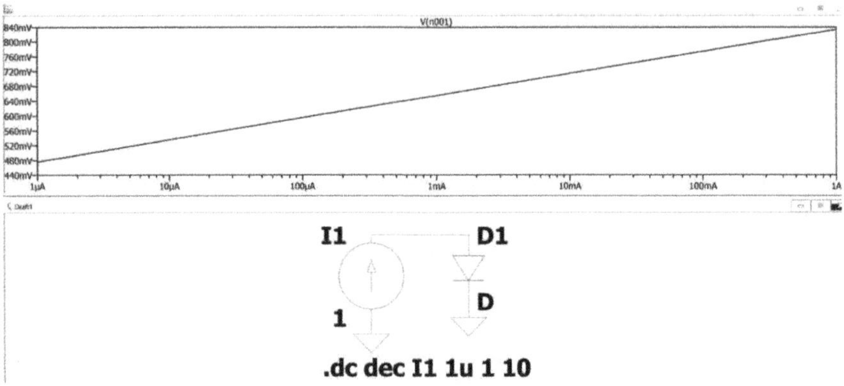

Figure 4.1 LTspice ideal diode I–V curve.

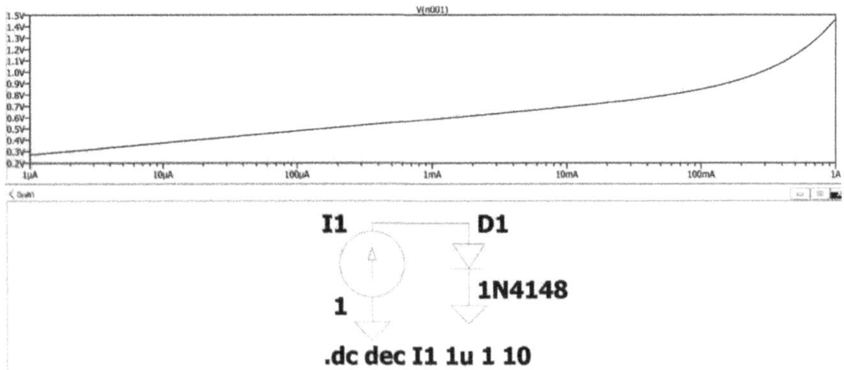

Figure 4.2 LTspice 4148 model I–V curve.

typical diode, like the ideal one they teach you about in school, and is suitable for, say, OR'ing two signals together.

However, if we care about the forward voltage in the circuit, we might want to replace it with the specific component value. For example, let's think about the 4148 in a pulse current application. Replacing the ideal diode in this simulation with the LTspice 1N4148 model (see Figure 4.2), we see that it is quite a bit different. Down at 1 µA it's at 270 mV rather than at 480 mV for the ideal diode; while it stops being a pure exponential somewhere around 50 mA and goes into its resistive region, with V_f rising to 1.45 V at 1 A, rather than 830 mV in the ideal model.

We put the comparison between the LTspice 1N4148 model and the datasheet in Table 4.1, using the datasheet's "typical" curve for V_f. Generally, the datasheet is 100 mV lower than the model at the most popular currents. Of course, the engineer who built the model (or the machine that tested the device) might just have

Table 4.1 Model vs. datasheet 4148.

Parameter	Model	Datasheet
V_f at 1 mA	712 mV	600 mV typ.
V_f at 10 mA	803 mV	710 mV typ.
V_f at 100 mA	804 mV	915 mV typ.
Breakdown voltage	>1000 V	100 V min.

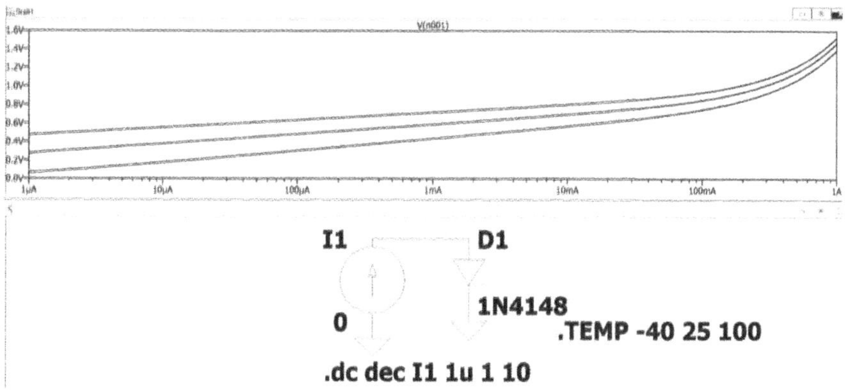

Figure 4.3 Temperature variation of 1N4148 V_f model.

used a different lot of parts than the datasheet we're using; or more likely, there are variations from manufacturer to manufacturer. Since the difference between the LTspice model and the datasheet is on the order of 10–15%, you shouldn't take the power dissipation too seriously; but then, this device isn't intended for applications where the power dissipation (or the exact V_f) is important.

We haven't discussed the temperature dependence of V_f for the 1N4148. As the diode (operating in its normal range) gets hotter, its V_f goes down. As shown in Figure 4.3, there is a temperature dependence of the V_f built in to the LTspice model.

The one last thing we haven't mentioned about diodes is their reverse voltage. Naturally, you should check the simulation that the reverse voltage limit of the diode you've selected isn't exceeded in the actual circuit. But assuming that's true, the actual reverse voltage rating of the model doesn't matter: LTspice models don't break! See Figure 4.4 for a sweep of the reverse voltage of the 1N4148 voltage. So this isn't something you need to be concerned about in your simulation, only in real hardware.

We now turn to the second diode we'll be using in our robust designs, the 1N4005, a 600 V standard rectifier which we'll use in the bridge of off-line SMPS. As it turns out, LTspice doesn't have a model for this; but what they do have is a

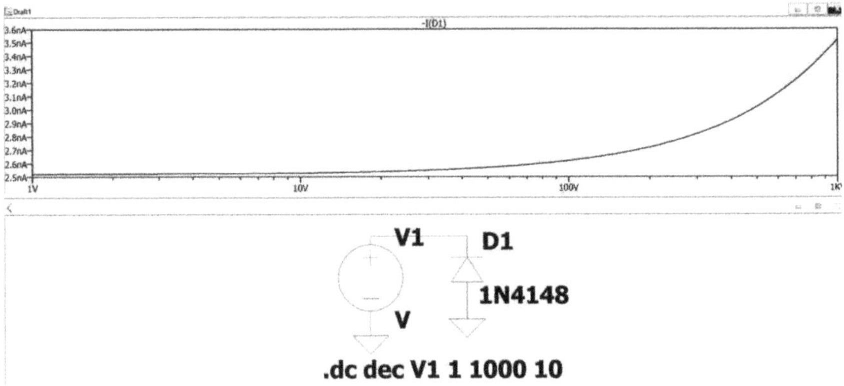

Figure 4.4 LTspice 1N4148 leakage as a function of reverse voltage.

Table 4.2 LTspice model vs. datasheet 4005.

Parameter	Model	Datasheet
V_f at 100 mA, 25 °C	758 mV	790 mV typ.
V_f at 1 A, 100 °C	954 mV	820 mV typ.
Breakdown voltage	>10 kV	600 V min.

model for the 1N4007, which is the same device but rated at 1000 V. The datasheet has a single set of curves for both devices, so we can satisfactorily use this model.

We can do the same sweeps as for the 1N4148, and end up with the following characteristics vs. datasheet in Table 4.2.

For the actual operating mode, the forward voltage of the diode at 100 °C is the most important. The table shows that the model's V_f is 15% higher than the datasheet. So the power dissipation that LTspice calculates is also going to be 15% higher. Since in high-efficiency SMPS the bridge diodes have a significant effect on the efficiency, you have to account for this error if you're using your model to approximate the real efficiency. Of course, it's good that the V_f is too high rather than too low, so that your calculated efficiency is a lower bound; what this really tells you, again, is that you have to build the circuit.

It is to be noted once again, LTspice's 1N4007 model doesn't break in the reverse direction. You have to verify (using the model's probe) that the voltage applied never exceeds the 600 V (or 1000 V) rating of the part you're using (for example, during lightning).

Our final diode is the 860, an 8 A, 60 V Schottky, used in the output of a medium-current, low-voltage SMPS. We choose a through-hole version because of the power dissipation; even without knowing any details of how it's going to be used, we can guess that the power dissipation will be something like

6 A × 500 mV = 3 W, too much for a surface-mount version, and probably requiring a heat sink even in the through-hole version.

LTspice has a part not vastly different than this, the RBQ10T45ANZ, which is 45 V at 10 A. We again run the same sweeps in LTspice, and get Table 4.3. We see that V_f is again about 20% higher in the model than on the datasheet. Also, observe that *this* model has a maximum reverse voltage; above 45 V it conducts substantial current.

Now, here is the measurement of the equivalent capacitance. This is important for the Schottky, because, even though it doesn't have any reverse recovery time, the capacitance provides a path for current to go in the reverse direction.

We build the circuit shown in Figure 4.5. We have a forward current of 10 A, which turns off at the same time that a reverse voltage of 20 V is applied (within 10 nsec).[6] Peak reverse current is seen to be 1 A. We calculate the capacitance from $CV = It$, giving 500 pF. This is in the same ballpark as the datasheets 195 pF. Without details of the datasheet's measurement technique, we can't hope to get closer. It's enough to know that the Schottky's capacitance is indeed modeled in LTspice's model.

Table 4.3 LTspice model of RBQ10T45ANZ vs. datasheet 860.

Parameter	LTspice model of RBQ10T45ANZ	Datasheet of 860
V_f at 1 A, 25 °C	438 mV	390 mV typ.
V_f at 10 A, 100 °C	672 mV	560 mV typ.
Breakdown voltage	45 V	45 V min.
Capacitance	500 pF	195 pF typ.

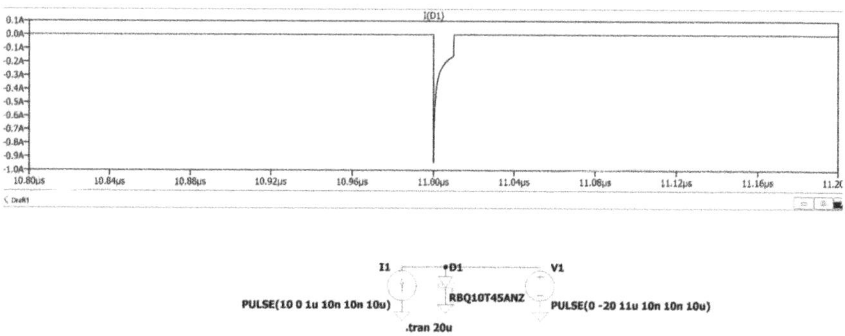

Figure 4.5 RBQ10T45ANZ LTspice model capacitance.

6 We set Reltol to 0.0001 to get this smooth result; see Section 4.11 on Controlling Spice for details.

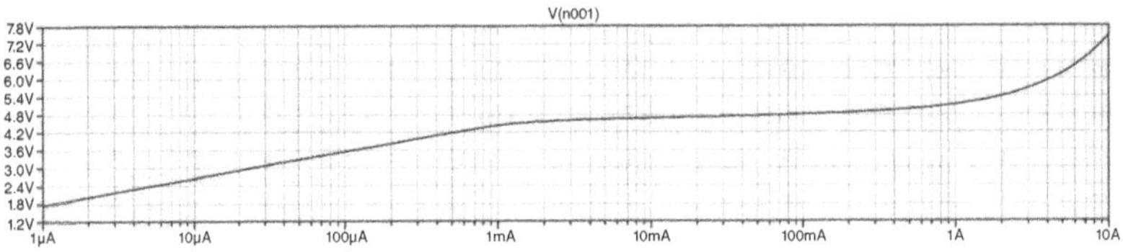

Figure 4.6 *V*–*I* curve of 1N750 Zener.

4.5.2 Zeners

Our only use for zeners in this book is as voltage clamps. So, while there are quite a number of characteristics of zeners that could be evaluated, the only one that matters for us is their breakdown voltage vs. current.

As a representative example, we take the 4.7 V 1N750. We build the obvious circuit, passing a ramped current through it, and look at its V–I curve, as shown in Figure 4.6. It starts to regulate at 1 mA, below which it just has some nonregulated voltage. It continues regulating up to perhaps 400 mA. Above that, the voltage starts to rise, and above 1 A, it's not working anymore—although it may be observed that it doesn't positively *break* even at 10 A, where it would be dissipating 72 W and a real device would certainly have failed. The real problem with the actual use of this part is that its nominal current is 20 mA. We would, of course, use something that doesn't require so much current for its rated voltage.

4.5.3 LEDs

Since LEDs are only used as indicators' lights in this book, we haven't bothered to build a model for them; there don't seem to be any LED indicator light models in LTspice, though strangely enough, there is one for a neon bulb!

4.5.4 Transistors

Just as for diodes, what model you want to use for transistors depends on what information you want the model to give. We'll be looking specifically at the performance of the BJT 2222 and the BJT complementary pair 3904 and 3906.

There are a number of models for transistors in LTspice. In practice, we will always use specific devices for the transistors, both bipolar junction transistors (BJTs) and MOSFETs, and never the ideal models, because these require selecting a number of parameters that aren't in datasheets. Pursuant to this, whenever we want a MOSFET we will choose either the "nmos" or the "pmos" model, never the "nmos4" nor the "pmos4." And similarly, for the BJTs, we will always pick the "npn" or the "pnp" models, never the models with digits in their names.

We start then with the 2222, a 40 V, 600 mA NPN, widely used for signal processing. As with the diodes, we want to check LTspice's model against the datasheet. We start with breakdown voltage, V_{CEO}, with the circuit shown in Figure 4.7. As with the diodes, the 2222 model in LTspice doesn't have a breakdown voltage. We do a similar setup for base-emitter breakdown voltage, V_{EB}, which is typically about 5 V and find that it isn't modeled at all; this won't matter in most applications, but you need to remember to look at the simulation result to check that you aren't exceeding the datasheet limit.

We next turn to on-state voltages. We check base-emitter voltage at 15 mA base current and a gain of 10 as shown in Figure 4.8. The value is close to the typical in the datasheet. We use the same simulation to find the $V_{\text{CE,sat}}$, which is also close to the datasheet value.

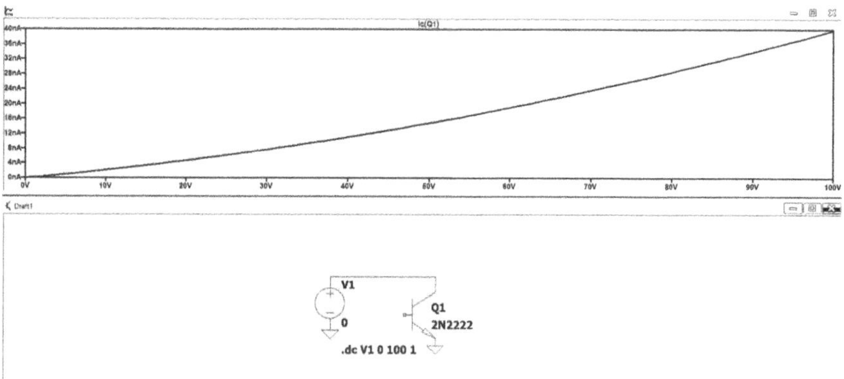

Figure 4.7 LTspice model of 2222 V_{CEO}.

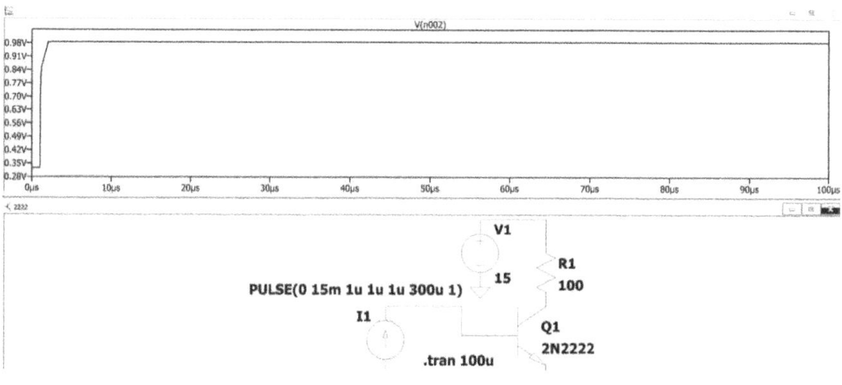

Figure 4.8 2222 $V_{\text{BE,sat}}$ LTspice model.

In one of our examples, we will be using the 2222 to pull current from a nominally 250 mV source. In that example, we can heavily overdrive the base of the 2222, so for future use, we also list the $V_{CE,sat}$ at 200 μA and 2 mA with unity gain.

Next, we look at the gain of the 2222, h_{fe}. We use the circuit in Figure 4.9. When the voltage on the collector is 10 V, the current through the resistor will be 10 mA; so we can sweep the base current and look at the V_{CE}. The result is that the base current is 50 μA, so the gain is 200, which is very close to the datasheet number of 210 (Table 4.4).

Finally, we look at the timing of the 2222, which we'll certainly want to know for use inside a control loop. We use the circuit shown in Figure 4.10, which differs somewhat from the datasheet circuit, but more closely resembles the circuitry we'll

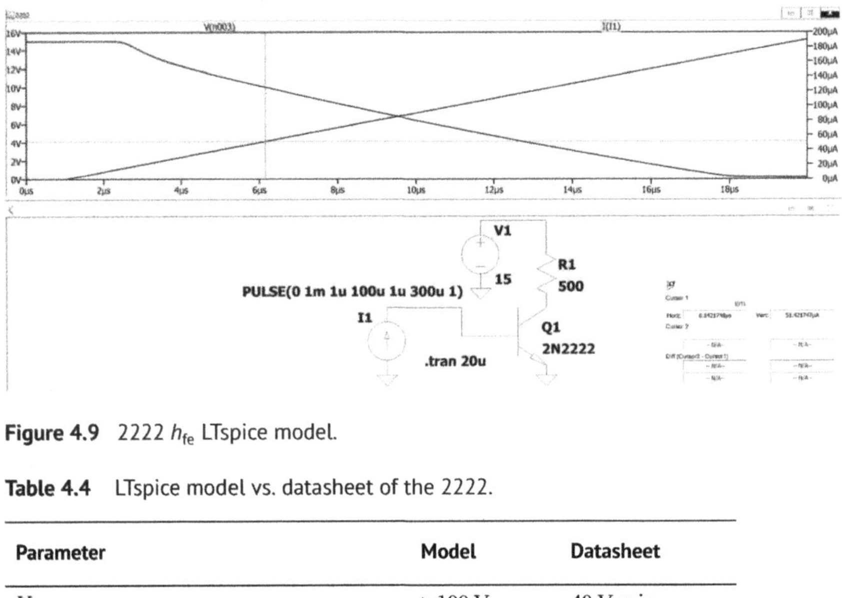

Figure 4.9 2222 h_{fe} LTspice model.

Table 4.4 LTspice model vs. datasheet of the 2222.

Parameter	Model	Datasheet
V_{CEO}	>100 V	40 V min.
$V_{EBO,max}$	>10 V	6.0 V
V_{BE} at $I_B = 15$ mA, $I_C = 150$ mA	980 mV	880 mV typ.
$V_{CE,sat}$ at $I_B = 15$ mA, $I_C = 150$ mA	130 mV	150 mV typ.
$V_{CE,sat}$ at $I_B = I_C = 200$ μA	13.6 mV	–
$V_{CE,sat}$ at $I_B = I_C = 2$ mA	15.2 mV	–
h_{fe} at $I_C = 10$ mA, $V_{CE} = 10$ V	200	210
Delay time	11 nsec	10 nsec max.
Rise time	100 nsec	25 nsec max.
Storage time	660 nsec	225 nsec max.
Fall time	12 nsec	60 nsec max.

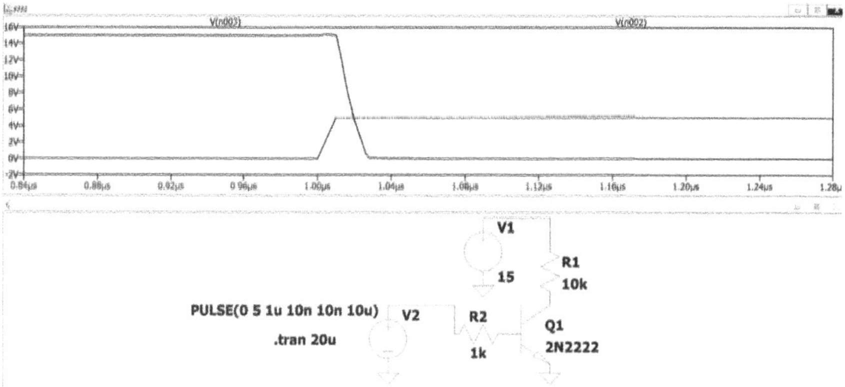

Figure 4.10 2222 timing LTspice model.

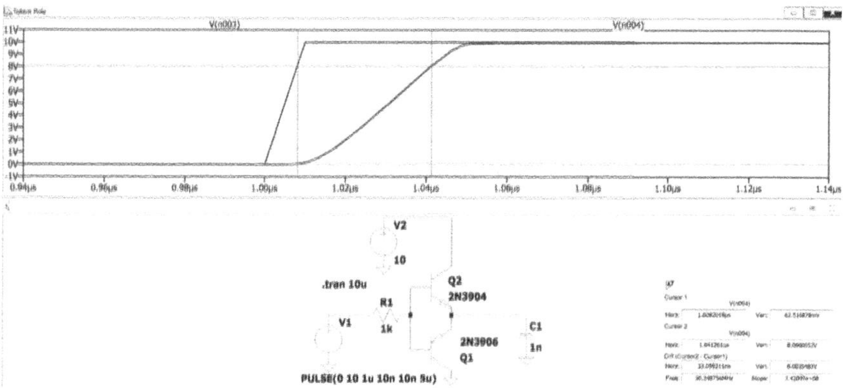

Figure 4.11 LTspice model of 3904/3906 totem-pole timing.

actually be using. Using cursors, we can measure that the delay time is 11 nsec and the fall time is 12 nsec, and looking at the other edge, the storage time is about 660 nsec and the rise time is about 100 nsec. Except for the delay time, these are all substantially longer than the numbers listed in the datasheet, probably largely due to the larger value collector resistor; but these are the numbers we should expect to be seeing in the actual application.

We turn now to the 3904 and 3906 BJTs which we'll use for a totem-pole pair to drive a (switching) gate. Since this is their only use, the parameters in which we will be interested will be the delay times for the rise and fall, and how close the output gets to the rails.

Using the circuit in Figure 4.11, we have a 1 kΩ resistor as a common base drive, and a 1 nF capacitor as a load to represent a gate capacitance. The delay from when the input drive reaches 8 V to when the capacitor reaches the same voltage is about

33 nsec; on the other edge, the delay to 1 V is about 39 nsec. The load capacitor goes up as high as 22 mV below the rail, and as low as 26 mV above ground.

When we get to power MOSFETs and BJTs, we're going to do the same sort of thing we did with the diode modeling. For MOSFETs, if the device we want is present in the LTspice library, we're going to look for a MOSFET that has roughly the same $R_{DS,on}$ that the MOSFET we're using has. For BJTs, we're going to look for one that has roughly the same h_{fe}. The V_{DS} and V_{CE} ratings don't matter because, as we've seen, LTspice model transistors don't break. If you're using a special MOSFET with (say) low gate threshold voltage, rather than looking for such a part in the simulation library, you need to monitor the V_{GS} during the simulation; a model is only going to show you typical voltages, so you can't rely on it. (What's more, it's difficult to identify which devices have low $V_{GS,th}$ from the LTspice selection list. You could look up datasheets one at a time, or just try it and see if it works, if the question actually matters for your design.)

The other thing we frequently care about in switching models using MOSFETs is the gate capacitance, or better, the total gate charge. This is important in SMPS because the gate drive impedance and the gate charge will determine the switching time of the MOSFET, which can be a significant portion of the power dissipation of the MOSFET. So, in addition to finding the approximately correct $R_{DS,on}$, you should also look for a device with approximately the same gate charge.

Finally, we've been discussing using device models. What about just using an ideal switch, **SW**, instead of a MOSFET? If all you care about is that sometimes it's connected, and sometimes it's not, an ideal switch could work. However, it requires you to make up parameters for it (how much resistance does it have when it's off? How much when it's on, what voltage does it transition at and with how much hysteresis?), as well as adding in a command line. And then it has rather odd behavior in how it transitions between states. It even seems to crash more often than MOSFET models do. So we never bother with this primitive. Why not just pick something that's closer to what you really want?

4.5.5 Op-amps, Comparators, and Voltage References

Op-amps, comparator, and voltage references are going to be the most commonly used ICs in our designs. So modeling them appropriately is key. Once again, what kind of model you use depends on what information you want the model to give you. Following our discussion in the previous chapter, for op-amps, we'll be looking at the 358 and the TLV9062; for comparators at the 339; and for voltage references, the 431 and the 4041. We will also, at the end of this subsection, look at using the 431 as an error amplifier.

Voltage references are easiest. In our section on components, we said there were only two of them that we were going to be using, the 431 and the 4041. Aside from being careful in your design to account for tolerances (and if you're buffering it with an op-amp to lower the impedance, input offset voltage and bias current), there's no reason you shouldn't just use a spice ideal voltage source as a model. Of course, in the real design, you have to make sure it's getting its minimum required current.

In one of the examples, we will use the LT1004 as a voltage reference because its model is available in LTspice. It is a fixed 2.5 V and is not adjustable. It has a nominal tolerance of $\pm 0.8\%$.

Op-amps and comparators require more care. You could build them out of a **B**-source just by writing the correct transfer function. But this has the issue that they would then produce exactly whatever voltage they're commanded to—even if it's millions of volts! But you could even limit their output to the supply voltage and ground, perhaps by using the "if" function. This would doubtless take some experimentation.

Another option would be to use one of the ideal op-amp models (though you should observe that there aren't any ideal comparator models in LTspice). They even have power supply pins, so they won't go too high or too low. But you need a special command on the schematic to get them to run. In order to do this, you have to select a model level, which means you need to know which parameters you want to use.

If you're going into all this trouble, why not just use the op-amp or comparator that you're actually going to be using anyway? The parameters that are going to be important for our designs are the slew rate, the minimum and maximum V_{CC}, the supply current, output voltage limits both high and low, gain-bandwidth product (GBW) (for the op-amp only), and the delay time from input to output. We'll check our LTspice models for these parameters and compare them to the datasheets. For most of our designs, input offset voltage and input bias current aren't going to matter, so we won't measure those.

Let's start out with the multisourced op-amp LM358, which has a GBW of 700 kHz and a slew rate of 500 mV/μsec. This used to be in the LTspice library, but for some unknown reason isn't anymore. However, the LT1013 is pretty similar; if your circuit works with this, it will probably work with the 358, and conversely. The LT1013 will be our most common op-amp. Or for only slightly more work, you can download the 358 model from the web.

Let's see how well the parameter set of the LT1013 matches up with that of the LM358. A summary of the results is shown in Table 4.5. Our basic simulation model is shown in Figure 4.12, a unity-gain configuration with a V_{CC} power supply whose voltage we sweep. The LM358 is supposed to work with a V_{CC} from 3 to 30 V. The LT1013 model starts outputting the 1 V source at about 2.6 V, which is reasonable, but continues working past 50 V. Just like the diodes and transistors, apparently the LTspice model of the LT1013 can't be broken by a high supply voltage.

Table 4.5 LT1013 model vs. LM358 datasheet.

Parameter	LT1013 model	LM358 datasheet
Minimum V_{CC}	2.6 V	3.0 V
Maximum V_{CC}	>50 V	30 V
Supply current at 5 V	350 μA	300 μA
Output high (from rail)	960 mV	1.35 V
Output low	33 mV	100 mV
Slew rate	413 mV/μ sec	500 mV/μ sec
GBW	1.3 MHz	1.2 MHz

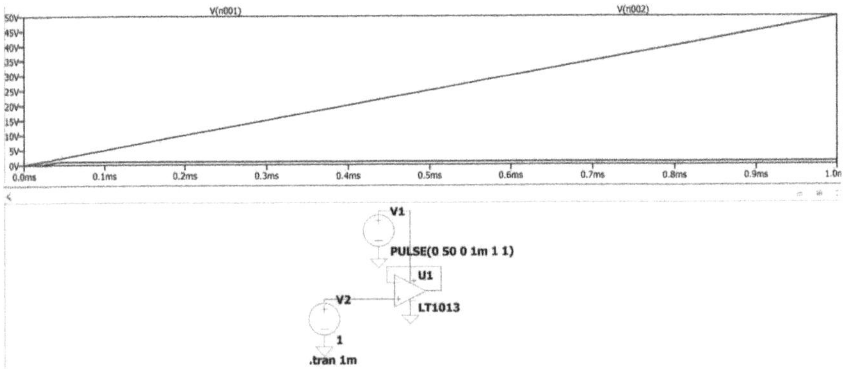

Figure 4.12 LT1013 LTspice model VCC.

With this same model, we can determine the supply current by setting the V_{CC} to a constant and finding the operating point, which, for the LM358 according to its datasheet, is supposed to be typically 300 μA at a supply voltage of 5 V; for the LT1013 model it's 350 μA, which is good enough.

We can further use this model to determine how close to the rails the model can get by sweeping the input voltage rather than V_{CC}; for the LM358 at no load, this is supposed to be 100 mV and V_{CC}—1.35 V typical. For the LT1013 model, the output goes down to 33 mV and up to 960 mV below V_{CC}. These numbers are fairly comparable as well.

For the slew rate, we can simply make V_1 a step and look at the output, as shown in Figure 4.13. The LM358 datasheet shows a typical slew rate of 500 mV/μsec, while the model shows 413 mV/μsec.

For the GBW, we can change the input to a swept AC model (don't forget that it needs a DC level as well), and look at the transfer function (output divided by input). Since it's a unity-gain, the GBW is the frequency at which the gain is down by a factor of 3 dB (=70%). The LM358 datasheet shows a GBW of 1.2 MHz; as shown in Figure 4.14, it measures 1.3 MHz, very close to the LM358.

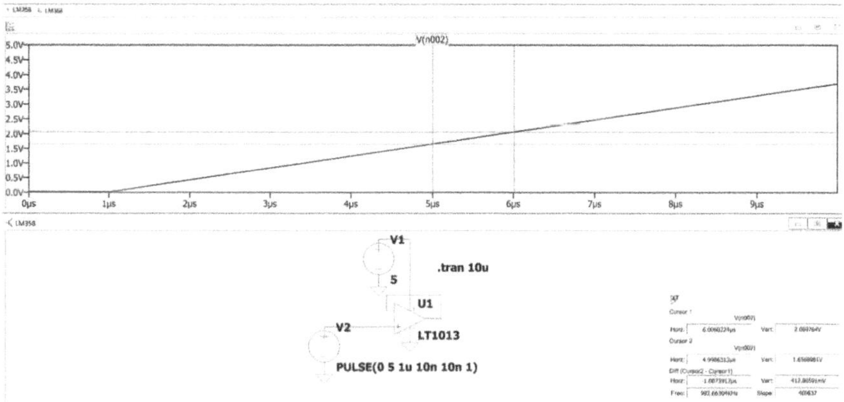

Figure 4.13 LT1013 model slew rate.

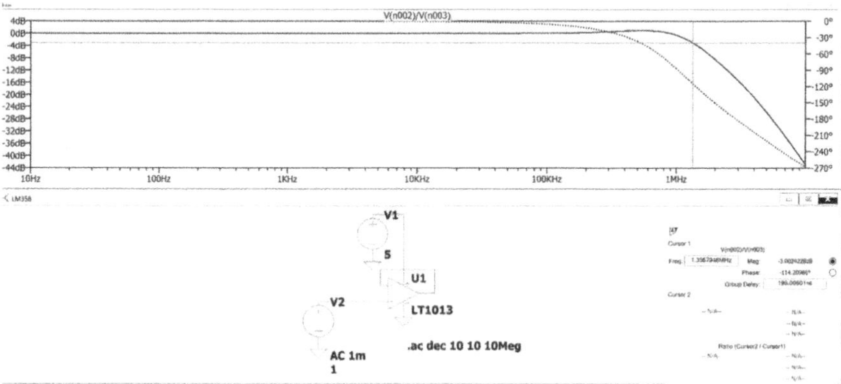

Figure 4.14 LT1013 model gain-bandwidth.

Overall then, the LT1013 model makes an entirely suitable model replacement for the LM358—as long as your design doesn't try to power it from greater than its rated voltage.

For higher speed work, we'll be using the multisource TLV9062. This has a GBW of 10 MHz and a slew rate of 6.5 V/μsec. As we mentioned in Chapter 3 on Components, there are many parts by various manufacturers that meet or exceed its specifications. For modeling, we can use the LTspice model of the MAX40077 (the MAX40079 is the single version), with a GBW also of 10 MHz and a slew rate of 3 V/μsec. We compare the MAX40077 with the TLV9062 datasheets, with the result shown in Table 4.6.

The main difference in our simulations is going to be that the MAX40077 draws 2 mA more current than the TLV9062.

As for the comparator, we usually start with the 339. This, too, isn't in LTspice, and it seems that they gave up making conventional comparators long ago. For a

Table 4.6 MAX40077 model vs. TLV9062 datasheet.

Parameter	MAX40077 datasheet	TLV9062 datasheet
Minimum V_{CC}	2.7 V	1.8 V
Maximum V_{CC}	5.5 V	5.5 V
Supply current at 5 V	2.5 mA	538 μA
Output high (from rail, 10 kΩ)	45 mV	20 mV
Output low	40 mV	20 mV
Slew rate	3 V/μsec	6.5 V/μsec
GBW	10 MHz	10 MHz

Table 4.7 MAX9095 model vs. LM339 datasheet.

Parameter	MAX9095 model	LM339 datasheet
Minimum V_{CC}	2 V	2 V
Maximum V_{CC}	6 V	30 V
Supply current	57 μA	200 μA typ.
Vios	3 mV	2 mV
I_B	30 pA	25 nA
V_{out} at 4 mA, low	117 μA	150 mV
Response time	100 nsec	1.3 μsec typ.

reasonable replacement model, we can try the MAX9095.[7] The most demanding comparator application we will have is in the bang–bang controllers. We compare the most important parameters for this application for the 339 and the MAX9095 model in Table 4.7.

We start by measuring the V_{CC} of the MAX9095 model. We configure the input so the output (open-drain) should be low, and sweep V_{CC} until it goes low, and then continue up. As we see in Figure 4.15, the output isn't fully functional until V_{CC} reaches 6 V; it works up to 6 V, and then the model fails.[8] Since we will generically be using 5 V as V_{CC}, this is acceptable, but you have to be wary of swapping this model in without ensuring that V_{CC} isn't too high.

7 Most of the other LTspice comparator models have complementary metal oxide semiconductor (CMOS) outputs rather than open-collector; this is another reason to choose the MAX9095. When different comparator characteristics are needed, we'll convert an LTspice comparator to an open-collector version by swapping the inputs and then following the comparator with an NPN inverter.

8 Note that this is almost the first device we've tested that has an upper voltage limit; the conclusion is that if it matters to your design, you have to independently determine it for every component.

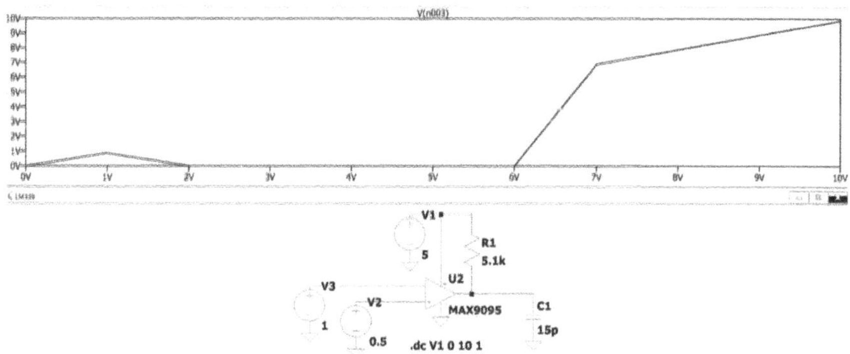

Figure 4.15 LTspice model of MAX9095 VCC.

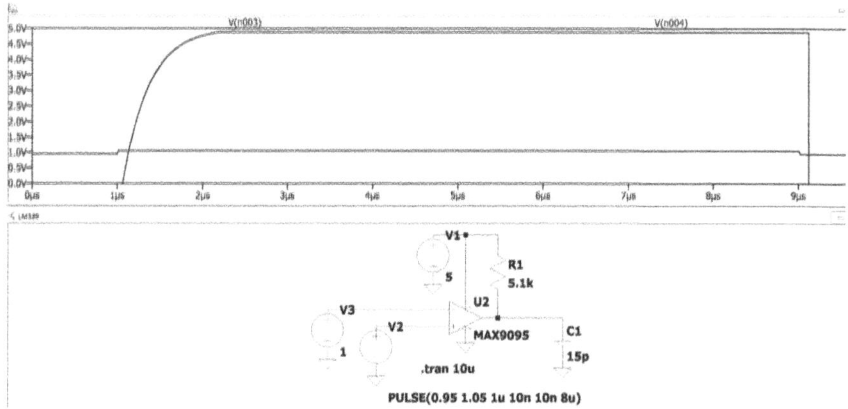

Figure 4.16 MAX9095 model response time.

We can use this same model to find the supply current at 5 V by removing the pull-up resistor. It turns out to measure 57 μA, rather lower than the LM339. This shouldn't matter.

Output low voltage at 4 mA can be measured by changing the pull-up resistor to 1.25 kΩ; it turns out to measure 117 μA, the difference being that the MAX9095 is open-drain, whereas the LM339 is open-collector. This too shouldn't matter.

We measure the response time by forcing the output low (since it's open-drain), as shown in Figure 4.16. Looking at the falling edge of the output (the rising edge depends in part on the pull-up resistor and the load), the response time is 100 nsec, substantially faster than the LM339.

Finally, we measure the input offset voltage by sweeping one of the inputs and measuring at what voltage (difference) the output is half of the pull-up voltage, as

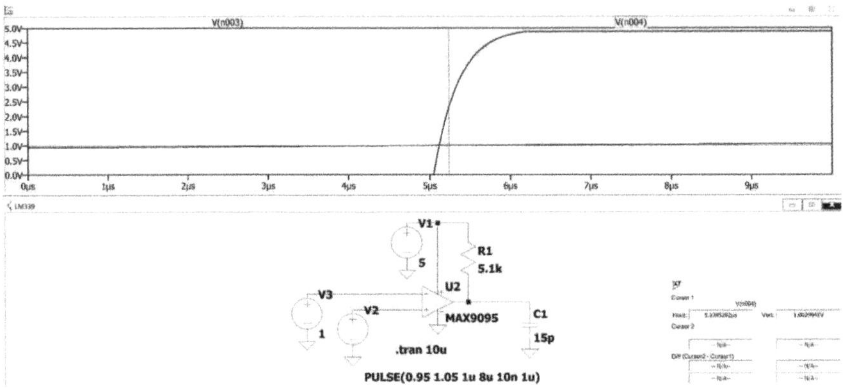

Figure 4.17 MAX9095 model input offset voltage.

shown in Figure 4.17. This happens at 1.003 V, and so the input offset voltage is 3 mV, adequately similar to that of the LM339. The input bias current is 30 pA, which doesn't matter for our circuits.

Finally, we turn to the 431, which as we already discussed above can be used as a voltage reference, but can also be used as an error amplifier. It is commonly so used on the secondary side of an isolated converter.

There's no model of a 431 in LTspice, although when used as a reference, such a model isn't necessary. Caution: *Neither the LT1009 nor the RH1009 function the same way as the 431*, even though their schematic symbol is identical; thus *they cannot be used as error amplifiers*![9]

To get around this serious deficiency, the author has constructed the following LTspice model that functions quite closely as a (1.24 V) 431. Looking at Figure 4.18, you apply the feedback voltage to the REF pin, and the ANODE pin sinks current.

The model finds the difference between the REF pin voltage and 1.24 V and amplifies it by 100. This difference is filtered, and controls a current sink. The voltage source in series with this is to prevent the ANODE voltage from dropping below 1.24 V; at 1 mA, the ANODE goes as low as 1.37 V, good enough for our models. Injecting a 1 kHz sine wave superposed on 10 VDC into the external supply, as shown in Figure 4.19 produces a 6 mVpp ripple, corresponding to 44 dB of rejection at this frequency, good enough for our models.

9 Perhaps the authors of the datasheets for the parts should not be too much blamed, as the symbol for the comparator and the operational amplifier are also the same, and yet are not used interchangeably.

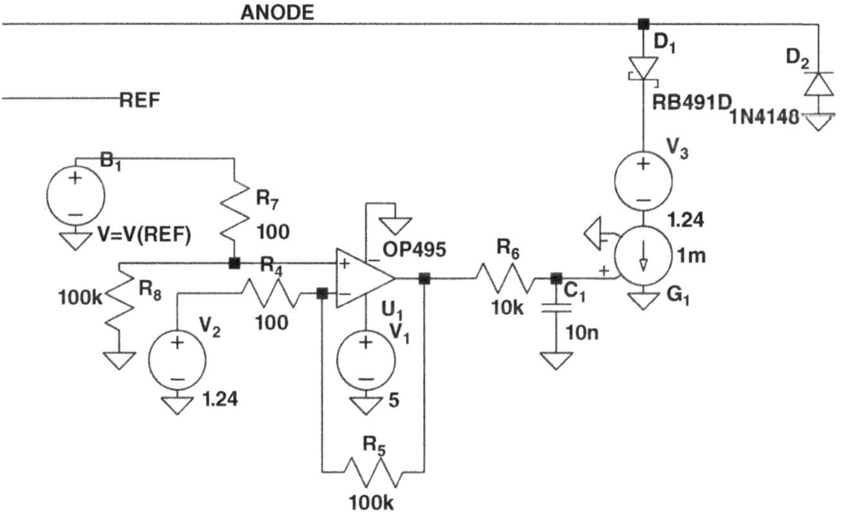

Figure 4.18 LTspice model of a 431.

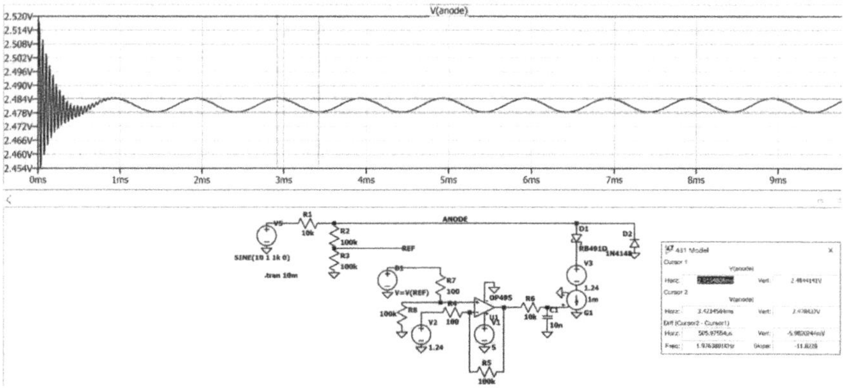

Figure 4.19 LTspice model of 431 rejection at 1 kHz.

The reader may note that the op-amp used as an instrumentation amplifier is powered from an internal 5 V. This is a convenience which the designers of the real part didn't have, but doesn't matter for our purposes.

The interested reader can put this into an LTspice model with the appropriate symbol. As mentioned above, the author prefers to keep the schematic on-schematic so that you know what you're getting.

4.5.6 Opto-couplers

As discussed in Chapter 3 on Components, we will most often be using the PC817A opto-coupler for crossing the isolation barrier in an isolated converter model, since LTspice doesn't have a 6N137 model. The three important parameters in actual usage are the gain, speed and temperature; the last is not something we look at in this book. In one particular application, we'll take a look at the opto-transistor $V_{CE,sat}$.

To test the opto-coupler model in LTspice, we build the model shown in Figure 4.20. A 1 mA source is pulsed on very quickly through the opto-diode, and the response is measured by looking at when and how much current is drawn from a 5 V source on the opto-collector to the opto-emitter, which latter is grounded.

Results are shown in Table 4.8. We don't compare the results with the datasheet, since the PC817 isn't an opto-coupler we would actually use. Its only importance is that it's a model that LTspice actually has. It can be seen that its gain is close to unity, and its response time is pretty slow. This will limit the speed with which the secondary in a bang–bang controller in an isolated converter can signal the primary, and thus limit the achievable peak-to-peak output voltage ripple. In practice, we will use the 6N137 opto-coupler, and we will show its datasheet parameters. Some of the parameters aren't specified on the datasheet, so we've used numbers from the same family on different datasheets. The $V_{CE,sat}$ test is run 10:1, that is, the opto-diode is run at 1 mA and the opto-transistor is run at 100 μA.

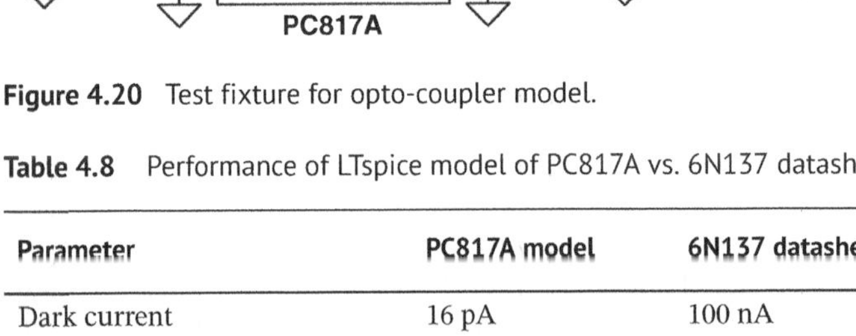

Figure 4.20 Test fixture for opto-coupler model.

Table 4.8 Performance of LTspice model of PC817A vs. 6N137 datasheet.

Parameter	PC817A model	6N137 datasheet
Dark current	16 pA	100 nA
Gain at 1 mA	0.89	7.0
Turn-on propagation delay	7.2 μsec	75 nsec
Turn-off propagation delay	2.5 μsec	75 nsec
$V_{CE,sat}$ at 100 μA at 10:1	107 mV	1.2 V

4.5.7 Logic Gates

A few types of logic gates are included in LTspice. The reader needs to be aware that their inputs and outputs are (literally) either 0 V or 1 V, so for many purposes this will have to be multiplied to get to the typical 0 V and 5 V output. (Inputs can be 5 V, as anything above 500 mV is considered a high.)

It turns out that the simulation of gates in LTspice is fairly slow, but that's just the way it is.

4.5.8 The 555 Model

We're going to be using in our examples three controllers, the 555, 3845, and 9910. The first one is easy: LTspice has an NE555 model. Since we will be using the 555 in many different SMPS designs, we need to compare the model's performance with that of the datasheet. This will include the frequency relationship set by the resistor-capacitor network, the rise and fall time of the output, VCC current and the output drive current. A summary of the results is shown in Table 4.9.

We start by looking at the switching frequency. Recall that we're using 555s that switch at least at 1 MHz, and the values we try are those mentioned in the datasheet as typically generating a 1 MHz switching frequency. It turns out that the NE555 model in LTspice does (unexpectedly!) run with these values, but the frequency is 2.5 times too high! Increasing the resistor values to those shown in Figure 4.21 (rather than increasing the capacitor value, in order to decrease the current drawn) reduces the frequency to 1 MHz. So, when you use the actual IC, use the values per the datasheet; when you model it, check to see that it's running at the correct frequency.

Using this simulation also allows us to see the rise and fall times, 32 nsec and 37 nsec respectively, somewhat faster than the actual device. Since our SMPS switching frequencies will be in the hundreds of kilohertz, the simulation should be OK.

Table 4.9 555 LTspice model vs. 555 datasheet.

Parameter	NE555 model	555 datasheet
Frequency 470 Ω, 270 Ω, 200 pF	2.5 MHz	1 MHz typ.
Output rise time	32 nsec	75 nsec typ.
Output fall time	37 nsec	75 nsec typ.
Supply current	1.6 mA	40 μA typ.
Output source current	250 mA	800 μA typ.
Output sink current	250 mA	3.2 mA typ.

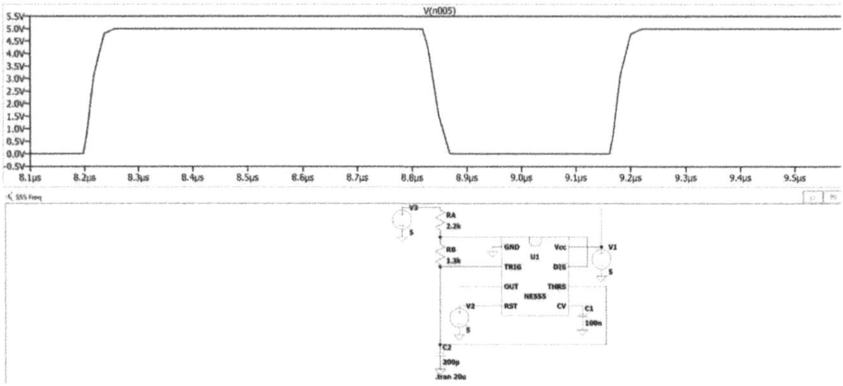

Figure 4.21 LTspice model of 555 astable frequency check.

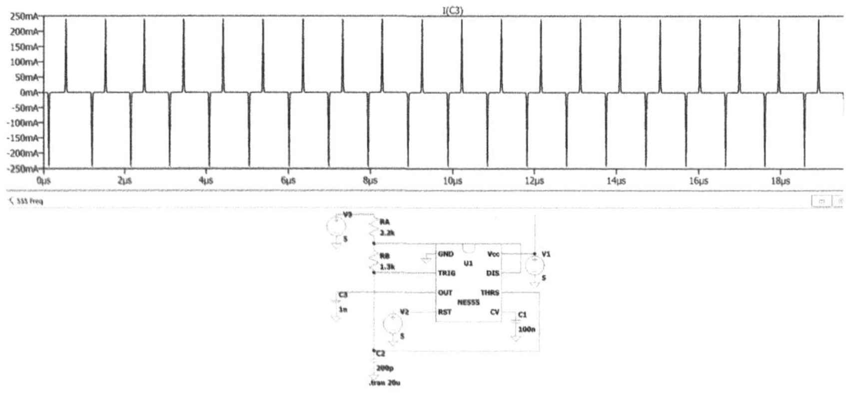

Figure 4.22 LTspice model of 555 drive current.

The simulation also allows us to measure the IC current, which is an average of 1.6 mA. This is hugely more than the datasheet, but in the context of SMPS, it really usually doesn't matter. In case it does, you can power the 555 model from a separate supply voltage, and add the 40 μA to the actual supply to emulate the current draw of the real part.

Finally, we look at the sink and source capability of the 555 output, as the specification is so low that it will routinely require a driver to make it suitable for our applications. We add a 1 nF capacitor load on the output pin so that it momentarily resembles a short. As seen in Figure 4.22, both the source and sink on the LTspice model reach 250 mA! This is wildly more than the datasheet specifies for the real part; be aware of the actual part's drive limitations when designing the gate driver.

4.5.9 The 3845 Model

The next one is straightforward too—if you know where to look. LTspice has the LT1242/3/4/5, which are close to the 3842 family. This multisource family of devices can be used for SMPS at less than 75 W, where PFC isn't required.

The four members of the 3843 family are the 3842, 3843, 3844, and 3845. The difference is the startup and operating voltage for the IC, and the maximum duty cycle each will support. Briefly, the 3843 and 3845 require only 9.0 V maximum to start (and a minimum of 8.2 V to operate), whereas for the 3842 and 3844, these numbers are 17.5 and 11.5 V. The maximum duty cycle of the 3842 and 3843 is 48%, while that of the 3844 and 3845 is 97%. The same numbers are true for the LT124x family.

We will be using only the 3845. It turns on by 9 V V_{CC}, and has a maximum duty cycle of 50%. Since it's an old part, its maximum switching frequency is only 300 kHz, whereas the 555 can exceed 1 MHz. It's also to be noted that it is explicitly a current-mode controller, though there is a way to run it in voltage mode.

We use a test circuit close to that shown in the 3842 datasheet, as shown in Figure 4.23. A summary comparison of the most significant of the model's and the datasheet's parameters for our designs is shown in Table 4.10.

The switching frequency is almost exactly the same (bearing in mind that the parts of this family that have 50% maximum duty cycle do so by cutting the frequency in half). Rise and fall times are much faster than the datasheet, but this won't matter for our circuits. Supply current is radically less than the datasheet, though both numbers are low enough to not have a significant effect on efficiency for most converters; it may, however, be important if you need to source the current from the AC line, in which case you can add the extra current from your source. Finally, to measure sink and source current, we add a capacitor at the output. Peak current is 3.2 A, though, of course, this is not a recommended operational mode.

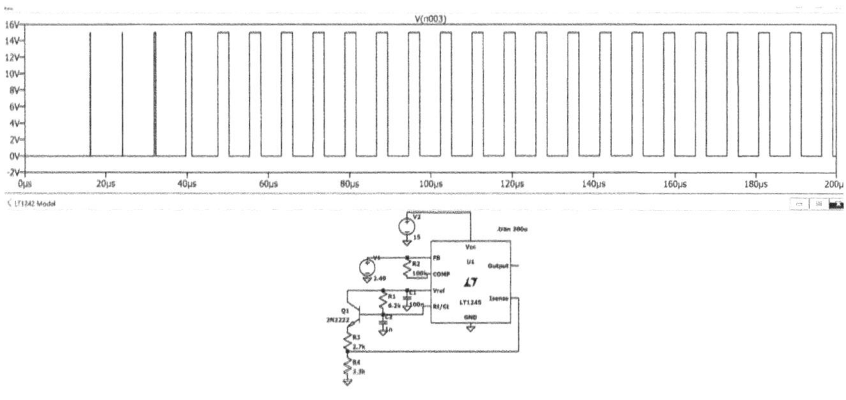

Figure 4.23 LTspice LT1245 model.

Table 4.10 LT1245 LTspice model vs. 3845 datasheet.

Parameter	LT1245 model	3845 datasheet
Frequency 6.2 kΩ, 1 nF	128 kHz	125 kHz typ.
Output rise time	5 nsec	50 nsec typ.
Output fall time	8 nsec	50 nsec typ.
Supply current	237 μA	11 mA typ.
Output source current	3.2 A	200 mA typ.
Output sink current	3.2 A	200 mA typ.

Considered overall, the LT1245 seems a reasonable model for the 3845.

4.5.10 Synchronous Gate Driver

We chose the IR2213 as a typical synchronous gate driver for a buck converter using both a high-side and a low-side MOSFET. As a substitute part for modeling, we propose the LTC7063 model in LTspice.

The big difference between the LTC7063 and the IR2213 is, of course, their boost voltage rating, the former doing only 150 V and the latter 1200 V. For some purposes, this won't matter. For high voltage inputs, though, this is a crucial failure in the available LTspice models, as, unlike other LTspice models, the LTC7063 doesn't work beyond 150 V. We will, in the appropriate spot, show a kludge to get around this.

Otherwise, the characteristics of the LTC7063 model are close enough to those of the IR2213. The peak drive currents of the two are somewhat different, but this relates to transition losses; since our frequencies are not super-high, this won't matter too much.

One thing that is worth paying attention to in the LTspice model is that the MOSFET model apparently includes the drain-gate capacitance of the MOSFET and the finite impedance of the LTC7063 driver. As seen in Figure 4.24, which is the LTC7063 datasheet test figure, the very high-speed drive of the LTC7063 produces a spike of current through the drain of the MOSFET. Although it has an amplitude of some 30 A, it's present for only a couple of nanoseconds. The reason this is important is that at higher voltages, the average current lost this way can be 15 mA or more; this can contribute non-negligibly to the input power to the converter. If the driver that you end up using has a slower rise time and lesser drive current, the losses in your circuit may be less.

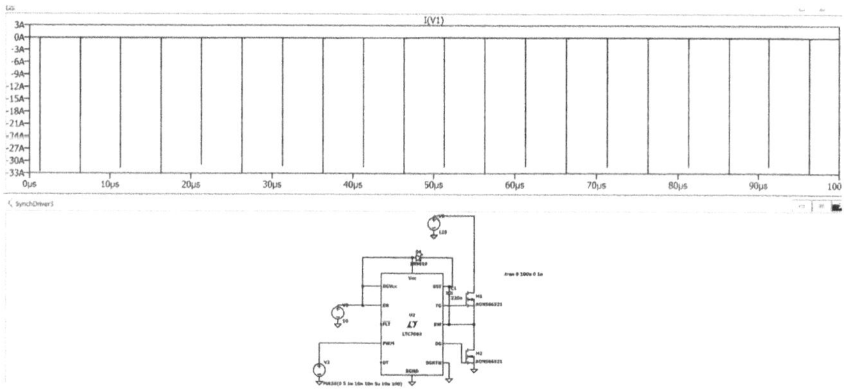

Figure 4.24 7063 model test fixture.

4.6 Modeling the 9910

We will be using the 9910 to build a PFC power supply. However, LTspice has no model of a 9910. We thus have to build our own, and that's why this IC is in a separate section from the other ICs. The construction of a model for the 9910 will give insight into how to build a functional-level model of an IC, without knowing or needing to know transistor-level details.

The first thing to know is that the intention is to build a minimal equivalent circuit of the 9910, good enough for our purpose of using it in a SMPS. The components we're going to use don't need to resemble what's actually inside the IC (which we don't even know), indeed they don't need to be real components at all. All that is needed is that the model act *functionally* the way the 9910 does in our application, and hopefully not simulate too slowly.

As the block diagram and the typical application of the 9910 from its datasheet show in Figures 4.25 and 4.26, the 9910 is a constant frequency IC. The frequency is set by the RT pin, with the formula

$$f(\text{kHz}) = \frac{25{,}000}{R(\text{k}\Omega)} \tag{4.2}$$

So the output is turned on at a constant frequency. It is turned *off* when the current sense pin reaches 250 mV. There is a small blanking time to prevent the gate current from being perceived as drain current by the current sensor. And finally (for now), the gate drive is at 7.5 V; later, we will deal with the other control pins. We're going to ignore the high-voltage LDO inside the 9910, that doesn't affect the operation.

So let's look now at the first-cut LTspice model we've built of the 9910, shown in Figure 4.27. The 9910 itself is shown inside a double box (which could ultimately

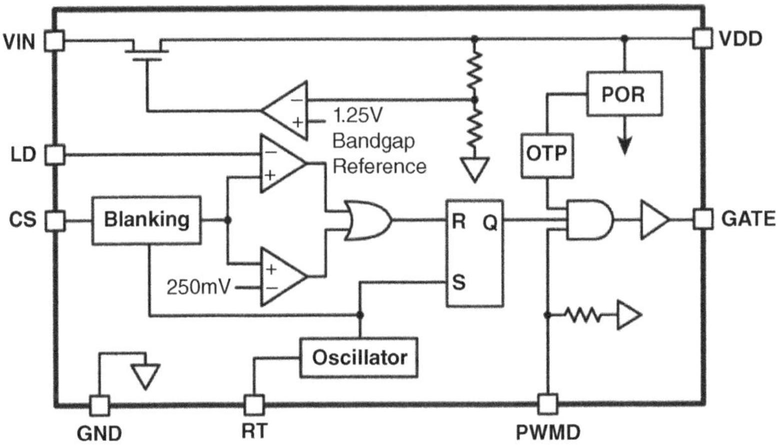

Figure 4.25 9910 block diagram.

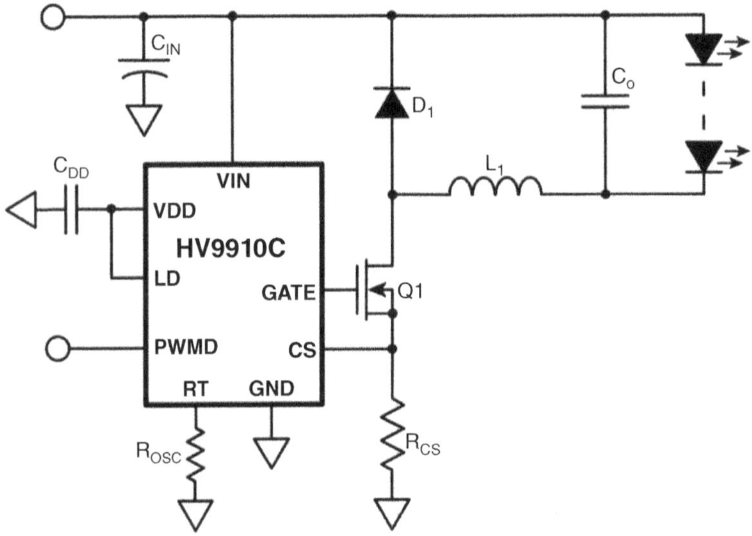

Figure 4.26 9910 typical application.

be encapsulated in a hierarchical model, though we don't intend to do so). The power stage is to the right of the box.

The external power stage is sourced by V_2, arbitrarily set here to be 10 V. This feeds a standard boost converter consisting of a 33 μH inductor, a MOSFET and a Schottky. None of these are robust parts; it's just a text fixture for our 9910 model. The source of the MOSFET goes to a current sense resistor, which feeds back to

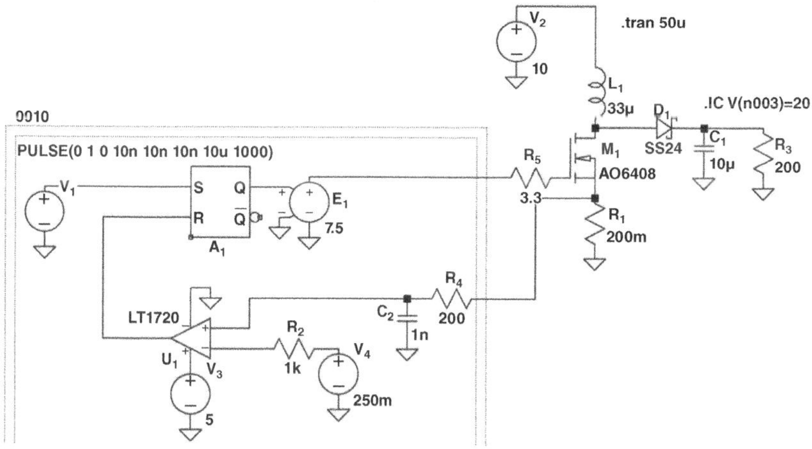

Figure 4.27 9910 model, first-cut.

the 9910 pin, and the gate of the MOSFET is driven from the IC through a gate resistor (necessary both in real life and in simulations to prevent high-frequency ringing). The output of the diode is a smoothing capacitor, plus a resistive load.

Now we can turn to how this power stage is driven and controlled by the 9910 model. A switching cycle is started at a regular period of 10 μsec by V_1. We haven't bothered to use an external resistor to set the frequency, it's just "hard-wired" in to V_1. It would be straightforward to include the frequency-setting resistor with a **B**-source.

V_1 is programmed to be a pulse generator with fast rise and fall times and a short on-time, as its function is to turn on the SR flip-flop A_1 without affecting the minimum on-time.

The Q output of the flip-flop is multiplied by 7.5 by the voltage-controlled voltage source E_1 because the outputs of LTspice digital components are 1 V. This corresponds then to the 7.5 V gate drive voltage of the 9910.

When the MOSFET turns on, inductor current starts to ramp up. (We are here assuming discontinuous-conduction mode for explanatory simplicity, though continuous-conduction mode would also work.) The inductor current flows through the MOSFET and then through the current-sense resistor to ground.

The voltage from the current-sense resistor is filtered by an RC with a 200 nsec time constant, corresponding more or less to the blanking time of the 9910. (A real blanking time could of course be implemented, but the RC seems to work well enough.) This filtered current-sense voltage is compared to a 250 mV reference with a very high-speed comparator. (The resistor between the reference and the comparator is for later when we add in the dimming pins.) The comparator is chosen simply by finding an LTspice model that is fast enough not to add significant propagation delay in the loop, it doesn't need to match what's inside the 9910.

Finally, the comparator feeds the Reset of the SR flip-flop. So the cycle, to recapitulate, is: the flip-flop is turned on by the pulse generator, which through the gate driver turns on the MOSFET and causes current to ramp up. The current is sensed and filtered, and when it reaches 250 mV, the comparator goes high. This turns the flip-flop back off, until the next switching cycle.

Does it work? We look at Figure 4.28, showing the current-sense resistor current and the Schottky current, which added together, make up the inductor current. The rising part of the current is the inductor current; the falling part is the diode current. The negative spike between the two is 50 nsec long and corresponds to the turn-off of the MOSFET and the turn-on of the Schottky. The current spike at the turn-on of the inductor ramp is something like 35 nsec, and corresponds to the gate drive current.

While we're on the topic of the gate drive, you'll notice that there's no explicit impedance of the gate drive in the model. The datasheet implies a resistance of something like 7.5 V/165 mA = 45 Ω, but this seems much higher than performance with the actual part suggests. For the time being, we simply leave it with no internal resistance, and limit the current by an external gate resistor. If the rise and fall times turn out to be too slow on the actual hardware, you could a totem-pole gate driver to the output.

We still need to add in the two types of dimming control, both of which we will be using in our SMPS. The linear dimming (LD) pin works by reducing the current-sense threshold. If it's set to some value lower than 250 mV, that overrides the internal set-point; if it's set higher than 250 mV, it has no effect. We can model this

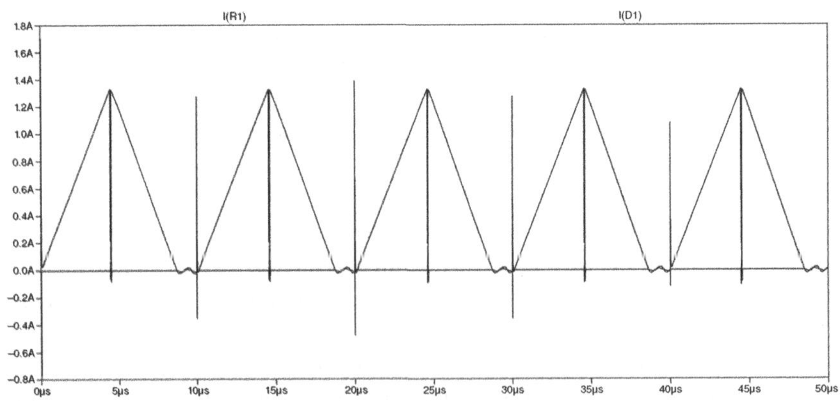

Figure 4.28 9910 model inductor and current-sense currents.

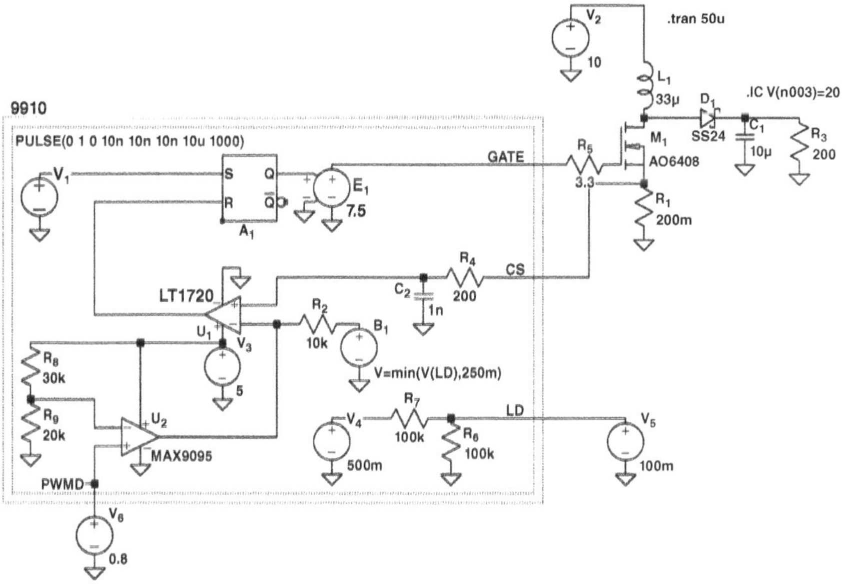

Figure 4.29 9910 model, final, with representative load circuit.

by changing the 250 mV reference to a **B** source with a "min" statement, which then senses the voltage applied to an internal 100 kΩ resistor as per the datasheet. The 500 mV voltage source and additional 100 kΩ resistor produce 250 mV on the LD pin if the LD pin is left open. The extra resistor is to prevent the two voltage sources from fighting each other, or, in the actual usage case of an op-amp controlling the LD pin, from it being broken.

Finally, the pulse width modulation dimming (PWMD) pin either turns the current sense completely off or has no effect, and is intended to be a TTL signal with "high" being normal operation. We use a (open-drain) rail-to-rail comparator set at 2 V to pull down the 250 mV reference, which turns off the gate drive. (We're not bothering with the low-end of TTL, since an intended operation will pull it to the ground.) The final 9910 model is shown inside the rectangular box in Figure 4.29.

4.7 Isolation

LTspice has a native model of a transformer. You can put multiple windings on it, put the connection dots on to get the proper polarity, and even set the coupling constant in order to simulate leakage inductance. You can put resistors in series with the windings to find the windings' copper loss. The only thing missing is the power loss in the core—which no simulator yet really has a handle on, anyway. What's not to like?

Since we thus already have a transformer model, how about isolation? Given that LTspice is, after all, SPICE, it shouldn't be surprising that you can only have one ground. But (at least in the modern version of LTspice) you can have a completely isolated circuit, not connected to ground in any way. So, this way, you can have an isolated secondary of the transformer. Of course, you can still add in an isolation-barrier crossing capacitor, as is frequently necessary for common-mode noise.

The other thing that crosses the isolation barrier is the feedback, which is done with an opto-coupler. We are using the 6N137, which isn't in LTspice, but as we mentioned in the chapter on components, the gain isn't critical in the design; if it was, the design might not work properly in practice either. So we can just use one of the opto-couplers that are already in LTspice. Our preference is the PC817A, as it doesn't have the base of the opto-transistor pinned out, though you might instead choose one of the series with a gain range close to what you are actually using in hardware.

One more note about isolation: if you have a separate ground system, or if you connect the secondary ground system to the primary through a high-value resistor, the simulation can go *very* slowly. This is especially true if the secondary side is complex. So, while in our examples, we will sometimes show the secondary ground as isolated from the primary, other times we will simply use the primary ground, even though the two grounds are actually going to be isolated. In these cases, we still use the opto-coupler across the isolation barrier, and so the simulation proceeds just as it would if the actual isolated ground were being used.

4.8 Modeling AC Power

For most of the AC power we will be using in this book, we use a two-phase system, which is two-wire, hot and neutral. Both of the wires carry current; what goes in on the hot comes back out on the neutral. (Bearing in mind, of course, that the current is AC.) Refer to Figure 4.30 for an overview of the different types of AC power.

However, there is also such a thing as a three-wire two-phase system, with the addition of an earth wire. Power is still carried by the hot and neutral; earth is there for safety and doesn't, under normal circumstances, carry any current.[10] Don't be confused between neutral and earth! Earth is ultimately supposed to be a length of copper buried in the actual planet, nothing to do with power.

In the examples of this book, you won't see an earth. Let's be honest, that's often the way things are actually built. The earth (green) wire, if present at all, goes to the box and the PCB of the power supply, but not to any components, or at most only to an MOV.

---

10 In fact, it's a sign of a serious problem if Earth *does* carry current; that's called a ground fault.

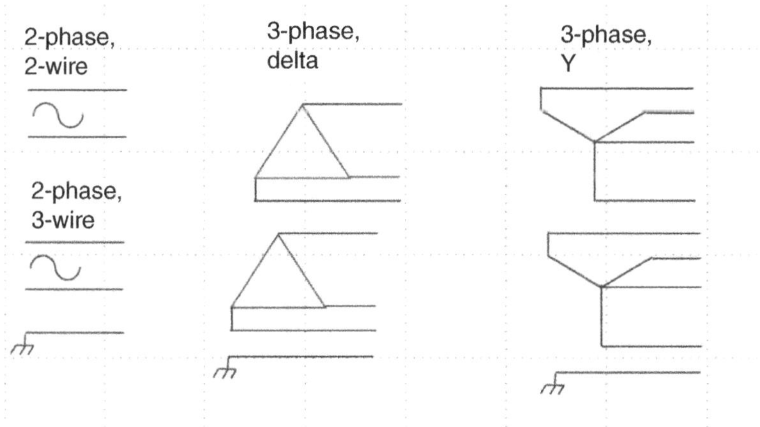

Figure 4.30 Different AC power systems.

At higher power levels you'll come across three-phase systems (we have an example using three-phase power in the PWM chapter). There are two different kinds, "Y" and "Delta" (refer again to Figure 4.30). The "Delta" system has three wires, each of which carries power 120° out of phase with the others. If there is an earth, then this is a four-wire system.

The "Y" type of three-phase power has, in addition to the three power lines, a neutral, the midpoint of the three power-carrying phases. Even though this is the same name, this is different than neutral for a two-wire system, because it is not supposed to be carrying current. But it is not the same as earth, either. So this is a four-wire system also, unless earth is present, in which case it is a five-wire system.

4.9 Modeling EMI

Although details of modeling electro-magnetic interference (EMI) are presented in the relevant examples (in Chapter 8 [PFC Converters]), some general comments are in order here.

There are two types of EMI: conducted and radiated.[11] Radiated depends very strongly on the physical layout of your components, the box they're in, the wiring etc. So in this book, we are only interested in conducted EMI, which can be controlled with discrete components (and physical layout, once you get that far).

11 Actually, there's also a third, EMV, or electromagnetic vulnerability, which is how resistant a unit is to EMI generated by other devices. We won't deal with that in this book.

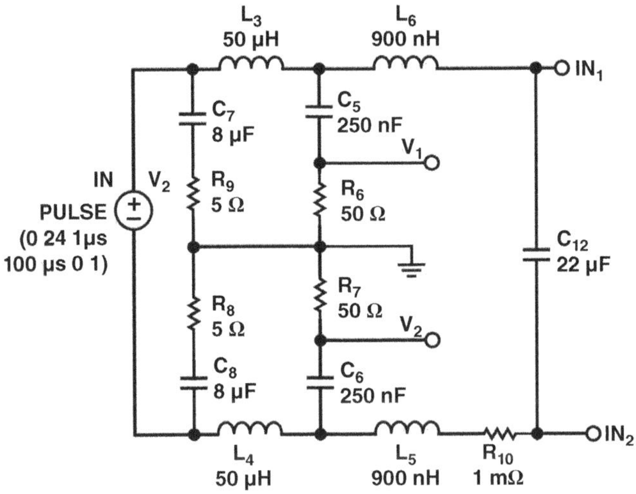

Figure 4.31 LISN model.

Conducted EMI in turn has two types: common-mode and normal-mode. Normal-mode conducted EMI is noise that is present on one of the power lines but not the other (or others plural, in the case of a three-phase system). Common-mode conducted EMI is noise that is present identically in both of the power lines with respect to neutral (or earth). If there are only two power wires coming in, there *can't* be any common-mode noise by definition.

When measuring conducted EMI, labs use something called a LISN. Correspondingly, when we want to simulate conducted EMI, we need a model of the LISN. Such a model is shown in Figure 4.31, following [17]. To be absolutely clear, the actual input (120 VAC or other) should replace V_{in} in this model; In_2 and In_2 are the hot and neutral lines that go to the input of your power supply, presumably the EMI filter and MOV. The voltage between V_1 and V_2 is the EMI measurement.

This measurement has to pass EMI specifications, which are shown in Figure 4.32, again from [17]. Since the line impedance stabilization network (LISN) model shows an earth ground, this figure shows common-mode and normal-mode noise separately. For the examples in this book, there is no earth, and consequently (while the limits are the same) the thing to plot is 1,000,000*$V(v_1,v_2)$.

In the US, EMI requirements start at 150 kHz, while elsewhere they start at 10 kHz, so what frequency you start at is determined by the input voltage range. Furthermore, in practice, noise above a few MHz is typically caused by radiated noise getting on the power lines. Thus, in this book, we will be looking at the noise spectrum only from 150 kHz to 1 MHz from 120 VAC and 277 VAC lines, and from 10 kHz to 1 MHz from 230 VAC and universal input lines.

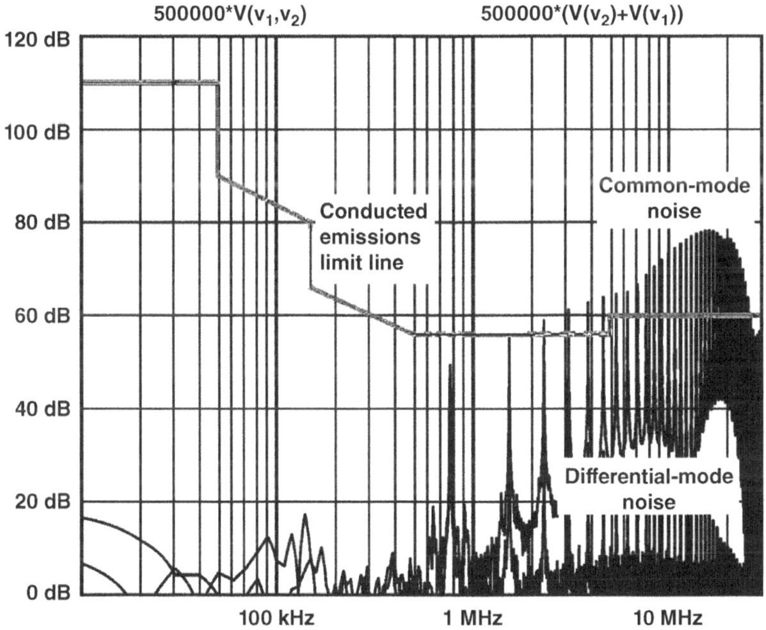

Figure 4.32 EMI limits.

4.10 Spice Directives

So far in this chapter, we've been discussing the models you should use to get the data that you want about your circuit. However, there are some remaining challenges in the area of actually running a simulation.

Let's start by talking about LTspice Directives, which are additional information you have to put on the LTspice schematic (beyond the information about what type of simulation you have to run). The most common of these is of the form ".IC V(n003)=5," which tells LTspice that there is an initial condition, namely that the voltage at node 3 is 5 V when a transient simulation is started. You can put multiple such statements in a single .IC. The author finds it convenient to place the .IC statement close to the node to which it refers. It's considered a good practice to give the net a name rather than "n003," so that the .IC directives can refer to the net that way, and don't have to change when you add additional nets.

We should also mention "K" which gives the coupling between two (or more) windings on a transformer as in "K1 L1 L2 1." Putting this statement on the schematic is the only way to tell LTspice that the various inductors cited (L1 and L2 in this case) are on the same core; it also conveniently shows the winding dots on the transformer schematic.

4.11 Controlling Spice

This is the ugly section, about the Settings used to control the simulation. SPICE has always had esoteric controls that were never brought out of the realm of the specialist. And even all of these years later, LTspice still has a settings control panel with ten tabs, each full of options. How on earth are you supposed to choose?

A good starting point, of course, is just to accept the default settings. This works well for a lot of analog circuits ... but sometimes not so well with SMPS, because of their stiffness. A simulation that has devices that change state very rapidly can cause accuracy errors. And simulating things very accurately for a long simulation time can cause errors to creep in.

The first step to address such issues is to choose carefully the parameters of your ideal devices. It seems unlikely that your voltage source needs to turn on and off in 1 ps; surely 10 ns is both adequate and more realistic? And whatever it's running, can it respond that fast? Taking care of things like this has a significant impact on the stiffness of your model, and thus on the overall simulation time and generated file size.

Once such things are taken care of, the most important controls in the Settings are on the "SPICE" tab. It's tempting to start changing things: maybe the waveforms will look better if I decrease Gmin by a factor of 1000? Maybe Gear will converge better than Modified trap? But the reality is that for almost all circuits, it's the author's experience that if there's something that causes simulation problems, it's more likely to be your circuit and its parameters than the simulator. The defaults are usually just fine.

There is one exception to this in the Settings. If the simulator seems to produce waveforms that, when zoomed in on, don't seem smooth, it may be that you could try changing the Reltol setting, perhaps dividing it by 10. This usually won't slow the simulation down *too* much. Don't bother with decreasing Reltol by lots more orders of magnitude. If you're still having issues, again, it's probably your circuit and its parameters.

There's one final thing you can try to improve the waveforms. When you enter the Configure Analysis page and go to the Transient tab, there's an option to set the Maximum Timestep. If you pick something ridiculous like 10 ps, your simulation will take weeks. But maybe something like 10 ns might help you get a little better resolution.

So far we have only talked about getting more accurate simulation results. But many of the things we mentioned will result in larger file sizes, which seems natural if you want better resolution. And file sizes are just naturally going to be huge for simulations that have lots of fast devices running for a long time, which is to say for power supplies.

On the same Transient tab, there is an option where you can tell LTspice when to start recording data. If, for example, you have a start-up transient, you can tell LTspice to wait until it's finished before it starts saving the data. While this will reduce file size, it's been the author's experience that when you do this you always end up wanting to know what happened *before* the file saving started, and you end up re-simulating the whole thing.

One last control thing: there's (at least sometimes) no need to run more than one switching cycle. Setting the simulation up to run for only the Stop Time that you need to get the information you want can save a lot of your time.

4.12 Things to Watch Out for When Modeling

We've already mentioned some of the potential issues with modeling in general and with LTspice in particular. To recapitulate:

- Not all models are created equal. Using anything other than a primitive (**V** and **B** and so on) requires some care. Even ideal models (**L** and **R** and so on) have potential issues. In particular, the **L** model has a default winding resistance of 1 mΩ. If you're planning on passing 100 A through the inductor, you have to be sure that you change it to a better value. The strongly preferred method is to *always* set it to zero and to add a series resistor.
- Device models come with no guarantees. MOSFET models frequently don't break when their drain-source voltage is higher than the datasheet claims is possible, and the same for diodes. IC models sometimes work at higher voltage than the datasheet allows. In general, if you're going to use a device model, you need to test its performance on the parameters you care about before using it. It's a nuisance, but the alternative is having a simulation that doesn't match the lab bench.
- If you're trying to measure power, be aware that LTspice (and presumably other simulators) measure voltage times current. This sounds right, but that also includes capacitors. So, if, during the simulation, your capacitor ends up at a different voltage than it started at, this will be reflected by more input power from the source during the simulation. This can mess up your efficiency measurement. You can, of course, simply type into the x-axis the desired start and finish times. But given the difficulty of picking precisely the right amount of time to measure power, the recommended plan is, instead, to measure the power losses in the components that can dissipate power (resistors, MOSFETs, etc.) and add them up. It's a nuisance, but beats getting the wrong answers.

4.13 Notes for the Hobbyist

The hobbyist can do no better than to download a copy of SPICE, preferably a free one. We consistently use LTspice in this book, and to the author's taste, it has the best user interface. If you're trying to design a power supply, this should be your first step.

Or rather, based on a lot of experience by the author, it should be your *second* step. Your first step ought to be to write a specification. Not, of course because you don't know what you want, but because even professional engineers often get it wrong. Writing down what you want makes you check the details. You want 12 V to run the op-amp? If you write that into your specification, it should cause you to go look at the op-amp's datasheet, and perhaps discover that it only runs on 5 V.

Having written down a few things about what you want your power supply to do and downloading SPICE, your next step is to put a design into the SPICE GUI. The list of problems commonly found when building a simulation model in the earlier section on Robust Simulation is well worth reading; here, the author presents it again:

1. Don't try to throw a schematic directly onto the simulator's GUI. Draw it out by hand on a piece of paper first, complete with values, just the way you would if it were going to a schematic capture person. And actually *design* it—don't just throw in a value with the thought that you'll figure out what it should be as a result of what your simulation shows.
2. Don't put off entering the values—do it for each component as you place it, rather than all at the end.
3. Place the components in a reasonable fashion: avoid lots of line crossings. Sometimes, the GUI will connect lines together that aren't supposed to be connected. And in any case, it's confusing.
4. Try to avoid using models you found on some non-vendor website. Who knows how much they've been tested (if at all)?
5. While it probably sounds antediluvian, use primitive models instead of vendor models when possible. Vendor models are typically encrypted, so you can't see what their limitations are. Spice primitives certainly work, so if all you want to do is make sure that the comparator switches states, use an ideal comparator rather than the vendor-provided high-speed, low-offset comparator that has dual outputs and bias pins. In fact, why not use a **B** source? Using lower complexity parts will also simulate way faster when you have 200 of them.

Let's add a few more, now, about the actual simulation.

1. Don't mess with the Settings on the "SPICE" tab. It's a reasonable possibility that if your simulation crashes, you've done something wrong, not the simulator.

2. If the simulation takes a long time and generates huge files, it may be that your circuit is stiff (things happen on very different time scales). Try a state-space average model instead. Of course, the first thing to check is that you haven't done anything wrong.
3. Look at all of the node voltages and currents. You should know what you expect to be there, and you should be seeing approximately what you expect. If not, there's something wrong.
4. Finally, remember that what you simulate isn't necessarily what the circuit actually does. You *have to* go build it.

The only way to test your model is with an oscilloscope. The old ones from the 1960s are almost worthless; don't even bother getting one from a flea market (even if it's in working condition, which is doubtful). There are now brand-new oscilloscopes with decent performance available on the market for under US$100. You're going to need it; just break down and get one.

What happens if you need to measure voltage, that's higher than your oscilloscope and/or probe will allow (for example the output of a PFC)? You can do the same thing your circuit is doing, and measure it divided down by a resistor divider. Voltage too small (for example, ripple voltage on a 1 V bus)? You can include an op-amp with a gain of ten on the board and measure that. Just be sure it has adequate GBW for what you expect to measure.

There don't seem to be any network analyzers (machines used to test control-loop bandwidth and phase margin) at an affordable price. This can be a problem even for professionals. One option might be to rent one. They're typically 10% of the purchase price for one month, though that's still quite expensive.[12] A better plan is to build a step-load test fixture. If it doesn't oscillate, it's pretty much OK and has decent phase margin. With a little effort, you can actually estimate the bandwidth by looking at the step response on your oscilloscope.

12 Unless you plan to eventually buy it anyway. You may be able to get the rental price deducted from the eventual sales price.

5

Low Dropout Regulators

Linear regulators are circuits intended to provide a constant duty cycle (DC) voltage from a (higher voltage) source, which can be either DC or DC with an AC component. **L**ow **d**rop**o**ut regulators (LDOs) are a particularly common variety of linear regulator, where, as the name suggests, the required minimum voltage (headroom) between the input and output is small (typically 2 V or less).

There are a number of robustly available IC LDOs. In this chapter, we will focus on constructing LDOs for situations in which robustly available LDOs are not available, either because there are no multiple sources for them or else because there are no sources at all. In particular, in our examples in this chapter, we will design a series of robust LDOs, including ones that: operate with only 50 mV of required headroom; operate at an output voltage of only 500 mV; operate at 76 V output; and one that regulates a negative voltage.

The basic idea of an LDO is to compare the actual output voltage with the desired output voltage and adjust the current going through a transistor to the output in such a way that the difference between the actual and the desired voltage is zero.

In our designs, the desired output voltage will be set by a voltage reference, typically a precision reference. The difference between the desired and actual output voltages will be determined by an op-amp circuit (known as the error amplifier), configured as a subtractor with a certain (possibly frequency-dependent) gain. Finally, the current will be adjusted by controlling the gate voltage of a metal-oxide semiconductor field-effect transistor (MOSFET) (or the base current of a [bipolar junction transistor] BJT) by means of the output of the op-amp and potentially a low-impedance gate/base driver.

The application for LDOs is typically to produce a modest amount of current at a specified voltage from a higher input voltage. The limitation on the current is because the LDO's power dissipation is proportional to the current; at too high a load current, the LDO will throw away too much power, and a switch-mode power supply (SMPS) may be a better choice.

One other application in which LDOs are occasionally used is as a post-regulator for a SMPS. The SMPS has switching noise, which shows up as voltage ripple and spikes on the output capacitor. Since LDOs don't switch, they are far less noisy, so sometimes a SMPS is designed at a voltage slightly higher than the load requires, and then an LDO is used to linearly regulate the voltage down to the desired voltage with the noise removed.

5.1 LDO Specifications

LDOs have numerous possible specifications, as detailed in the subsections of this section. In many cases, only a subset of these is required for a particular application.

5.1.1 Input Voltage Range

The first consideration when designing (or selecting) an LDO is what the input voltage range is.

The maximum input voltage is important because this determines the type of components used. If the input voltage goes up to 12 V, for example, you can't use an LDO that is rated for 5 V input maximum. In the case of discrete LDOs, this maximum voltage determines what voltage the transistor should sustain input-to-output. The components internal to the LDO also need to be rated to take this, unless the voltage is very high, in which case there may be an internal regulator (such as a 431 shunt regulator) to power the internal devices.

5.1.2 Output Voltage

Obviously, the output voltage needs to be specified as well. This is usually a fixed voltage, but it may be that the circuitry being powered by the LDO hasn't decided exactly what voltage is required, or it may change as new circuitry is developed, and so it may want to be changed later. In this case, the LDO needs to be designed to be adjustable until a decision is reached.

A less common possibility is that the LDO is supposed to be adjustable even in production, in which case some mechanism must be provided to either change the reference voltage or the feedback divider.

5.1.3 Tolerance

For fixed-output LDOs, the specification should state the tolerance of the output. ±2% is not uncommon and doesn't require any special design consideration. We show an example of this.

Tolerance as tight as ±1% can be achieved without heroic efforts. However, it doesn't hurt to question the end user about whether a tight tolerance is actually necessary, as sometimes it is simply written down so the user doesn't have to think about the effects of a wider tolerance.

5.1.4 Output Current

The output current of the LDO has to be specified as well. Furthermore, in addition to a resistive load, there is often a capacitive load, at least that of the LDO itself. Sometimes the capacitive load can be quite large, and this should be in the specification, as it could affect stability or startup. Once in a while, there is even an inductive load.

Usually the minimum current for an LDO is zero, though not always; even when the specification states a nonzero minimum, it is probably prudent to design the LDO so that it doesn't have a large output voltage when it is unloaded, even if it doesn't necessarily meet all of the other specifications in that condition.

The maximum current also has to be specified, as this, along with the voltage drop across the LDO, determines the power dissipation and thus the size of the pass transistor. Of course, if the output current is large, a SMPS should be considered instead.

What happens if the load current exceeds the specification's maximum? While it requires a little bit of extra circuitry, it may be desirable to add in a current limit, perhaps 20% higher than the maximum specified current. Some hysteresis should be added as well so that the output doesn't just sit at 20% too high current. The final chapter of this book discusses the design of an overcurrent limit for discrete LDOs.

Finally, what happens during turn-on of the LDO? If there is an output capacitor, especially a large one or ones external to the LDO, the LDO will try to quickly charge them up. If there is a current limit on the LDO, it will simply hit the current limit for a little while until the capacitors are charged. If there isn't a current limit, a soft-start circuit can be used instead. A typical way of doing this would be to let the internal reference voltage come up slowly, with an RC setting the speed.

5.1.5 Power Loss and Efficiency

Power loss and efficiency are typically not *required* as separate specifications, as they follow from the input–output voltage difference and the current. However, the maximum power loss has to be calculated in order to select the appropriate package for the pass transistor, and both it and the efficiency are worth putting into the specification so that everyone can consider the effects of using an LDO rather than a SMPS.

5.1.6 Temperature Rating and Tolerance

Another consideration is the temperature rating of the LDO. In addition to the self-heating of the pass transistor, the ambient temperature in which it operates may be elevated as well, and these two have to be added together to find the maximum temperature of the components.

Indeed, more sensitive than the pass transistor (which is likely to be rated at least 125 °C), the other components inside the LDO have to be rated to take the ambient temperature. Of special concern are the other active components, such as the reference voltage and the op-amp. You need to leave some temperature margin on these components to ensure adequate mean time to failure (MTTF) (which is to say, a 70 °C rated part should not be operated at 65 °C).

5.1.7 Stability

There are two types of stability, both of which need to be considered in the design of an LDO.

Small-signal stability of an LDO (and also of other systems such as SMPS) refers to the LDO not breaking into sustained oscillation when there is a change of operating point. It also determines over- and undershoot when the operating point changes. It is measured by injecting a small (ideally infinitesimal) AC signal into the control loop and then measuring the response when the frequency of the signal is swept.

A typical specification will be the bandwidth of the LDO (which is where the rejection gain is 0 dB) and the phase margin (how much phase there is at the bandwidth frequency). The latter should typically be 45 °C, or 30 °C in the worst case.

It can be the case that large capacitive loads lead to lower phase margin and eventually oscillatory instability. You should check the design with the maximum possible output capacitance to be safe.

5.1.8 Step Response

The other type of stability is large-scale. This refers to the operation of the LDO reaching limits, such as the error amplifier reaching one of its rails or being slew-rate limited. This can be checked by placing a step load on the output of the LDO, typically switching between zero and full load. If the LDO doesn't oscillate, it's large-scale stable. If the LDO is large-signal stable, it is also small-signal stable, but not vice versa. If it takes a long time for the LDO to settle back into stability after a load step, the small-signal phase margin may be inadequate.

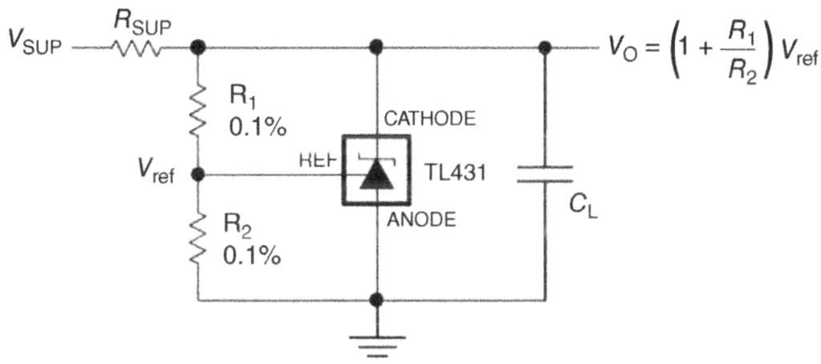

Figure 5.1 431 LDO typical application.

5.2 Shunt Regulators

Before learning how to design LDOs, let's mention two shunt regulators that can be used as LDOs, the 431 and the 4041, both of which we listed as multi-source.

A typical application is shown in Figure 5.1 from the 431 datasheet; the 4041 is similar. The best way to think about this is that the current through the resistor R_{sup} splits into two parts: the amount that is needed for the output and the rest into the 431 (the feedback resistors' current is typically negligible). The resistor thus needs to be sized such that the total of the two currents can be sourced by the input minus the output voltage. This means that this amount of current will *always* be going through the resistor, determining its power dissipation even when there's no load. And when there is no load, all of that current will go into the 431, so that it must be able to handle that current (and power) as well. Obviously, this severely limits the current rating of this type of LDO.

5.3 LDO Design

In this section we're going to look in detail at the general design of a discrete, robust LDO, and at the selection of the components for it to ensure that it is robust against supply chain issues. We'll follow this by examples later in this chapter.

5.3.1 Basic Idea

The basic circuitry used for a discrete LDO is shown in Figure 5.2. It consists of a number of distinct components: A pass transistor, usually a MOSFET (or BJT), which, though usually a p-channel (or PNP), may be n-channel (or NPN); a voltage reference; an op-amp configured as a difference amplifier that acts as the error amplifier, along with its gain components; possibly a gate drive (not shown) to interface the output of the error amplifier to drive the pass transistor; and finally

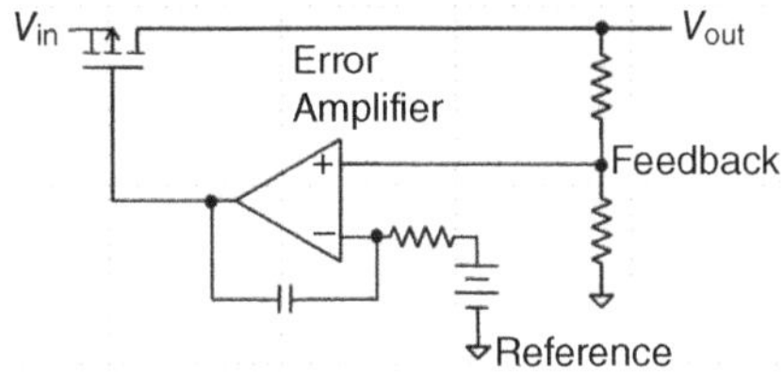

Figure 5.2 Basic LDO concept.

some output capacitance (also not shown), which may either be present externally to the LDO or included as part of the design or both. The following subsections give specific guidance on how to select each of these components for various LDO applications, followed by design examples in the sections below.

Let's take a moment to observe how this works. Normally, one expects to see the feedback going to the inverting pin of the error amplifier. However, in this case, it goes to the non-inverting pin. This is because the p-channel MOSFET acts as an inverter compared with the usual n-channel. When the output voltage goes too high, the output of the error amplifier goes high, which reduces the gate-source voltage of the p-channel MOSFET. This in turn reduces the current through the MOSFET, and so this is the correct polarity for the feedback to the error amplifier. However, the *gain* of the error amplifier still needs to have negative feedback for stability, so the capacitive compensation goes from the output of the error amplifier to the reference voltage, which is made to have finite impedance with a resistor buffer. In fact, the RC sets the frequency at which the zero cuts in.

5.3.2 Pass Transistor

The pass transistor in an LDO is the central component in the design: by controlling it, the output current is controlled in such a manner that the output voltage is regulated. The pass transistor also dissipates the bulk of the power needed to drop the input voltage to the output voltage.

There are several possibilities for what device to use for the pass transistor, each with its own pros and cons.

- **N-channel MOSFET:** There are tens of thousands of n-channel MOSFETs available, and as a result, it's easy to find second sources that will function well in an application. However, n-channel MOSFETs have a problem in this application: in order to turn them on, you generally have to apply at least 2 V on their gate to source. There *are* n-channel MOSFETs that have lower gate-voltage thresholds, but these are much less second-source available. If your input-to-output voltage drop is large, needing a big drop to the source may not be an issue. But unless the output current is low, this will result in a lot of power loss, and you may be better off with a SMPS. Of course, you could design a method of generating a voltage higher than the input, but this could be more complex than the rest of the LDO!

- **NPN BJT:** NPN BJTs have the same issue as n-channel MOSFETs, but less so. The base voltage has to be about 700 mV higher than the emitter. So, if the input-to-output voltage drop is about a volt or more, an NPN may be a reasonable choice. Of course, you still have to find one with adequate h_{fe}, or else you'll need a driver to push adequate current into the base. An NPN Darlington could in theory be used, but then the base emitter drop is closer to 1.4 V. And don't forget that *all* the parameters of BJTs are temperature dependent, making their design more complex.
- **P-channel MOSFET:** P-channel MOSFETs don't have the big problem that n-channels do: to turn them on, you pull the gate *down* from the source, so a voltage higher than the input isn't required. And there are many types from many manufacturers. However, for high-input voltage LDOs there can be a problem: the maximum gate-to-source voltage is usually limited to 20 V, so if your input is above 20 V, you will have to design a circuit to limit the source-gate voltage, although it might be something as simple as a zener.
- **PNP BJT:** As with p-channel MOSFETs, PNP BJTs also don't require a voltage higher than the input. And similar to the NPN, their base voltage only has to be about 700 mV less than their emitter, making them good for low-voltage inputs. As with NPNs, you, of course, control them by regulating their base current, and so you don't have to worry about exceeding their voltage rating. Their one downside is that there isn't a huge selection of them anymore, especially at higher currents. Still, many applications can find suitable multi-source parts. There are even p-channel Darlingtons, if the occasion arises.

In summary, our first choice for a pass transistor is going to usually be a PNP. If the input voltage is >4 V, and especially if the current is relatively high or the headroom very small, we may try a p-channel MOSFET first. We'll see examples of such design choices in the following sections.

For most LDO designs, the power dissipated by the pass transistor will be 1 W or less. So for most designs, the pass transistor package can be small and surface-mount, either a SOT-23 or a DPAK, and no heatsink is required. If the power dissipation goes much above 1 W, serious consideration should be given to using a SMPS, or to the possibility of lowering the input voltage.

5.3.3 Reference Voltage

The reference voltage in an LDO is a key element of circuit design. Its accuracy (including temperature effects) determines the accuracy of the output voltage.

For many purposes, it's adequate to use a zener, typically one with a ±2% tolerance. If your input is above 5 V, you can power a 5.1 V zener. This has a minimum temperature coefficient compared to other zener voltages; make sure to bias the zener at the recommended operating current (microamp zeners are available).

If the input is below about 3 V, a zener won't work for the reference, as zeners don't go that low in voltage and there will be no way to power it. In this case, a multisourced alternative is a 431 IC. They come in a variety of tolerances, though ±1% is inexpensive. And although they often require 1 mA of bias current, lower-current devices are available. The other important thing is to select one with a nominal voltage of 1.24 V rather than 2.5 V; most LDOs will be functioning with an input voltage higher than this, and so the device will be suitable.

If tolerance is required to be tighter than the 1% of the 431 plus the 0.1% of the resistor divider (plus input offset voltage and input bias current of the error amplifier), the 4041 is multi-sourced available down to 0.2%. Make sure the application really requires this tight tolerance before designing it.

5.3.4 Error Amplifier

As discussed in Chapter 3 on Components, there are a large number of specifications for an op-amp; that's why there are so many of them!

For most LDO designs, the most important specification of an op-amp is its operating voltage. Just as with the other components, when possible the op-amp needs to function from the input voltage, so that no additional circuitry is required to power it. It is also nice (though not essential) that the output be rail-to-rail, so that the pass transistor can be driven directly from the error amplifier; if it's not, the output of the op-amp has to be buffered with a gate (or base) driver.

Most op-amps are fast enough for most LDOs, so that gain-bandwidth product and slew rate are usually not important in op-amp selection. And again, since most LDOs produce outputs of volts, input offset voltage and input bias current are usually unimportant in op-amp selection.

It is considered good practice to bypass the op-amp's VCC pin with a 100 nF ceramic capacitor, to keep down its susceptibility to VCC noise.

5.3.5 Gate/Base Drive

In some higher-power applications, the output of the op-amp may not be enough to drive the pass transistor. This is especially the case with BJTs, as they are current driven rather than voltage driven. (Darlingtons require less base current, but then the voltage required to run them is larger.)

If you need to have a considerable amount of base drive current, you can use a totem-pole, as shown in Figure 5.3. Don't put the totem-pole as a follower to the

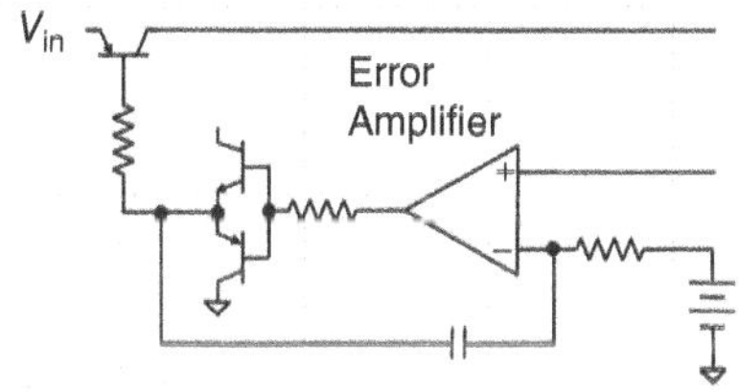

Figure 5.3 LDO base drive with a totem-pole.

op-amp; include it in the feedback loop of the error amplifier to enhance controllability. Make sure you choose transistors with sufficiently high gain (particularly the PNP, which does most of the work), and a sufficiently small value base resistor that the output of the totem-pole goes close to the input voltage, so that the pass-BJT can truly be turned off. Of course, here a Darlington pass transistor could be an advantage. We'll have an example of a design of an LDO with a totem-pole base drive below.

5.3.6 Compensation

Compensation of the LDO feedback is done using a network of capacitors and resistors around the error amplifier. The most common compensation is just a pole at the origin to roll off high-frequency effects, which is to say, a capacitor from the inverting input of the op-amp to its output; this will need to work against a resistor to set the pole position. As a rule of thumb, the response time of the LDO will be approximately equal to the RC time constant.

If you don't know or don't want to deal with the pole position, it's easy to just put a large capacitor in the compensation. If it's large enough (say a time constant of 10 msec with the reference voltage feed resistor), the LDO will be stable. You can then decrease the size of the capacitor step at a time until it's fast enough without having too much over- or undershoot during load steps. We will show below how to simulate the LDO so that the step-load response can be verified to be adequate.

5.3.7 Capacitors

Though often not a part of the LDO circuit proper, the performance of the LDO will be enhanced by including moderate-valued capacitors at the input and output; in many circumstances, these capacitors are already present in the source feeding the LDO and in the load that, in turn, it has to feed.

It's important to know the values of the external capacitors being used in the load, as excessive capacitance can destabilize the control loop. (Though it is straightforward to compensate for external capacitance, *not knowing* what the value of the output capacitance is can result in a stability issue.) Similarly, having an input capacitor will help with the transient response of the LDO, particularly if its input source is not immediately next to the LDO. Ceramic capacitors are desirable for both the input and output, though of course the load's design may include electrolytics. If they do, make sure to include the ESR of these capacitors when you do a stability check.

5.4 Example: Very Low Dropout LDO

As a first example of the design of a robust LDO, in this section, we're going to design one with very low dropout voltage. As mentioned above, the "L" in LDO stands for low, but in commercially available LDOs this may be as high as 2 V, and few are less than 100 mV. What we want here is to design a robust discrete LDO that has a dropout of less than 50 mV at 1 A. To be concrete, suppose we need to design a 4.95 V nominal output from a 5 V nominal input, both with 1% tolerance. The minimum 5 V input is 4.95 V, and with 50 mV drop at 1 A, the output will be a minimum of 4.90 V, which is 4.95 V − 1%. The commercially available parts at this sort of rating are all single-source, which is why we need to design our own.

We assemble the specifications for our design into Table 5.1, which will act as a prototype for all of the succeeding examples in this chapter. We've made up some

Table 5.1 Specification of a very low dropout LDO.

Parameter	Specification	Design
Input voltage min.	4.95 V	
Input voltage max.	5.05 V	
Output voltage min.	4.90 V	
Output voltage max.	5.00 V	
Output current min.	100 mA	
Output current max.	1 A	
Max. ambient temperature	25 °C	
Min. ambient temperature	25 °C	
Bandwidth	1 kHz	
Min. phase margin	30°	
50% step recovery time	100 μsec	
50% step max. undershoot	100 mV	
50% step max. overshoot	50 mV	

easy-to-meet numbers for the small-signal performance of the LDO (the bandwidth and the phase margin) and for the large-signal performance (the step recovery time along with over- and undershoot).

5.4.1 Robust Design

Using the basic design in Figure 5.2, we start the design by selecting the pass transistor. In this case, the input is 5 V, so either a p-channel MOSFET or a PNP could work. However, with only 50 mV of headroom, a suitable PNP is going to be difficult to find, especially at 1 A. So, we choose a p-channel MOSFET for the pass transistor.

Worst-case power dissipation is going to be (5.05 − 4.90 V) * 1 A = 150 mW, which is too much power for a SOT23 package; we should choose a SOT-223, a DPAK or possibly a D2PAK. Since the input is only 5 V, pretty much any drain-source voltage rating will be fine, and the gate-threshold voltage can be up to 4 V since we have that much (plus margin) on the source. The main selection criterion thus comes down to the $R_{DS,on}$, which has to be less than 50 mV / 1 A = 50 mΩ, including temperature effects; to be safe with the increase of the $R_{DS,on}$ with temperature, we select for those with $R_{DS,on} \leq 40$ mΩ. We end up with over 200 parts from some 20 vendors that meet all of these specs; selecting down to only DPAKs, we still have 159 choices from 176 vendors, so we pick the cheapest one, the G450P04, and choose parts from other vendors to be added to the APL for robustness.

Having chosen the MOSFET, let's think about the voltage reference. The design specification explicitly says that the output is ±1% tolerance. So, we need our reference voltage to be better than this, since we also have to account for the resistor divider tolerance, as well as the input offset voltage and input bias current of the error amplifier.

For this purpose, we pick a 431 with an output of 1.24 V with a tolerance of 0.25% or better, for which there are multiple sources. Typical operating current for multiple manufacturers is available at 200 μA or less, so our feed resistor should be less than (5 − 1.24 V) / 200 μA = 18.8 kΩ. We can choose a 10 kΩ, which gives 376 μA.

Next, we choose the op-amp. Op-amps that run on 5 V are common, and since nothing is specified about the dynamic performance of the LDO (other than our placeholders for small- and large-signal stability), we can pick almost any device. We choose the multi-source 358, which has two op-amps in a single eight-pin package and is available with rail-to-rail output. Perhaps the second op-amp can be used in a second LDO.

We can now choose the resistors for the feedback network. They should be relatively low value, so that the input bias current of the 358, which most vendors

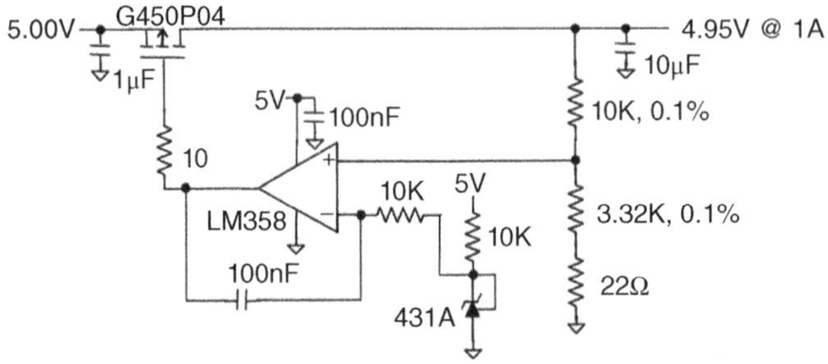

Figure 5.4 Very-low dropout LDO.

have as less than 27 nA, isn't significant; 10 kΩ gives only 270 μV error, while presenting an insignificant load on the LDO. Selecting this as the top resistor at 0.1%, to get a nominal output of 4.95 V, the bottom resistor should be 3.342 kΩ. This, of course, isn't available; we choose the largest 0.1% resistor less than this value, 3.32 kΩ, and put in series a 1% resistor[1] of 22 Ω, which makes up the rest of the resistance.

The remaining components are the input, output, and compensation capacitors. Since there are no real dynamic performance specifications, we can choose the input to be 1 μF, the output to be 10 μF (ceramic) and the compensation to be 100 nF; all we have to do then is to verify that the LDO is small-signal stable.

Our preliminary design with these parts is shown in Figure 5.4. We've added in a gate resistor to prevent very-high-frequency ringing (to which MOSFETs are subject), and a resistor from the voltage reference to the inverting input of the error amplifier for the compensation to work against. Next, we have to build a model and check it against the specifications.

5.4.2 Modeling

Let's go through the model component selection in the same order we selected components for the LDO design. Starting with the MOSFET, then, the specific device we chose, the G450P04, is, of course, not in LTspice. But when we chose it, all we wanted was a 40 mΩ p-channel. Looking through the LTspice library, we see there are a variety of choices with this characteristic. Since the G450P04 has the high gate charge of 25 nC, we pick the Si9434DY, which is a comparable 30 nF.

The capacitors are ceramic, so we can just use ideal models (we ignore the 1 μF input capacitor and the 100 nF bypass capacitor on the error amplifier since we'll feed both of these from an ideal voltage source). The resistors are ideal too, except

1 1% is fine for this trim resistor because the 22 Ω is less than 1% of the total resistance already, and 1 of 1% is negligible.

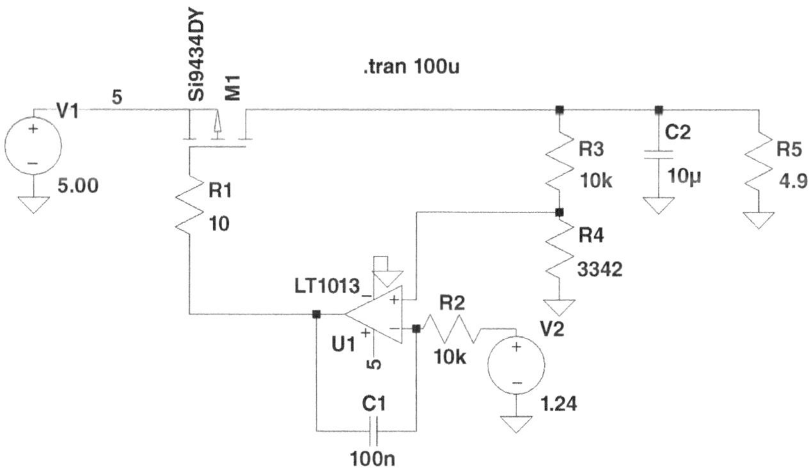

Figure 5.5 Model of very-low dropout LDO.

that we'll combine the two resistors in the bottom of the divider for simplicity. The 431 is just acting as a 1.24 V reference, so we can just use a voltage primitive. The only remaining item to select is the op-amp. In the chapter on modeling, we found that, although LTspice doesn't have an LM358, the LT1013 is reasonably close, and so we use this.

So, our starting point is the LTspice model shown in Figure 5.5. We've added a 4.9 Ω resistor to be a 1 A load. An initial run looks good—the output is 4.951 V. Now, we need to start testing to the specification.

We can do quite a bit with the model just as it is. We change the input voltage to minimum, 4.951 V, and find the output to be 4.913 V, which is within spec. We similarly increase the input voltage to 5.05 V, and verify that the output remains unchanged at 4.951 V.

To find the minimum output voltage, we change the reference voltage down by 0.2%, and decrease the top resistor of the feedback divider by 0.1% while increasing the bottom resistor by the same percentage. We find the output to be 4.913 V, which is within spec; and taking everything to the opposite limits while decreasing the load to its minimum of 100 mA, the maximum output voltage is 4.968 V, which also is in spec.

We don't test the temperature performance of this unit, since it is only specified at room temperature.

We can continue on to look at the dynamic performance of the design. We start with measuring the small-signal stability. To measure the bandwidth and phase

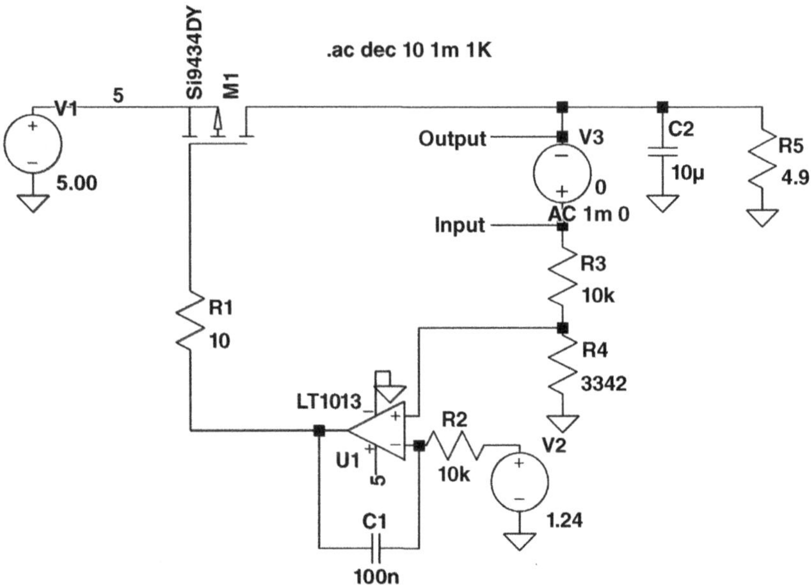

Figure 5.6 VLDO model configured for AC sweep.

margin, we need to add in an AC sweep model, just like we would do on the actual device.[2]

We show the model in Figure 5.6.[3,4] All parameters can be left at nominal, since their tolerances are negligible when measured in dB, except for the load, for which we will need to sweep at both minimum and maximum, as its range is 20 dB (100 mA to 1 A).

The closed-loop transfer function of the LDO is the output of the AC sweep divided by its input, as labeled in Figure 5.6. Sweeping the AC, then, we find (as might have been expected from the size of the compensation capacitor) that the BW is only 500 mHz! The transfer function is linear, so to get to 1 kHz BW, we reduce the error amplifier's capacitor by a factor of 200–470 pF and its feed by a factor of 10–1 kΩ for a product of 2000, which gives the plot shown in Figure 5.7. It has a bandwidth (that is, where the gain is zero dB) of 878 Hz, and a phase margin of 90°. Repeating this now with the load reduced to 100 mA, the BW is 526 kHz with a phase margin of 43°, so that the specifications are met at both ends of the load range.

2 For a detailed description of how to make this measurement on the lab bench, refer to [9].

3 Observe that since there is no switching in the LDO, no state-space average model is required.

4 Actually, the AC sweep voltage can be left in the model for DC operating point and Transient simulations. With its DC specified to be zero, it has no effect except when an AC sweep is run.

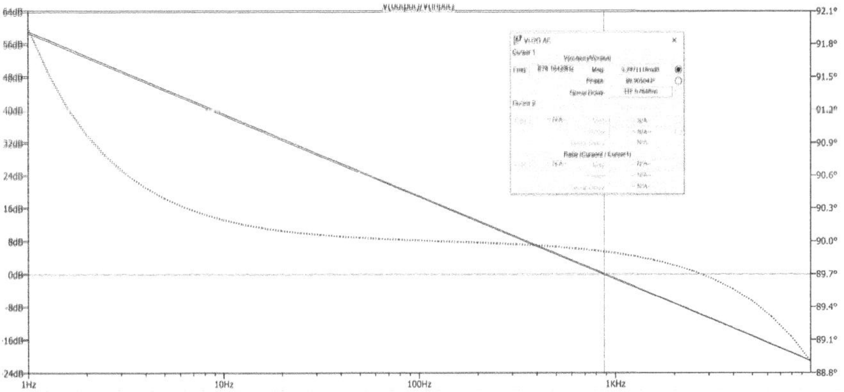

Figure 5.7 VLDO AC sweep.

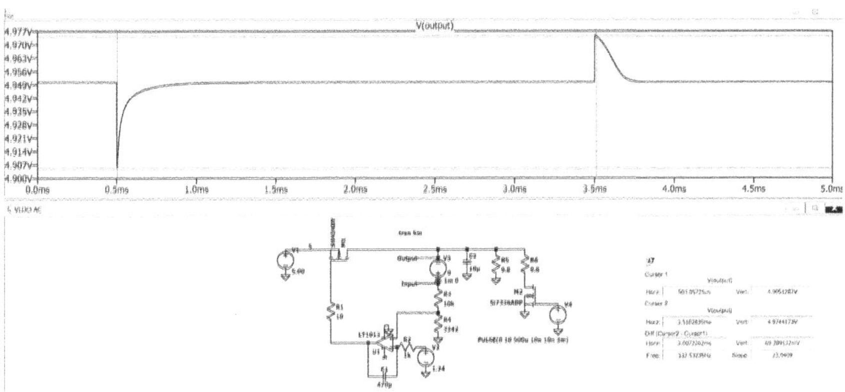

Figure 5.8 VLDO step response.

Our final step is to modify the model for a step load change of 50%, which we do as shown in Figure 5.8. 50% of the load is static; the other half is switched in by the MOSFET. The particular MOSFET we use doesn't matter at all, as long as its resistance is small compared to the load it's switching; what we're actually making is a model of a test fixture.

As the figure indicates, recovery on the up-step is something like 250 μsec, and that on the down-step is something like 350 μsec. Maximum undershoot reaches about 4.905 V, which is 44 mV below nominal, while maximum overshoot reaches 4.974 V, which is 25 mV above nominal.[5] If we were concerned with

5 It takes a very long time to recover because the LDO can only source current, not sink it.

Table 5.2 Specification and performance of a very low dropout LDO.

Parameter	Specification	Design
Input voltage min.	4.95 V	4.94 V
Input voltage max.	5.05 V	>5.1 V
Output voltage min.	4.90 V	4.913 V
Output voltage max.	5.00 V	4.968 V
Output current min.	100 mA	<100 mA
Output current max.	1 A	>1 A
Max. ambient temperature	25 °C	25 °C
Min. ambient temperature	25 °C	25 °C
Bandwidth	1 kHz	878 Hz
Min. phase margin	30°	43°
50% step recovery time	100 μsec	3350 μsec
50% step max. undershoot	100 mV	44 mV
50% step max. overshoot	50 mV	25 mV

simultaneous worst-case tolerance and step response, we could go back and increase the output capacitor, though, of course, then we would need to rerun the AC sweeps to recheck the stability.

Finally, we end up with the completed specification, shown in Table 5.2. This design meets all of its specifications and is robust since all of its components are multi-sourced. It's time to build a prototype and verify its performance in real life.

5.5 Example: Very Low Output Voltage LDO

In Section 5.4.2, we made our own LDO because an integrated circuit with its specifications isn't multi-sourced available, namely one with 5 V input and an extremely low overhead. In this section, we take on another LDO example, this time one where the overhead isn't quite as low, but the output voltage is *quite* low. Let's take as a specific example, then, an LDO with 500 mV ±2% output, coming from a 900 mV ±5% input, and that can provide up to 3 A of current. Its external load will be resistive, plus four 10 μF and three 1 μF capacitors, all ceramics and all in parallel.[6] An external 5/12 V supply is available for powering the internal circuitry.

6 That is, these capacitors are attached to the load, rather than being a part of the LDO. We'll add additional capacitance internal to the LDO.

Table 5.3 Specification of a very low output voltage LDO.

Parameter	Specification	Design
Input voltage min.	855 mV	
Input voltage max.	945 mV	
Output voltage min.	490 mV	
Output voltage max.	510 mV	
Output current min.	2 A	
Output current max.	3 A	
Max. ambient temperature	50 °C	
Min. ambient temperature	0 °C	
Bandwidth	100 kHz	
Min. phase margin	30°	
33% step recovery time	100 μsec	
33% step max. undershoot	20 mV	
33% step max. overshoot	20 mV	

A quick check shows that there are parts that do this—but the cheapest one is over US$4.50 at 1000 pieces and there's only one manufacturer. In order to be robust, we ought to design our own. We are looking for the following specifications, Table 5.3.

We have included an operating temperature range of 0–50 °C, whose only significant effect will be on the pass transistor.

5.5.1 Robust Design

To start, then, we consider that with less than 2 V as input, an ordinary p-channel MOSFET, requiring 2–4 V of V_{GS}, isn't going to work. But with only 900 mV of emitter voltage and needing 3 A output, the V_{EB} of PNP BJTs may be too high as well. The best choice, then, seems to be an n-channel MOSFET, driven by the external 12 V supply. (The 5 V is too low, because the 358 op-amp would swing up to only 5 − 1.35 V = 3.65 V, which might be below the gate threshold voltage of the MOSFET with 500 mV output. And most MOSFETs can take 20 V V_{GS}, so 12 V won't harm them.)

We need an $R_{DS,on}$ of less than (855 − 510 mV) / 3 A = 115 mΩ. Since the specification says the temperature goes up to 50 °C, we look for MOSFETs with $R_{DS,on} \leq 100$ mΩ. As for power dissipation, we have (900 − 500 mV) * 3 A = 1.2 W, and so we should choose either a DPAK or a D2PAK. There are a variety of manufacturers of suitable devices.

For the voltage reference, we need less than the minimum 1.24 V that the 431 can provide, but the tolerance is 2%, so we can pick a 1% version of the 431 and power it from the external 5 V with a 10 kΩ resistor as in the previous design. Since the reference is greater than the desired output, we don't need a resistor divider in the feedback; instead, we'll divide the output of the 431 down to 500 mV with 0.1% resistors: 10 kΩ on top and 6757 Ω on bottom, which we synthesize from a series combination of a 6.73 kΩ at 0.1% and a 27 Ω 1%. We put a series 1 kΩ from the output to act as an impedance against which the feedback capacitor works; it doesn't require buffering because it is the input to an op-amp.

5.5.2 Modeling

We build out our model using these parameters. The op-amp is the LT1013 again, powered by the external 12 V. Since, unlike the p-channel MOSFET, the n-channel MOSFET is not inverting, we have to remember to invert the inputs (and the compensation network) compared to the op-amp from the previous schematic.

Finally, we'll increase the output capacitor to 47 μF because the current is now 3 A, and everything else will be the same as in the previous design (recall that the 43 μF of output capacitors are external). We end up with the schematic shown in Figure 5.9. We can double-check that we got the polarity of the error amplifier right by considering that if the output voltage is too high, the error amplifier goes low, turning down the conduction of the n-channel MOSFET and thus reducing the output voltage.

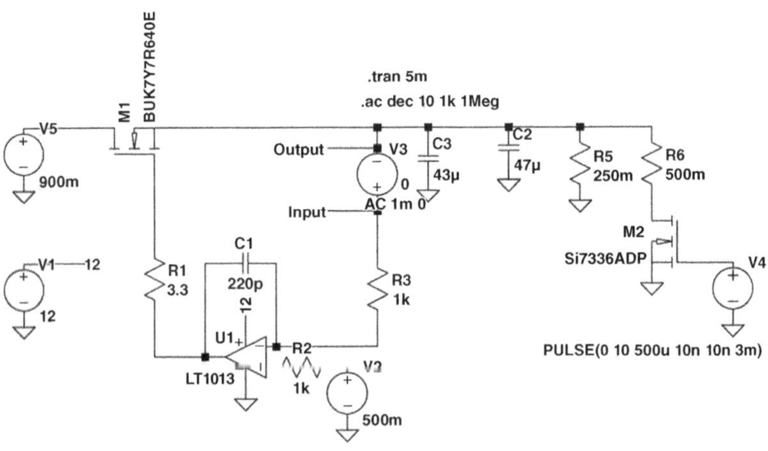

Figure 5.9 Very low output LDO schematic.

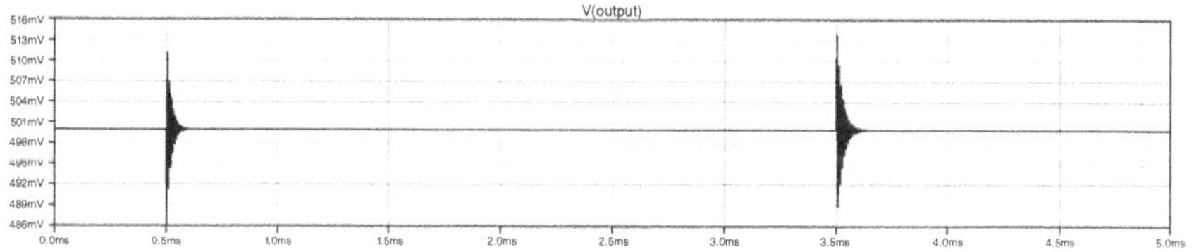

Figure 5.10 Very low output LDO step response.

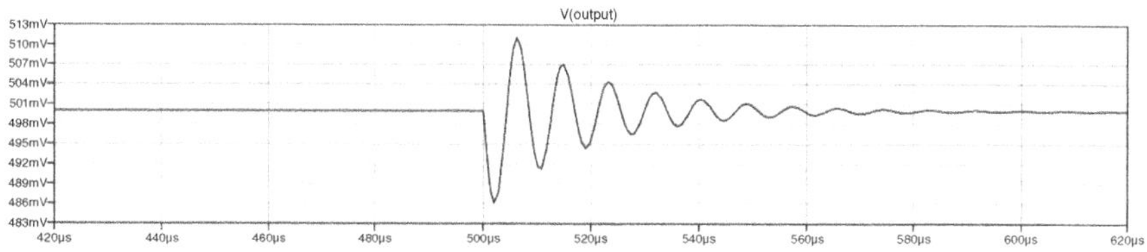

Figure 5.11 Very low output LDO step response zoom.

We've already done the compensation. You may note the schematic already includes the AC source for the sweep.[7]

The output during the step load is shown in Figure 5.10. This very good-looking result meets the specification, as the maximum deviation goes to 514 mV and the minimum to 486 mV, both less than the 20 mV limit. Zooming in on the rising step in Figure 5.11, we can measure that the ring is at 116 kHz and is not particularly well damped. If this actually mattered (remember that it already meets the specification), it could be fixed either by a little tweaking of the error amplifier's compensation, or more easily, by an RC filter at the output before the load, with a small resistor value to limit power dissipation and voltage drop, and a small capacitor value to reduce the 116 kHz noise, depending on how much reduction was desired (Table 5.4).

We haven't measured the final phase margin (though it's probably quite close to 30° given the ringing). We also haven't really looked at the temperature rating, though there are no electrolytic capacitors, and the MOSFET's $R_{DS,on}$ was picked so that it would work at 50 °C.

You might be curious as to whether the speed of the LT1013 model was contributing to the ringing. Replacing it with MAX40077 doesn't seem to improve the ringing much, so there seems to be no reason to go to a more expensive op-amp.

Given the high simulated performance of this LDO, this is clearly a case where you have to build it in order to verify that it matches with the simulation.

7 You might notice that there are *two* LTspice commands present; when you put a second one in, LTspice blues out the first. You can then restore the first by deleting the second. A better way, actually, is to change the first one so that its initial period is a semicolon, which makes it a comment. You can then alternate which one is the comment without having to delete either.

Table 5.4 Specification and performance of a low output voltage LDO.

Parameter	Specification	Design
Input voltage min.	855 mV	<850 mV
Input voltage max.	945 mV	>900 mV
Output voltage min.	490 mV	494 mV
Output voltage max.	510 mV	506 mV
Output current min.	2 A	<qty1 A
Output current max.	3 A	>3.3 A
Max. ambient temperature	50 °C	
Min. ambient temperature	0 °C	
Bandwidth	100 kHz	116 kHz
Min. phase margin	30°	
50% step recovery time	2 μsec	2 μsec
50% step max. undershoot	20 mV	14 mV
50% step max. overshoot	10 mV	14 mV

5.6 Example: High Voltage LDO

Now let's consider the other extreme, a very high-voltage LDO. Suppose that we need an LDO with an output of 76 V at 24 mA, from an input of 84 V. There are not many LDOs that operate at this high a voltage, none, of course, that are at this specific voltage, and none even of the adjustable type that are second-sourced. We might, of course, be tempted to build a SMPS for this application, but LDO efficiency is going to be 76 / 84 V = 90.5%, and this might be comparable to the losses of a SMPS at a rather lower price.

5.6.1 Robust Design

We thus design our own. We start with a specification, Table 5.5.

Our first consideration, as usual, is what to use for the pass transistor. Were we to use a p-channel MOSFET, the gate voltage would be limited to 20 V below the source. This would necessitate somehow generating a gate driver that would be sure not to pull the gate any lower than 60 V. While this is possible (perhaps with a zener), it seems like a lot of trouble for a 25 mA output.

A PNP is much easier to deal with. It doesn't care about the base voltage, which is going to be something like 700 mV below the emitter no matter what. All you have to do is to pull the current out of the base at a controllable level.

Table 5.5 Specification of a very high voltage LDO.

Parameter	Specification	Design
Input voltage min.	88 V	
Input voltage max.	80 V	
Output voltage min.	74 V	
Output voltage max.	78 V	
Output current min.	0 mA	
Output current max.	25 mA	
Max. ambient temperature	25 °C	
Min. ambient temperature	25 °C	
Bandwidth	1 kHz	
Min. phase margin	30°	
100% step recovery time	100 μsec	
100% step max. undershoot	1 V	
100% step max. overshoot	1 V	

The emitter-collector voltage needs to withstand 100 V, and since there are plenty of such parts, we've picked one for our model, the 2SAR372P5.

As for the base drive, the 358 can only take 40 V and so it can't turn off the PNP. Instead, we pull current from the PNP with an NPN, which in turn is driven by the 358. Again, there are many choices, and we choose a 100 V part, the BSS64A. Of course, we still need some voltage to run the 358. Since it only takes 300 μA, we can just pull it from the input, dissipating 25 mW. Unfortunately, LTspice's only 5 V zener, the 1N750, takes 20 mA (sic), but we'll just ignore the power loss in the resistor in the model.

A final issue turns out to be the loop gain. The gain of the PNP is over 100, and the gain of its driver, the NPN, is over 30, giving a gain of over 3000. With a capacitive feedback, the gain is so high that the LDO oscillates. In fact, the gain of the two transistors is high enough that we simply make the error amplifier gain unit, and this regulates well enough, as we will see.

We also pick our base resistor at this point. To avoid dissipating a lot of power, we choose 100 kΩ for the PNP base (giving 70 mW in the resistor), and then saturate its driver NPN with another.

We end up with the schematic shown in Figure 5.12. We've added a step load that takes the output from zero to maximum current.

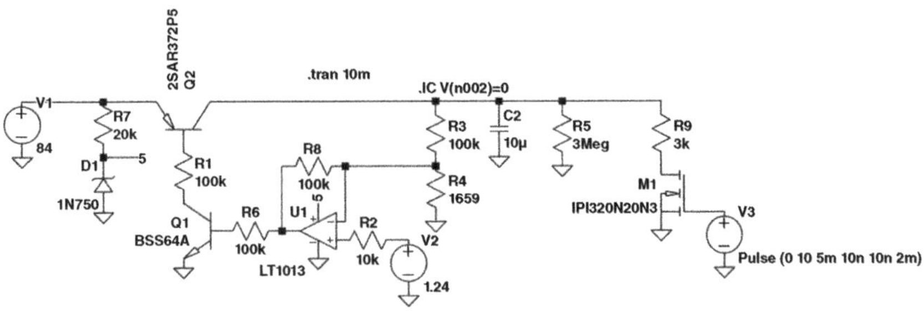

Figure 5.12 High voltage LDO model.

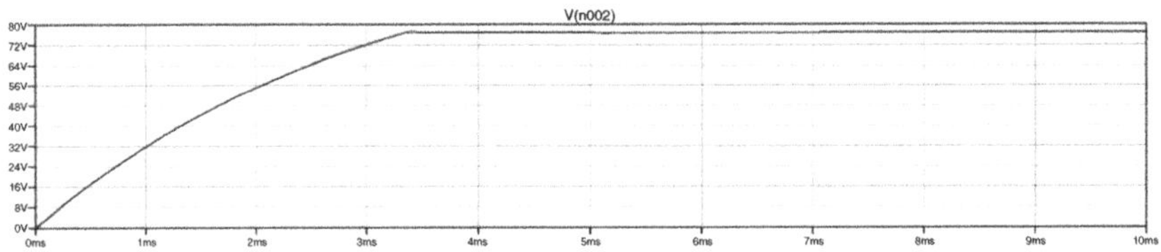

Figure 5.13 High voltage LDO startup.

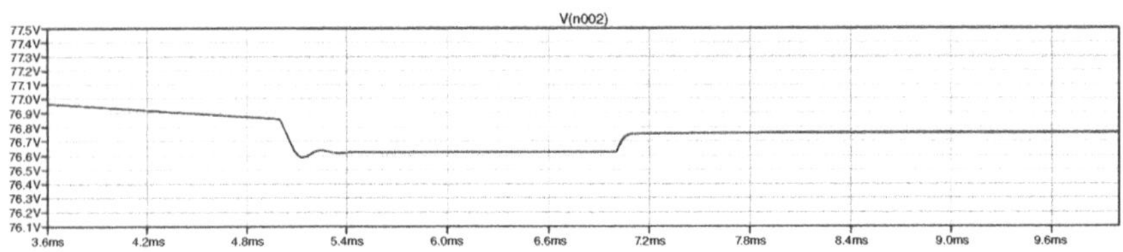

Figure 5.14 High voltage LDO step response.

5.6.2 Modeling

Running the model, we see in Figure 5.13 that it reaches steady state about 3 msec after startup. Its steady-state voltage is 76.88 V, so perhaps a little more gain on the error amplifier would have been OK.

Looking at the step response in Figure 5.14, we see that going from no load to full load drops the output voltage 260 mV with recovery after about 350 μsec, and that there is no overshoot on recovery to zero load. This could be improved by increasing the output capacitance. In summary then, the specifications seem to be approximately met, with a clear path to improvement if needed.

5.7 Example: Negative Output LDO

Negative output LDOs are considerably less common than positive output ones, no doubt because negative voltages are used a lot less than positive. Nonetheless, they are definitely used *sometimes*,[8] and we will design one to, among other things, remove the confusion surrounding which direction the current flows.

5.7.1 Robust Design

As a specific example, let's suppose we have a −12 V source (coming, perhaps, from a 555 inverter as shown in Chapter 10 (Special Topics), or from a winding on a transformer) and we also need to get −5 V. Of course, this could be done with a resistor and a zener, but our load current is going to be 60 mA, which times 7 V is 420 mW, so the resistor and zener will have to be fairly hefty, and there will be a lot of power wasted when the load is light.

This can be designed very similarly to a positive LDO; we just have to get the flow of current in the right direction, and then the appropriate pass transistor, also in the right direction. To get this right, then, consider the diagram shown in Figure 5.15. It's backwards from the way we've drawn other LDOs, but conceptually it's the same.

We have a 7 V source that we want to regulate, and that can be pulled down by pulling current out of it. We compare this source with a 7 V reference generated from a 12 V supply. If the 7 V is too high, the error amplifier goes higher, turning

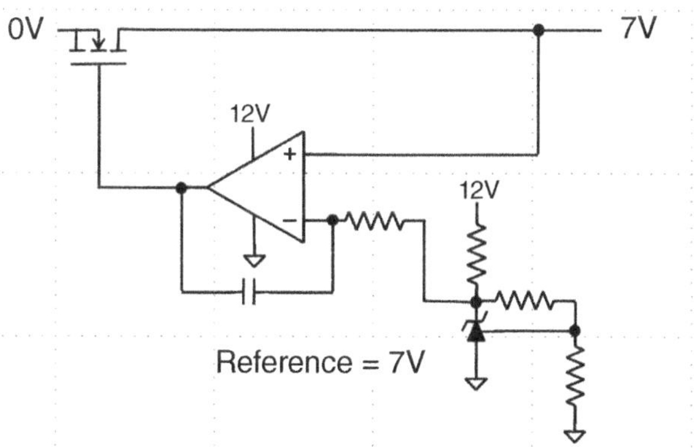

Figure 5.15 Conceptual derivation of a negative output LDO.

8 One usage is to smooth out the output of a SMPS for low-noise power to an op-amp which has poor PSRR at high frequencies.

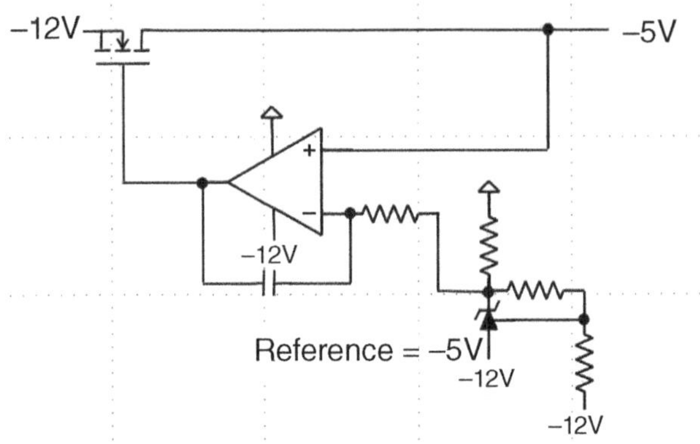

Figure 5.16 Diagram of negative output LDO.

the n-channel MOSFET on harder. This pulls more current from the 7 V, pulling it down. This is thus appropriate negative feedback.

Now let's take this diagram and subtract 12 V from everything, including ground. Remember from college that "zero volts" doesn't exist, all voltages are relative to other voltages, and so changing every voltage by the same amount has no real-life actual effect.

We get the diagram shown in Figure 5.16. Now, we have a −5 V source that we want to regulate, and that can be pulled down by pulling current out of it. We compare this source with a −5 V reference generated from a 0 V supply; zero volts is higher that negative five volts, that's why this works.[9] If the −5 V is too high, the error amplifier goes higher, turning the n-channel MOSFET on harder. This pulls more current from the −5 V, pulling it down. This is thus appropriate negative feedback, exactly the same as before.

Observe that there is nothing wrong with running the op-amp from ground (attached to its V_{CC} pin) and −12 V (attached to its GND pin). The V_{CC} pin is 12 V higher than the GND pin, which is what is required. Similarly, the cathode of the 431 is higher than its anode, and is powered from a voltage that is yet higher. Again, this is being operated correctly.

5.7.2 Modeling

It's now straightforward to build a model of this, see Figure 5.17. We take as the load −60 mA, which is 300 mW. The only thing you have to be a little careful about is that the negative supply of the op-amp is −12 V... but the way LTspice displays it, you can't really see the negative sign.

9 If this is still confusing, think of the number line. Twelve volts is higher than five volts because it's to the right of it. In just the same way, zero volts is higher than negative five volts because it's to the right of it.

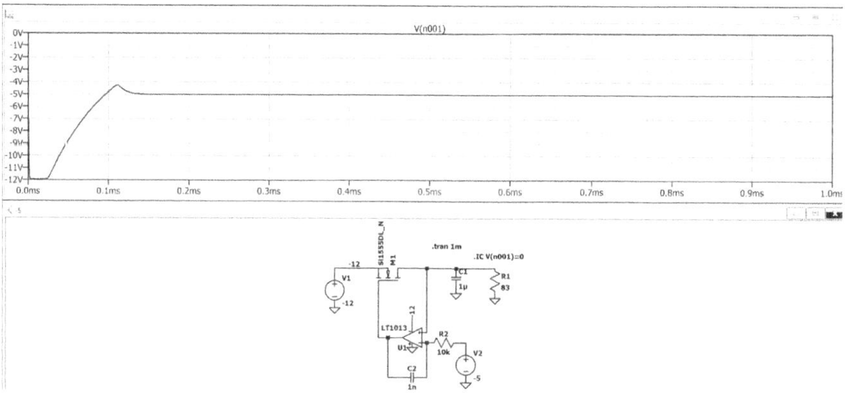

Figure 5.17 −5 V output model.

The output undershoots to −12 V for 25 μsec, and then overshoots very briefly, but settles in to a clean −5 V within 140 μsec. Doubtless, the compensation could be designed to avoid the undershoot.

5.8 Notes for the Hobbyist

For most purposes of the hobbyist, there won't be a call to make your own LDOs. 12 V in and 5 V out? Of course, you can just pick up a 7805. The times when you need something unusual is when you might consider a build-your-own. −5 V out at 2 A? Aren't many of those around. Of course, this might cause you wonder why it is that your design needs this.

The times when you need to make your own LDO are pretty much those suggested by the examples in this chapter: If the input-to-output voltage is very small; or if the output voltage is very low; or if the output voltage is very high. Negative output LDOs are also less common than positive. In these sorts of cases, the easiest thing to do is to copy one of the examples' schematics, and then modify it. Of course, after that you need to run all the tests that the Example ran, to make sure you haven't accidentally broken something.

6

Bang–Bang Converters

6.1 What a Bang–Bang Converter Is

Historically, bang–bang control is very old and dates back at least to the 1950s [18]. In words, a bang–bang converter is a SMPS that turns on a switch when its output voltage is too low and turns the switch off when its output is too high. Thus, it does not have a constant frequency, and in fact the only IC it needs is a comparator. While this is conceptually simple, the more difficult part is in the details.

As shown in Figure 6.1, the basic bang–bang converter typically uses a (p-channel) MOSFET as the switch. When it turns on, the voltage across the inductor is (say) positive, since the output voltage is lower than the input voltage in a buck converter, and so the inductor current ramps up. The current goes into the capacitor at the output (shown as an electrolytic, though it will often be a ceramic), causing the voltage to rise. When the voltage goes above the reference voltage, the output of the error comparator goes high, turning off the MOSFET. When the MOSFET turns off, the input to the inductor swings down in voltage until the diode turns on. The inductor current ramps down. This provides less current into the capacitor, the load causing the voltage to start to drop. When the voltage goes below the reference, the MOSFET turns back on, completing the cycle.

Observe that we are using an error *comparator* rather than an error amplifier. This is because (as suggested by the name), the output is either high or low—it isn't like an LDO, where the gate voltage is controlled linearly. For the bang–bang converter, the MOSFET is either on or off, and spends as little time as possible in the linear region, because it is a form of SMPS.

A key element of the bang–bang controller is the hysteresis, which is set by the hysteresis resistor around the comparator, working against the feedback resistor (in the divider, if there is one). The smaller the amount of hysteresis (which is to say, the larger the hysteresis resistor compared to the feedback resistance), the

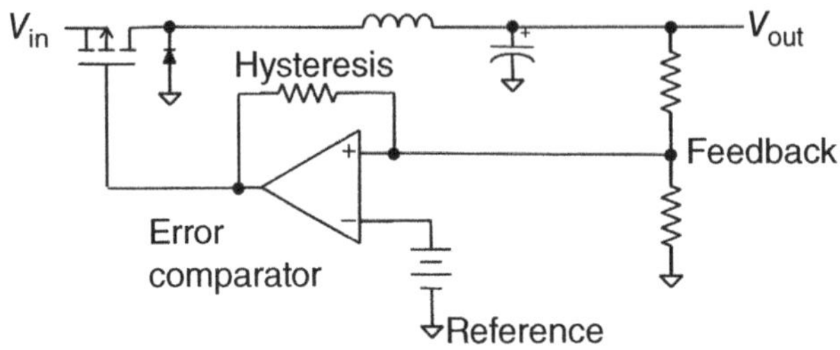

Figure 6.1 Basics of a bang-bang SMPS.

tighter will be the regulation ... and also the higher the frequency at which the transistor is switched. It will be essential to find a comparator that is fast enough for the maximum frequency of the converter.

Now it has to be admitted that bang–bang converters have a poor reputation. Some of their perceived issues are:

- They're old-fashioned. This is true in a sense, as the concept of bang–bang control dates back at least to the 1950s. And there was a time in the 1970s when there was a lot of tinkering with bang–bang control for power supplies as the first ICs were becoming commercially available. However, MOSFETs date from the same era, and no one thinks of *them* as being old-fashioned. It's not how long the concept has been around that dates a technology, but whether it can perform approximately as well as modern systems do. We'll demonstrate how to get modern performance from a bang–bang in this chapter.
- They're hard to design. Historically, again, this is true. Long ago, the author derived the equations governing the switching frequency of a bang–bang controller based on the input and output voltages, the inductance, the capacitance and the hysteresis, which is a lot of variables for an equation. And it didn't even include the capacitor's ESR! In the day, then, what engineers actually did was throw something together on the lab bench and then spend days swapping components to look for the right performance. In this book we'll, for the first time, give some rules of thumb to select components that are first approximations to the right values and then tune the values by simulating the performance.
- They're not very efficient. Once again, this relates to history. When MOSFETs were new, they didn't have very good performance, and so neither did the SMPS built from them. And many people didn't understand that their circuits had to be designed differently than those with BJTs. But today we have vastly improved MOSFETs, and can build bang–bang converters with efficiency as high as PWM converters, and at similar switching frequencies.

- They're physically large. They used to be built to switch at 10 or 20 kHz. But actually *all* SMPS used to be built to run at that frequency and were large. If you build one today to run at a couple of hundred kilohertz, bang–bang controllers can be similar in size to other SMPS.
- They have EMI issues. This is a supposition that, because their frequency "is all over the place" they will be hard to noise filter. Actually, though, bang–bang SMPS have very *good* EMI characteristics. The switching frequency varies with the load, and the heavier the load, the higher the switching frequency. This makes EMI filters for bang–bang converters potentially smaller than those for PWM converters, as the higher load (requiring more EMI filtering) is to some extent compensated for by the higher switching frequency (as filters are more effective at higher frequencies).

While there is certainly some truth to these objections, in this chapter we will show how to design a robust bang–bang SMPS with modern performance: high switching frequency, high efficiency, physically small and passing (conducted) EMI.

6.2 Component Selection

The bang–bang converter at its most fundamental has nine components: the switching transistor and its gate drive, a rectifier, an inductor, an output capacitor, a voltage reference, feedback resistors, a comparator, and the comparator's hysteresis.

The most complex component to select is the switching transistor. In bang–bang converters, the switching transistor is usually going to be a p-channel MOSFET. As in any SMPS, the particular component selected has a significant impact on the efficiency of the converter. Of course, its voltage rating should be greater than the input voltage, while its $R_{DS,on}$ should be low enough that at the rated current it doesn't lose significant power in conduction.

But it also needs to respond appropriately to its gate drive. And this is particularly important in the basic bang–bang converter, where the gate of the switching transistor is driven directly by the (open-collector) comparator, with the turn-off relying on a source-gate pull-up resistor. This simple drive means that transitions are going to be slow. To help keep the losses under control, it's desirable, then, to keep the gate capacitance low. Fortunately, MOSFETs with low $R_{DS,on}$ and low Q_g are readily obtainable from multiple sources.

The choice of the pull-up resistor is not critical. As a very general rule of thumb, choose a value such that the power dissipation is, say, less than 1% of the total power delivered to the output. Later in this chapter, we'll discuss two different ways of making better gate drives.

The rectifier diode is often the dominant loss component in the bang–bang converter. Thus, when possible, we choose it to be a Schottky of as high a current rating as possible in the package space we have, in order to minimize losses at the converter's rated current. Following our guidelines in Chapter 3 on Components, we want to select a voltage rating for the Schottky that is double the maximum input voltage. When this isn't possible, we look for an ultra-fast rectifier with, again, as high a current rating as possible, to minimize the forward conduction loss. Later in the chapter, we'll discuss replacing the diode with a synchronous rectifier.

Since the bang–bang converter will be running at a fairly high frequency, we won't be using millihenry-valued inductors! We can make an estimate of the sort of value we need. First, let's calculate the duty cycle. Suppose that the input voltage is V_{in} and the output is V_{out}, so that while the transistor is on there's $(V_{\text{in}} - V_{\text{out}})$ across the inductor, and while the transistor is off, there's a diode drop below ground of V_{f}, so the voltage across the inductor is $(V_{\text{out}} - (-V_{\text{f}}))$. This gives a duty cycle determined by

$$(V_{\text{in}} - V_{\text{out}}) \times t_{\text{on}} = (V_{\text{out}} - (-V_{\text{f}})) \times t_{\text{off}} \tag{6.1}$$

which (as was anticipated) is the duty cycle of a buck,

$$\text{DC} \equiv \frac{t_{\text{on}}}{t_{\text{on}} + t_{\text{off}}} = \frac{V_{\text{out}} + V_{\text{f}}}{V_{\text{in}} + V_{\text{f}}} \tag{6.2}$$

Beyond the duty cycle, there's one more consideration we must make in order to select an inductor value: Is the current in the inductor continuous or discontinuous, that is, are we operating in CCM or DCM? For low-powered applications, DCM can be good enough; there is a certain minimum size to inductors no matter what you do. But for any medium- or high-power application, CCM has a big advantage. Since the current in the inductor is triangular, to get a certain amount of average current through it is going to require a large RMS current if the current has to start from zero every cycle, as it does in DCM. In CCM, however, the current doesn't need to go to zero each cycle. The triangular part of the current can be small compared to the average, and the RMS current will be close to the average current. This difference between DCM and CCM, that the RMS (and peak) current is smaller in the latter than in the former, will be essential to keeping inductor size down. For this reason, for medium- and high-power bang–bang buck converters, we will always be using CCM.

So now, let the average output current for the converter be I_{out}. Then the peak current in the inductor in DCM is I_{out} / DC. If we want a switching frequency at full load of something like F_{SW} (period = $1/F$), we need to ramp up the current by I_{out} / DC in a time $1/f \times \text{DC}$ by applying $(V_{\text{in}} - V_{\text{out}})$ across the inductor, so that using V-sec balance on the inductor

$$\frac{V_{\text{in}} - V_{\text{out}}}{L} = \frac{I_{\text{out}}f}{\text{DC}^2} \tag{6.3}$$

which is

$$L = \frac{(V_{\text{in}} - V_{\text{out}})\,\text{DC}^2}{I_{\text{out}}f} \tag{6.4}$$

This is at least a good starting point for DCM converters; we will be using this below in our low-power example. For converters that are *not* operating at low power, we will want to use CCM. As a rule of thumb, we will want the peak-to-peak current of the inductor to be 20% of the average current. In this case, the inductor equation will instead be

$$\frac{V_{\text{in}} - V_{\text{out}}}{L} = \frac{20\%\,I_{\text{out}}}{\frac{1}{f}\,\text{DC}} \tag{6.5}$$

resulting in

$$L = \frac{5(V_{\text{in}} - V_{\text{out}})\,\text{DC}}{I_{\text{out}}f} \tag{6.6}$$

where the factor of five in the numerator comes from using one fifth of the average output current. The linearity of the term in DC, rather than quadratic as in DCM, is because we are forcing the peak-to-peak current to be 20% of the output current, so there's no extra term proportional to DC. The inductance is bigger, and that's what causes the converter to operate in CCM.

Naturally, having a low winding resistance for the inductor is desirable to keep the losses down.

The output capacitor is also going to be partially determined by the switching frequency, since as soon as the load has pulled down the capacitor voltage to the lower hysteresis limit, the switch has to turn on, and conversely, when the capacitor voltage reaches the upper hysteresis limit, the switch turns off.

Suppose then that the hysteresis is set to produce V_{pp}. Using the same numbers as above, the capacitance should then be at least

$$C \times V_{\text{pp}} = I_{\text{out}} \times \frac{\text{DC}}{f} \tag{6.7}$$

so that

$$C = \frac{I_{\text{out}} \times \text{DC}}{f \times V_{\text{pp}}} \tag{6.8}$$

for DCM, while for CCM

$$C \times V_{pp} = 20\% \, I_{out} \times \left(DC \times \frac{1}{f} \right) \tag{6.9}$$

so that

$$C = \frac{I_{out} \times DC}{5fV_{pp}} \tag{6.10}$$

the capacitor is one fifth of the value in CCM vs. DCM.

Obviously, the capacitor's voltage rating has to be higher than the output voltage. The largest value multi-source ceramic capacitor is 22 μF, and is available at up to 50 V. A few of these should be enough for most applications. Of course, you could always go to an electrolytic, but then when you build a model you need to remember to account for the ESR at the switching frequency.

The voltage reference is usually going to be a 431, and so it will be 1.24 V. What tolerance 431 is required is going to be set by the tolerance requirement of the output voltage, but also including the hysteresis.

Next comes the feedback. The top feedback resistor can usually be conveniently taken to be 10 kΩ. The bottom feedback resistor is then selected such that the resistor divider gives the same voltage as the voltage reference when the output is at the correct nominal voltage.

Finally, we have to consider the comparator and its hysteresis. Recall that in a generic, low-power bang–bang converter, we're relying on the comparator to directly pull the MOSFET gate low in order to turn it on. Thus, the comparator has to both be able to sink enough current from the pull-up resistor and be fast enough. And, of course, it has to be able to operate from the input voltage.

In general, we will be using the LM339 comparator. The generic part runs from 2–36 V, has a response time somewhat less than 1 μsec and can sink 16 mA, so this will work for most bang–bang applications. For higher sink currents, we will buffer the output with a small-signal transistor such as the 2222.

As for the hysteresis resistor, it determines the peak-to-peak ripple voltage on the output. A typical choice would be 1% of the output voltage, which corresponds to a 1 MΩ resistor for the hysteresis (given the 10 kΩ feedback resistor, since 10 kΩ / 1 MΩ = 1%). You should note that decreasing the ripple voltage can be done by increasing the hysteresis, but that there is a limit: you can't use a capacitor for "infinite" hysteresis!

Reviewing the component selections, we see that only a few specification values are needed in order to get to the right ballpark for component values. We need the input voltage, the output voltage, and its tolerance, the maximum output current, the desired peak-to-peak output ripple voltage and the desired maximum frequency. All of these would already be present in any specification of any SMPS, and thus the preliminary design of a bang–bang converter is no harder than that of a PWM SMPS.

Let's also take a quick look at the components that most determine the efficiency of a bang–bang converter. The generic bang–bang producing moderate output current is going to have a Schottky diode, and this is often the dominant loss mechanism in the converter. Of course, it could be replaced with a synchronous rectifier, and we'll show how to do this later in this chapter.

Depending on the details of the specifications, the switching transistor can also contribute significantly to the losses, although here it's easy to find a better device to keep the losses down. The inductor winding resistance usually dissipates only a small amount of power, and the pull-up gate resistor on the switching transistor is usually not that important either. If it is, changing MOSFETs or going to an active drive are good possibilities.

6.3 Example: 5–1.8 V at 2 A

For our first example, we're going to make a bang–bang SMPS to convert 5–1.8 V at a 2 A load. It's obvious that an LDO would be impractical, as the power loss would be (5 − 1.8 V) × 2 A = 6.4 W, which is to say an efficiency of 1.8 V / 5 V = 36%!

Let's start with a specification Table 6.1.

Table 6.1 Specification of 5–1.8 V at 1 A bang–bang converter.

Parameter	Specification	Design
Minimum input voltage	4.75 V	
Maximum input voltage	5.25 V	
Minimum output voltage	1.755 V	
Maximum output voltage	1.845 V	
Minimum output current	0 A	
Maximum output current	2 A	
V_{ripple}, max	100 mVpp	
50% step max. undershoot	50 mV	
50% step min. undershoot	50 mV	
Frequency, max	100 kHz	
Efficiency, 2 A	85%	
Size	–	

Several observations about this specification are in order.

- There is no minimum frequency, because at no load the transistor will simply be off, and not switching at all.
- There is no bandwidth or phase margin specified for a bang–bang converter because there are no poles or zeros in the control system; it's essentially single-cycle control. And finally,
- A target of 85% for the efficiency is really pretty good, considering that the output voltage is 1.8 V and we're using a diode rather than a synchronous rectifier!
- We've included a specification of the (maximum) size of the converter. We'll describe later on how to estimate the size.

6.3.1 Robust Design

We'll begin with Figure 6.1. The right place to start the design is with the calculations in the previous section. We have the duty cycle

$$\mathrm{DC} = \frac{V_{\mathrm{out}} + V_{\mathrm{f}}}{V_{\mathrm{in}} + V_{\mathrm{f}}} = \frac{1.8\,\mathrm{V} + 0.5\,\mathrm{V}}{5\,\mathrm{V} + 0.5\,\mathrm{V}} = 42\% \tag{6.11}$$

We have made an initial guess of a V_{f} of the Schottky of 500 mV, but the result won't change all that much if we're a bit off.

Next, we can make an estimate for the inductor. Since this is low power, we'll use DCM:

$$L = \frac{(V_{\mathrm{in}} - V_{\mathrm{out}})\mathrm{DC}^2}{I_{\mathrm{out}} f} = \frac{(5\,\mathrm{V} - 1.8\,\mathrm{V}) \times 0.42^2}{2\,\mathrm{A} \times 100\,\mathrm{kHz}} = 2.8\,\mu H \tag{6.12}$$

with peak current of

$$I_{\mathrm{pk}} = \frac{I_{\mathrm{out}}}{\mathrm{DC}} = \frac{2\,\mathrm{A}}{0.42} = 4.76\,\mathrm{A} \tag{6.13}$$

You can see that even though the output current is only 2 A, peak current is almost 5 A because we're using DCM. While this is readily achievable in a common device, as has been noted, if the output were much higher it would be advantageous to switch to CCM.

And finally, for the capacitor,

$$C = \frac{I_{\mathrm{out}} \times \mathrm{DC}}{V_{\mathrm{pp}} \times f} = \frac{2\,\mathrm{A} \times 0.42}{100\,\mathrm{mV} \times 100\,\mathrm{kHz}} = 84\,\mu F \tag{6.14}$$

Let's then choose actual parts for these three. We want a Schottky that at 5 A has a (room temperature) V_{f} of 500 mV or less. (As diodes warm up their V_{f} goes down.) In Chapter 3, we talked about the multi-sourced 860; at 5 A, its typical V_{f} is about 390 mV.

For the inductor, we look for a value less than that calculated, and try to find the standard value of 2.7 μH with a minimum 2 A rating and minimum 4.8 A saturation current. Since there are many choices, we look only for surface mount, and see that typical devices have 15 mΩ winding resistance or less, which gives us something like 60 mW of power dissipation at 2 A, since RMS current is

$$I_{\text{RMS}} = 4.76\,\text{A}\sqrt{\frac{0.42}{3}} = 1.78\,\text{A} \tag{6.15}$$

For the capacitor, we said that 22 μF is standard; we don't want to cut it too close, so we picked five of them to get 110 μF.

Now we can choose the switching transistor. Since the input is 5 V, we can use any V_{DS}; the tricky part is rather ensuring that the gate threshold voltage is low enough that we can turn the transistor on with only 5 V. Not only that, but we also have to ensure that the transconductance (how many volts per amp the transistor passes once the gate is above the gate threshold voltage) is high enough. This is going to be a serious challenge, as many transistors *don't list* their transconductance on their datasheet.

We also need to think about the efficiency, which means getting a sufficiently low $R_{\text{DS,on}}$. Output power is 1.8 V × 2 A = 3.6 W, so to get 85% efficiency, input power must be no greater than 3.6 W / 0.85 = 4.24 W, which is to say total losses must be no greater than 4.24 W − 3.6 W = 635 mW, and two thirds of this has already been eaten up in the diode and the inductor. Let's try limiting $R_{\text{DS,on}}$ losses to 10% of the total, 64 mW. We want

$$R_{\text{DS,on}} = \frac{P}{I_{\text{out}}^2} = \frac{64\,mW}{(2\,\text{A})^2} = 16\,m\Omega \tag{6.16}$$

maximum. Of course, this ignores transition losses. Since we're dissipating little power, the package doesn't matter, but at that low an $R_{\text{DS,on}}$, the smallest is a DPAK. There are many manufacturers and over a hundred choices, and we choose the G40P03 because it specifies the transconductance (an astonishing 52 S typical!) with a maximum gate threshold of 2.5 V. Second-source parts can, if necessary live with considerably lesser specifications and still work.

The gate capacitance of this device is less than 3 nF, so we'll start by just leaving the turn-off resistor at 330 Ω and see how the transistor's power dissipation ends up when we simulate it.

For the voltage reference, we'll use the 431, and the feedback resistor can be a 1 kΩ on top and a 2.2142 kΩ on the bottom, which the latter we split into a series 2.21 kΩ 0.1% and a 4.2 Ω 1%. (Why not 10 kΩ on top? As far as the comparator is concerned, this resistor is in series with the feed to the non-inverting input, limiting the hysteresis to 10 kΩ / 1 MΩ = 1%.)

Finally, we must consider the comparator and its hysteresis. Our starting preference is the 339, which can sink a minimum of 6 mA and has a 1 μsec response time. We'll try this in our model and see if it's good enough. As for the hysteresis, we'll try the nominal value of 1% and see if it meets specs.

6.3.2 Modeling

Our starting point is thus the schematic shown in Figure 6.2; now we'll build a model and see to what extent changes seem appropriate.

We want a 16 mΩ p-channel MOSFET; the Si4403 is close, and has a gate capacitance of 2.4 nF (we find this from the datasheet, not the model). We want a Schottky with about 390 mV V_f at 5 A; the RB215T-40 is close to this.

We remember to include a separate component for the 15 mΩ winding resistance of the inductor, and combine the five capacitors into one. We use an ideal 1.24 V voltage source for the 431, and combine the two bottom resistors of the divider into one. Finally, we use the component suggested in Chapter 4 on Modeling, the MAX9095 to represent the 339 comparator.

We now have the initial model shown in Figure 6.3. We've added in the load and a switched load for when we look at step load response. Also, observe that we've added in an initial condition for the output voltage to be at the expected regulation voltage. While this could be set at any voltage, *some* initial condition is absolutely necessary, because otherwise the simulator doesn't know whether to start with the MOSFET on or off.

Simulation using this initial model shows that the peak-to-peak output voltage ripple is way too high, and the switching frequency way too low. These are both remedied by making the hysteresis band smaller, which we do by decreasing the 10 kΩ feed resistor to 1 kΩ, for a total feed of 2 kΩ. This still isn't enough, and with a little probing of the circuit we find that the turn-off delay is too large for this switching transistor, so we change the gate-source resistor from 330 to 100 Ω.

The switching frequency is now 75 kHz, and peak-to-peak output voltage is 90 mV, which would meet specification except that its offset: peak is 1.864 V (+64 mV from 1.800 V) while minimum is 1.773 V (−27 mV). We adjust the top

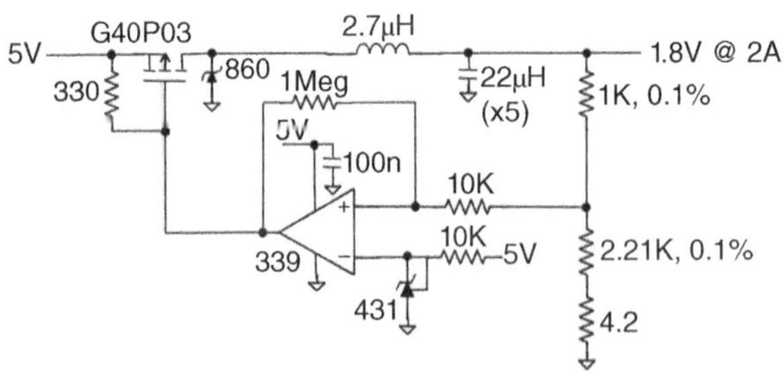

Figure 6.2 Initial schematic for 5–1.8 V at 2 A.

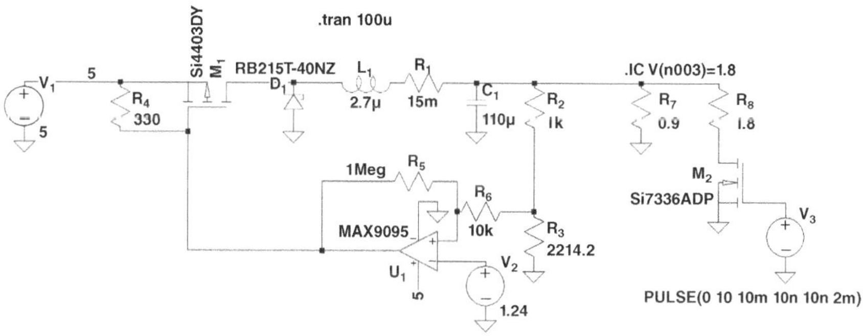

Figure 6.3 Model of initial schematic, 5–1.8 V at 2 A.

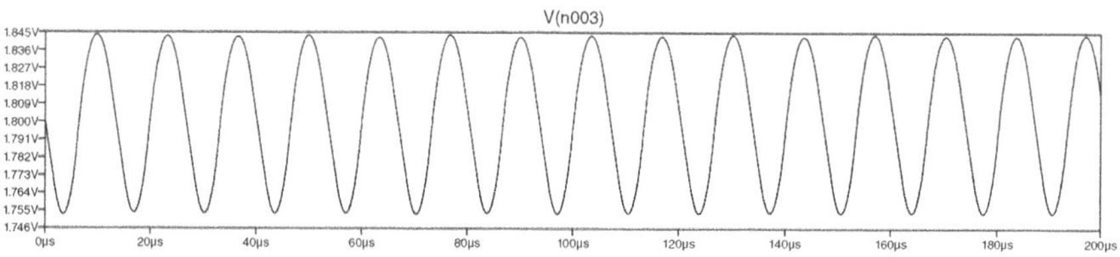

Figure 6.4 5 –1.8 V simulation, max load.

resistor down to 965 Ω, and now the voltage is centered on 1.799 V as shown in Figure 6.4.[1] Using this model, we can now fill in the specification Table 6.2 data for the performance. With 4.75 V input, the output is 1.798 V; with 5.25 V input, 1.803 V. Minimum output is 1.753 V and maximum is 1.844 V.

We next change the load resistor to 1 A and turn on the load step. Although there is no over- or under-shoot in the simulation, you should be aware that where the load step occurs in the switching cycle may influence the size of the voltage over- and under-shoot.

To evaluate the efficiency, all we need do is measure the average output power (with the load step back off again and load set back to maximum) divided by the average input power: $\eta = 3.60\ \text{W} / 4.459\ \text{W} = 81\%$. Note that we haven't included the power running the comparator nor the 431, because these are negligible.

To understand where the losses are, we compile a table of component losses, Table 6.3.

1 Note that to get this smooth-looking waveform in a switching model, we decreased Reltol to 0.00001.

Table 6.2 Specification and performance of 5 to 1.8 V at 2 A bang–bang converter.

Parameter	Specification	Design
Minimum input voltage	4.75 V	<4.75 V
Maximum input voltage	5.25 V	>5.25 V
Minimum output voltage	1.755 V	1.753 V
Maximum output voltage	1.845 V	1.844 V
Minimum output current	0 A	0 A
Maximum output current	2 A	>2 A
V_{ripple}, max	100 mVpp	90 mVpp
50% Step max. undershoot	50 mV	0 mV
50% step min. undershoot	50 mV	0 mV
Frequency, max	100 kHz	76 kHz
Efficiency, 2 A	85%	81%
CISPR25	–	
Size		

Table 6.3 5–1.8 V converter loss table.

Component	Power loss (mW)
MOSFET	258
Diode	428
Inductor resistance	102
Source-gate resistor	34
Total	822

As expected, the dominant loss is in the Schottky (52%), and there's not much that can be done about that without changing the design except getting an even higher rated current device to reduce further the V_f.

Second highest losses are in the MOSFET. Zooming in on the transitions, turn-on time is 800 nsec (sic) and turn-off is 1.05 μsec. This probably accounts for something like 200 mW of the losses in the MOSFET. And since the losses in the pull-up resistor are so small, it is probably worthwhile to try decreasing the pull-up resistance to try to minimize the sum of the two power dissipations. This is left for the reader. Other possibilities include having a comparator with a higher output drive current, or buffering the comparator output with a transistor inverter or a totem pole pair. Both of these last we will look at in later examples in this chapter. We will also look at synchronous rectification to more dramatically boost the efficiency.

Table 6.4 Size and cost of 5–1.8 V at 2 A bang–bang.

Component	Price	Size (mm^2)
Schottky, 40 V 20 A	\$0.25	30
p-MOSFET, 20 V 15 mΩ	\$0.12	30
Inductor, 2.7 μH 15 mΩ	\$0.23	36
LM339	\$0.08	32
431 1%	\$0.07	18
Total	\$0.75	146

6.3.3 Cost and Size

Before turning to efficiency improvement, we should take a brief look at the size and cost of this bang–bang converter. A cost estimate is fairly straightforward. For all the components other than standard surface-mount resistors and capacitors we look up their 1000-piece price at a distributor. If we use n of the same part type, we look at the $1000 \times n$ price. This gives a reasonable first guess at the production price at low to medium volume. Further, following the author's "rule-of-three," these prices then give an indication of how much the unit would cost in very high volume, by dividing the 1000-piece total cost by three. For the design in Figure 6.3, we show this in Table 6.4. Things such as PCB costs, pick-and-place costs, and so on are not included.

Also shown in the table are the sizes of the individual components. Naturally, this doesn't include the size of the traces that interconnect the parts. We may, as a zeroth-order assumption, suppose the number of traces to be roughly the same for every component of equal size. This table also doesn't include the possible extra copper needed for heat-sinking. But at the same level of assumption, this will depend on the efficiency of the converter, so that this will be roughly the same for differing SMPS handling roughly the same power level. To summarize, the approximation being used here, while not useful as a measure of the size of the PCB you're going to end up with, at least allows a comparison between differing designs by having a common basis.

6.4 Improving Efficiency

In the previous section, we've commented several times on the losses in the basic bang–bang converter. While we managed to achieve 81% efficiency on a 1.8 V, 2 A output, this is due to using high-performance (though robust) parts. With higher load currents and/or lower output voltages, finding such parts that are robustly available becomes increasingly difficult. In this section we're going to look at a variety of possibilities for improving the performance.

6.4.1 Pull-down Driver

In the previous example, MOSFET losses were 31% of itemized losses. So driving the MOSFET better would help, at least to some extent. The simplest choice for doing this is to buffer the comparator output. The comparator is open-collector and relies on its pull-up resistor for transistor turn-off. So we could buffer the turn-on simply by putting a bigger transistor at the output of the comparator.

We show how to do this in Figure 6.5. Several alterations have been made. The buffering transistor, the 2222 NPN, needs a pull-up resistor at its base, since the comparator is open-drain; we somewhat arbitrarily pick a resistor ten times larger than the source-gate resistor. Since the buffer is an inverter, we also have to flip the polarity of the comparator. And since the hysteresis has to go to the non-inverting input, we move the hysteresis resistor and its buffer resistor to the non-inverting side as well.

We now have the power losses specified in Table 6.5. Unfortunately, the remeasured power dissipation in the MOSFET is unchanged, because the turn-on is at zero current (we're operating in DCM); but the power dissipation in the Schottky is considerably reduced, presumably because the transition to "on" of the transistor

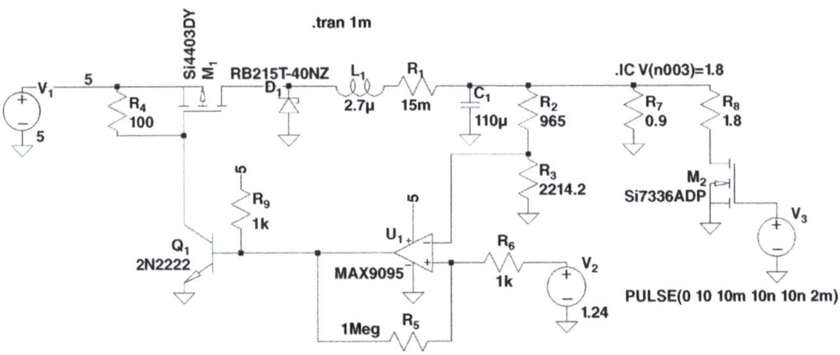

Figure 6.5 Model schematic with improved MOSFET turn-on.

Table 6.5 5–1.8 V loss table, improved buffering.

Component	Power loss (mW)
MOSFET	269
Diode	99
Inductor resistance	99
Source-gate resistor	75
Total	542

is much improved. Before it was 800 nsec, now it is 77 nsec. Total loss is reduced from 822 to 542 mW, saving 280 mW and increasing efficiency from 81 to 87%, for the space and cost of a SOT-23. Clearly, this would be a good choice for higher-power converters operating in CCM.

6.4.2 Totem Pole Driver

The losses in the source-gate resistor weren't large. But as we attempt to speed up the turn-off time, these losses will mount. And when we go to CCM, losses in the turn-on will also become important. For both these reasons, then, we should look at how to actively drive both the turn-on and the turn-off of the switching transistor.

A totem pole can accomplish this. As shown in Figure 6.6, when the base drive to a totem pole is high, the top NPN is on and the bottom PNP is off, the latter since the base is higher than the emitter. Conversely, when the base drive is low, the NPN is off, while the PNP is on. The totem pole is thus a non-inverting driver.

However, the totem pole as described works only if the base drive is high enough to turn the transistors fully on/off. In particular, suppose the common-emitter output of the totem pole needs to be 48 V in order to turn off a p-channel MOSFET whose source is attached to 48 V. Then the base of the totem pole also has to go up to 48 V.[2] If it's driven by 5 V, then, because of the base-emitter drop on the NPN, the output only goes to 4.3 V, which doesn't drive the switching transistor.

So we still need the transistor we added in the previous section to act as a voltage-level shifter. Although unnecessary for our 5 V input converter, we show the model in Figure 6.7 and the re-measured losses in Table 6.6.

Total losses are 587 mW, slightly higher than those of just the pull-down transistor driver. But, of course, it will be very much better when we need faster turn-off at higher currents.

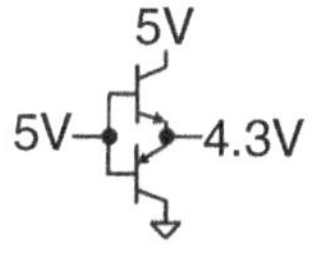

Figure 6.6 Totem pole.

2 If the input voltage is *really* high, it may be necessary to consider a gate-drive transformer. However, these have issues, and are best avoided if at all possible. Yet another possibility is a floating gate driver, in which a gate driver is floated close to the input voltage. But this then needs another whole power supply.

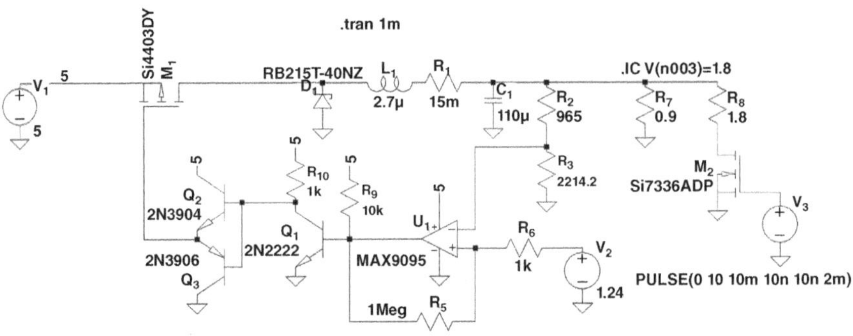

Figure 6.7 Model schematic with totem pole gate driver.

Table 6.6 5 V to 1.8 V loss table, totem pole gate driver.

Component	Power loss (mW)
MOSFET	62
Diode	436
Inductor resistance	89
Source-gate resistor	0
Total	587

6.4.3 Synchronous Rectification

Let's come to the real problem in all of these concepts: the dominant loss is in the rectifier. Fixing this would dramatically increase the efficiency. It's not just a matter of finding bigger and better Schottkies. While nowadays Schottkies are available at many hundreds of volts, their V_f at these stand-off voltages is often no better than that of ordinary (fast-recovery) diodes. For higher voltages and to achieve higher efficiencies, what is needed is synchronous rectification.

The concept is shown in Figure 6.8. When the input to the gates is high, the switching transistor is off, while the synchronous rectifier is on. When the gates are low, they are reversed, so that only one is ever on at a time, and there is only a single drive needed!

If you look carefully at the figure, you'll realize that the n-channel MOSFET used as the synchronous rectifier has a body diode that could in principle be used as the rectifier. This is undesirable for at least two reasons: the V_f of this diode is usually pretty high, and the reverse recovery is usually pretty long.

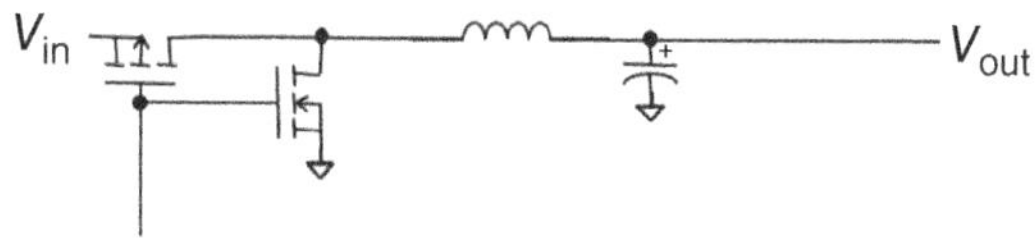

Figure 6.8 Concept of synchronous rectification.

There are a couple potential problems with this pleasant idea. Just as the p-channel MOSFET can't have its gate pulled more than 20 V below its source and thus requires some sort of voltage limit, the n-channel MOSFET can't have its gate pulled more than 20 V *above* its source. So it too needs some sort of voltage buffering. Thus, for low voltage inputs, say 15 V and less, Figure 6.8 shows a realistic connection. For higher voltage inputs, the two gates will need to be separately driven, albeit in phase.

Another problem involves shoot-through. If there were ever an occasion when both transistors were to be on at the same time, there would be a very-low resistance path from the input to ground. This would be a bad thing. However, if there were a short delay between the turn-off of one and the turn-on of the other and *mutatis mutandis*, a shoot-through wouldn't occur. So, in IC gate drivers, it is common to put a small delay between the two outputs.

6.5 Example: 4–1 V at 100 A

The idea of this section is to design a power supply operating from an auxiliary 48 V battery in an EV to run a (relatively) low-power AI chip which requires 1 V at 100 A. Clearly, with such a low output voltage, synchronous rectification will be essential. A diode, even a Schottky, is likely to have a V_f close to a volt, which being equal to the output voltage implies the efficiency would be close to 50%! It should also be clear that we will be using CCM to avoid excessive RMS current in the MOSFETs.

Let's start with some specifications in Table 6.7.

A few comments are in order. We've set the switching frequency low, because the duty cycle of the top switching transistor is going to be very small. This would make the on-time of the transistor difficult to control. We've set the target efficiency fairly low, since the output is such a low voltage at such a high current. And finally, since this is automotive this time, we will have to measure the conducted emissions.

Table 6.7 Specification of 48–1 V at 100 A bang-bang converter.

Parameter	Specification	Design
Minimum input voltage	38 V	
Maximum input voltage	52 V	
Minimum output voltage	0.975 V	
Maximum output voltage	1.025 V	
Minimum output current	20 A	
Maximum output current	100 A	
V_{ripple}, max	20 mVpp	
50% step max. undershoot	20 mV	
50% step min. undershoot	20 mV	
Frequency, max	20 kHz	
Efficiency	81%	
CISPR25 margin	6 dB	
Size		

6.5.1 Robust Design

As we did last time, we start by computing the basic operating values for the converter. We have the duty cycle

$$\mathrm{DC} = \frac{V_{\mathrm{out}} + V_{\mathrm{f}}}{V_{\mathrm{in}} + V_{\mathrm{f}}} = \frac{1\,\mathrm{V} + 100\,\mathrm{mV}}{48\,\mathrm{V} + 100\,\mathrm{mV}} = 2.3\% \tag{6.17}$$

We have made an initial guess of the $R_{\mathrm{DS,on}}$ drop of the synchronous rectifier of 100 mV, which at 100 A corresponds to 1 mΩ and may be a reasonable guess at this quite low V_{DS}. The duty cycle depends quite strongly on this value, since the output voltage is only 1 V, but since we're only looking for a starting point, the result won't change hugely if we're somewhat off.

Next, since we're operating in CCM, we have an estimate for the inductor,

$$L = \frac{5(V_{\mathrm{in}} - V_{\mathrm{out}})\mathrm{DC}}{I_{\mathrm{out}}f} = \frac{5 \times (48\,\mathrm{V} - 1\,\mathrm{V}) \times 0.023}{100\,\mathrm{A} \times 20\,\mathrm{kHz}} = 2.7\,\mu H \tag{6.18}$$

with peak current of

$$I_{\mathrm{pk}} = I_{\mathrm{out}} \times 1.1 = 110\,\mathrm{A} \tag{6.19}$$

the factor of 1.1, reflecting that 20% ripple, is 10% maximum above average.

And finally, for the capacitor,

$$C = \frac{I_{\mathrm{out}} \times \mathrm{DC}}{5 \times V_{\mathrm{pp}} \times f} = \frac{100\,\mathrm{A} \times 0.023}{5 \times 20\,\mathrm{mV} \times 20\,\mathrm{kHz}} = 1150\,\mu F \tag{6.20}$$

Starting with the inductor, while there are two vendors making inductors of 2.7 μH at 100 A, they are very pricey. Instead, we look for two inductors in parallel, each of which should be about 5.4 μH at 50 A. There are a number of such from a number of vendors, ranging from 4.7 to 6.8 μH with the less expensive ones having about 1.2 mΩ winding resistance, and having roughly similar footprints.

For the capacitor, this is going to be too much for a ceramic, even with such a low voltage. We can turn to aluminum-polymer capacitors, for which there are many vendors and many choices. Choosing surface-mount, a common capacitor is 1000 μF at 2.5 V, with an ESR of 15 mΩ, and a lifetime of 2000 hours at 105 °C. At the 25 °C rated operating temperature, this will have a life of

$$\text{life} = 2000\,\text{h} \times 2^{\frac{(105-25)\,^\circ\text{C}}{10\,^\circ\text{C}}} = 512{,}000\,\text{h} \tag{6.21}$$

We can choose through simulation whether to use a single one or two of these in parallel, or perhaps one of these with the remaining 150 μF in ceramic.

Now we need to choose the switching transistors. 60 V V_{DS} is adequate. For the top transistor, the duty cycle is only 2.3%, so the $R_{DS,on}$ is not very important; at 100 A, the power lost per milli-ohm is only 230 mW. (Recall that since we're using synchronous rectification, our top MOSFET can be n-channel.) But for the synchronous rectifier, which is conducting almost 100% of the time, at 100 A there will be 10 W lost for each milli-ohm, so getting very low $R_{DS,on}$ for the synchronous rectifier is critical to efficiency. There are a number of vendors with a number of parts meeting these specs; we choose one in an 8-DFNW package, and will use two, which will keep power per package down to 2.5 W, and total power loss 5 W, which is 5% of output power.

With a low-switching frequency, the 339 comparator will be fine, and we use a 1% 431 to achieve the 2.5% output voltage tolerance. We also want voltage supplies to run the 431, comparator and synchronous rectifier driver. We'll simply assume that somewhere in the EV there are 5 and 12 V already available, though it would be easy enough to design LDOs to generate these voltages directly from the battery.

6.5.2 Modeling

So our starting point for a model is shown in Figure 6.9. We have the (model of the) switching transistor at 8.9 mΩ, while the synchronous rectifier is 0.4 mΩ, corresponding to two 0.8 mΩ transistors in parallel.

The only other thing requiring mention isn't shown on the LTspice schematic. LTspice's native inductor model simply assumes that all inductors have a default winding resistance of 1 mΩ, which, of course, is rather larger than our actual winding resistance. To make sure we don't double count, we set the built-in inductor winding resistance low; we used 1 μΩ, which as we've commented before, doesn't show up on the schematic.

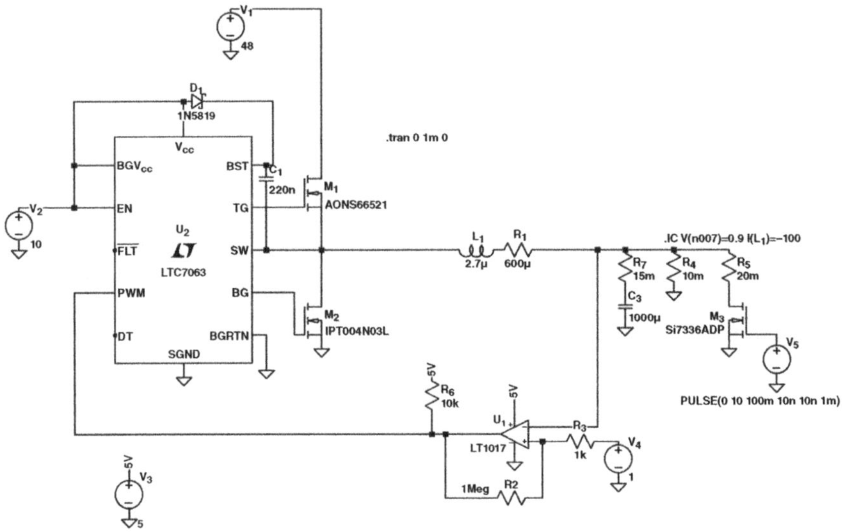

Figure 6.9 48–1 V at 100 A preliminary model.

We can now run a simulation to see where we are. If you try this yourself, you'll notice that the simulation is very slow; presumably this is because there are many things going on inside the IC model (at a guess, digital things, which LTspice is not too good at). The file size is also going to get very large for the same reason, so the general plan is going to be to simulate just a few switching cycles until we're at about the value set that gives us the performance we want, before running enough switching cycles to get good performance numbers.

This model has the output voltage between 1.24 and 0.96 V, which is both too large a spread and offset. Inductor current is between 140 and 95 A, also too large and offset. And the frequency of the switching is too low, just below 10 kHz.

Some examination of the circuit waveforms suggests that the issue may be the comparator speed. The output doesn't even reach 5 V before it starts coming back down. We replace the model with a faster comparator, the MAX9095. Now the output voltage is between 1.017 V and 998 mV, which is in spec, just slightly offset. The inductor current is now 103–99 A, which is great, and the switching frequency is 125 kHz. Everything is looking good, so we can calculate the efficiency as power in the load resistor divided by power from the 48 V source: 101.5 W / 121.0 W = 84%, very acceptable considering the very high current, the low voltage at the output and the switching frequency. The measured parameters are shown in Table 6.8.

We still need to think about EMI. CISPR25 conducted emissions requirements run from 150 kHz to 108 MHz in certain specific frequency bands. For our purposes, we will only be considering the two lowest frequencies, for two reasons: (1) our simulation isn't accurate enough to know what is happening at the very small timescales of the high frequencies; and (2) the high frequencies are usually strongly impacted by mechanical layout, not just electronic design, and are thus not adequately accounted for by an electrical simulation.

That said, then, the two bands about which we will be concerned are the LW band, from 150 to 300 kHz, and then the MW band, from 530 kHz to 108 MHz. The limits are 70 dBμV for the former and 54 dBμV for the latter.

We add in the model of the LISN network as shown in Figure 6.10, basically 5 μH and a 100 nF/1 kΩ divider for the measurement port. Inserting this network

Table 6.8 Specification and performance of 48–1 V at 100 A bang–bang converter.

Parameter	Specification	Design
Minimum input voltage	38 V	<38 V
Maximum input voltage	52 V	>52 V
Minimum output voltage	0.975 V	0.998 V
Maximum output voltage	1.025 V	1.017 V
Minimum output current	20 A	<20 A
Maximum output current	100 A	100 A
V_{ripple}, max	20 mVpp	19 mVpp
50% step max. undershoot	20 mV	30 mV
50% step min. undershoot	20 mV	4 mV
Frequency, max	20 kHz	125 kHz
Efficiency	81%	84%
CISPR25 margin	6 dB	15 dB
Size	–	–

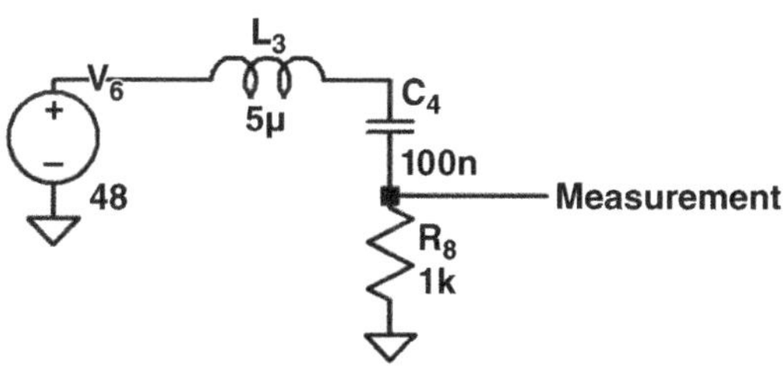

Figure 6.10 Artificial network for CISPR25.

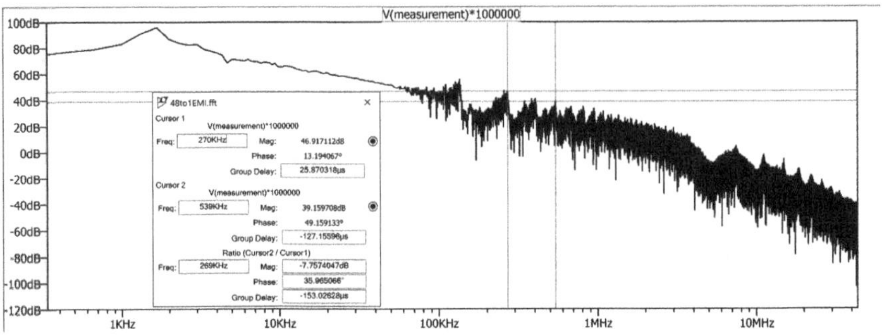

Figure 6.11 48–1 V at 100 A EMI spectrum.

into the model and doing the FFT[3] on the node marked "Measurement," we see that everything is 120 dBμV or more, as there are 7 kV (sic) spikes from the MOSFET turning on into an inductor. Obviously, we need an input capacitor.

Adding in a 100 μF, say in the form of five 22 μF ceramics, reduces the peak in the LW band to 85 dBμV and in the MW band to 67 dB. We need an extra, say, 20 dB of attenuation. Making the capacitor 10 times bigger would require going to an electrolytic; the other choice is having a discrete inductor in the input line, which is easy enough since input current is only 2 A. We choose 50 μH, with the result shown in Figure 6.11:[4] in the LW band the peak is 47 dBμV at 270 kHz, and in the MW band the peak is 39 dBμV at 539 kHz, both in spec. Of course, this is just a simulation; you have to go to an EMI lab and check for real.

We could consider a number of things at this point. For example, the frequency is higher than we expected, and this is presumably reflected in the output voltage and inductor current ranges being smaller than we targeted; we could try decreasing the hysteresis, on the presumption (not necessarily true) that this would improve the efficiency.

Another thing that could be looked at would be altering the dead time between when one transistor turns off and the other turns on. This actually has a control on the IC used in the model, the "FLT" pin. However, the common ICs that we would use for robust design typically don't have this control, so we could see what common dead-times are for those and adjust the FLT pin to match.

This design seems ready to go to a prototype and be tried out. Since the current is *so* high, the load needs to be very carefully located as close as possible to the converter, and adequate copper has to be used throughout the power stage to ensure there aren't large power losses. Perhaps busbars should be considered.

---

3 FFT is available in LTspice under "View."

4 The voltage at the measurement point is multiplied by one million so that the scale is in dBμV rather than dBV.

6.6 Example: Bang–Bang Boost

Both of our examples so far in this chapter have been buck converters, as this was traditionally the sole use of bang–bang converters. However, at this point it should be clear that there's no particular reason why other types of converters can't be built using bang–bang control, because it *is* fundamentally just a different type of control, and doesn't influence the power stage.

In this example, then, we're going to do a bang–bang converter with 5 V in and 12 V at 0.5 A out. The specifications are shown in Table 6.9.

Now the deterrent to controlling a boost with a bang–bang control is the right-half plane zero (RHPZ) inherent in a flyback. Look at Figure 6.12 to understand this. In a buck converter, when the transistor is turned on, the inductor current starts to rise, causing the voltage to start to rise.

Table 6.9 Specification of 5–12 V at 500 mA bang–bang boost converter.

Parameter	Specification	Design
Minimum input voltage	4.75 V	
Maximum input voltage	5.25 V	
Minimum output voltage	11.4 V	
Maximum output voltage	12.6 V	
Minimum output current	0 A	
Maximum output current	500 mA	
V_{ripple}, max	500 mVpp	
50% step max. undershoot	100 mV	
50% step min. undershoot	100 mV	
Frequency	100 kHz	
Efficiency, 500 mA	85%	

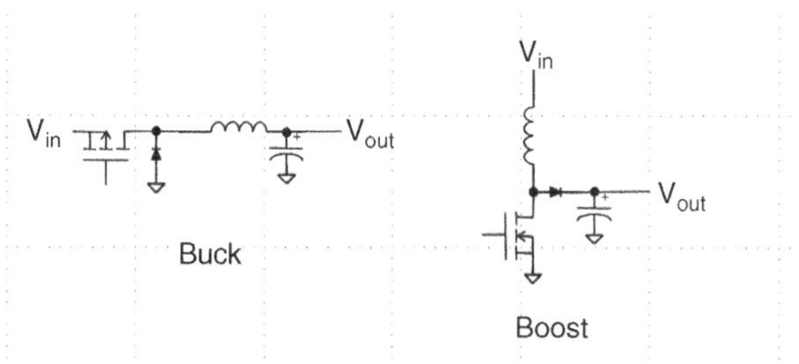

Figure 6.12 Buck vs. boost concept.

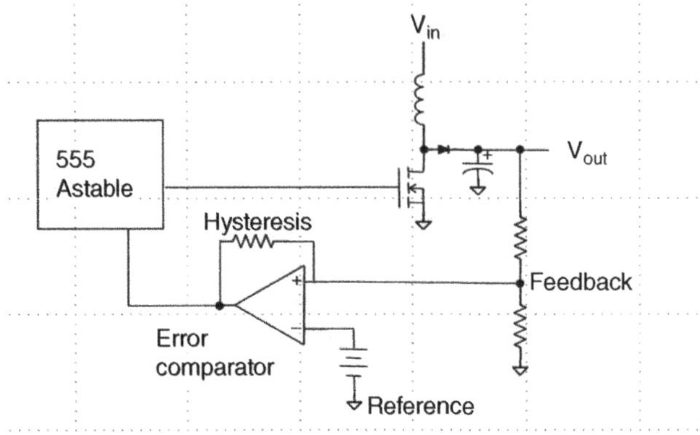

Figure 6.13 Concept of constant on-time bang–bang boost.

In the boost converter, however, when the transistor is turned on, the current also starts to rise, but the voltage on the capacitor doesn't start to rise *until the transistor is turned off*, because the current flows to the ground while the transistor is on.

This RHPZ, then, causes an issue for a simple-minded extension of the bang–bang control. When the output voltage is too low, the comparator would turn the transistor on—and then the voltage would *remain* too low because the voltage doesn't rise; the MOSFET would remain permanently on!

To circumvent this, the author introduces this novel, robust control scheme, which might be called a constant on-time bang–bang control. Looking at Figure 6.13, the transistor is controlled by an astable 555 timer set to have a constant frequency and duty cycle. The power is delivered to the output once every cycle, this is the constant on-time. When the output goes too high, the comparator shuts off the 555; this is the bang–bang control. When the voltage goes low again, the comparator re-enables the 555.

6.6.1 Robust Design

Let's put together some equations for this type of converter. The power delivered to the output is

$$P = \frac{1}{2} L I_{\text{pk}}^2 f \tag{6.22}$$

If we assume that the (maximum) power is given and the frequency is set by us to some convenient level, we can solve this for L and I:

$$L I_{\text{pk}}^2 = \frac{2P}{f} \tag{6.23}$$

We also know that the peak current is set by how long the MOSFET is on,

$$\frac{V_{\text{in}}}{L} = \frac{I_{\text{pk}}}{t_{\text{on}}} = \frac{I_{\text{pk}}f}{\text{DC}} \tag{6.24}$$

so that

$$LI_{\text{pk}} = \frac{V_{\text{in}}\text{DC}}{f} \tag{6.25}$$

Let's assume that we set the duty cycle also to some convenient value;[5] then we can divide the first equation by the second to find the required peak current:

$$I_{\text{pk}} = \frac{2P}{V_{\text{in}}\text{DC}} \tag{6.26}$$

and from this, we can now find the inductance,

$$L = \frac{V_{\text{in}}^2\text{DC}^2}{2fP} \tag{6.27}$$

For this design, P = 12 V × 500 mA = 6 W; frequency is 100 kHz; and let's set the duty cycle to 30%. Then

$$L \leq \frac{5V^2 30\%^2}{2 \times 100\,\text{kHz} \times 6\,\text{W}} = 1.8\,\mu H \tag{6.28}$$

with a peak current of

$$I_{\text{pk}} = \frac{2 \times 6\,\text{W}}{5\,\text{V} \times 30\%} = 8\,\text{A} \tag{6.29}$$

The other component is the output capacitor. For each switching cycle, the capacitor receives a total charge of

$$Q = I_{\text{avg}}t = \frac{I_{\text{pk}}}{2}\frac{\text{DC}}{f} \tag{6.30}$$

and since the charge is CV, if we want to keep the ripple voltage in this case to V_{pp},

$$C = \frac{Q}{V_{\text{pp}}} = \frac{I_{\text{pk}}DC}{2fV_{\text{pp}}} \tag{6.31}$$

with $V_{\text{pp}} = 100\,\text{mV}$, we have

$$C = \frac{8\,\text{A} \times 30\%}{2 \times 100\,\text{kHz} \times 100\,\text{mV}} = 120\,\mu F \tag{6.32}$$

readily available as five 22 μF ceramic capacitors.

---

5 Less than 50% since we are inverting the output of the 555, which has a minimum DC of 50%.

As for the transistor and diode, the average current is low (6 W / 5 V = 1.2 A), but the peak (for the diode) is high (8 A) and the RMS (for the transistor) is moderate, ($8\text{ A} \times \sqrt{\frac{30\%}{3}} = 2.5\text{ A}$). We can just pick a 10 mΩ $R_{DS,on}$, giving us power dissipation of $2.5\text{ A}^2 \times 10\text{ m}\Omega \times 30\% = 20\text{ mW}$, and the 860 type Schottky has V_f of 400 mV at 4 A during the 70% of the time it's conducting giving a power dissipation of 1.1 W.

6.6.2 Modeling

Putting together a model, we have Figure 6.14. We've done a few things here worth mentioning. Most significantly, we've ramped the reference voltage with a rise time of 3 msec in order to have a clean startup, and rather than trying to specify the initial condition, we just let the output start from 5 V, which it will be anyway, being a boost converter. And, since we're driving the MOSFET directly from the 555, we've picked one with a low gate charge in order to be able to turn it on and off quickly, improving the efficiency. We've put in a similar Schottky, since LTspice doesn't have the 860. We've also changed out the standard 358 for a faster comparator.

Once the converter reaches steady state, the voltage ranges from to 12.52 V, and should be shifted down by 280 mV to center it. Switching frequency is 8.3 kHz, and efficiency is eye-popping 6.22 W / 6.97 W = 89%.

Everything looks great, until we look at all the internal nodes' waveforms, see Figure 6.15. The gate drive is miraculously only 8.7 μsec / 125.8 μsec = 7%—and yet we know that the 555 has a minimum duty cycle of 50%, that's why it's always necessary to invert it. Except for this time. What's going on?

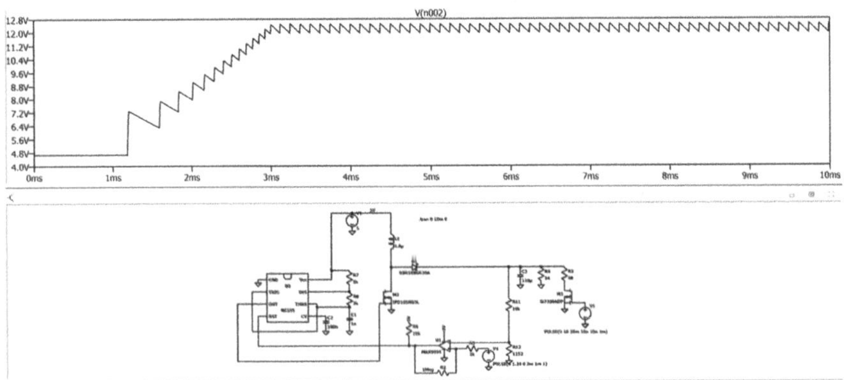

Figure 6.14 Bang-bang boost startup, 5–12 V.

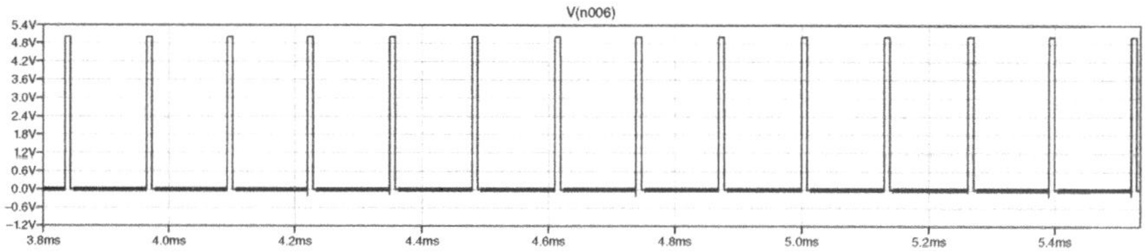

Figure 6.15 Constant on-time boost gate drive signal.

Table 6.10 Specification and measurements of 5–12 V at 500 mA bang–bang boost converter.

Parameter	Specification	Design
Minimum input voltage	4.75 V	<4.75 V
Maximum input voltage	5.25 V	>5.25 V
Minimum output voltage	11.4 V	11.8 V
Maximum output voltage	12.6 V	12.3 V
Minimum output current	0 A	0 A
Maximum output current	500 mA	>500 mA
V_{ripple}, max	500 mVpp	560 mVpp
50% step max. undershoot	100 mV	0 mV
50% step min. undershoot	100 mV	0 mV
Frequency	100 kHz	8.3 kHz
Efficiency, 500 mA	85%	89%

It turns out that we are truncating the pulse each time by the comparator asserting the RST pin 8.7 μsec after the output pulse starts. So the 555 turns on, and then suddenly it's reset. That's how we get less than 50% duty cycle on a 555!

Does it matter? Probably not. The converter works, producing a regulated 12 V. And it's stable in the sense that changing the parameters a bit doesn't affect the performance. Indeed, this might be considered a useful innovation: Now we know how to get less than 50% duty cycle on a 555!

Measurements are contained in Table 6.10.

6.7 Example: Multi-slice Bang–Bang

In a previous section, we designed a 100 A bang–bang buck converter, dropping the input voltage by a factor of 50 with 84% efficiency. Pretty impressive! But now what would happen if, instead, we needed even higher current? At some point, not too much higher in current, we're going to have to parallel up MOSFETs and inductors and capacitors, and then there will end up being difficult layout questions. And we already have a pretty high gate drive current. At some point, increasing the capability and number of MOSFETs is going to start slowing them down, resulting in decreased efficiency.

What we're going to propose in this section is the novel concept of a multi-slice bang–bang controller.[6] While multi-slice PWM control has been common for 20 years for powering microprocessors, the ability to do so using bang–bang control has never been explored.

Of course, the major challenge of a multi-slice converter is to ensure that the current is evenly shared between slices, for if one slice takes most of the current, the purpose of the multi-slice architecture is defeated. Multi-slice converters are usually also staggered in time, so that the net ripple voltage is reduced.

6.7.1 Extension to 200 A

What we specifically are going to design is the 48–1 V converter, but this time with double the output current 200 A. So the naive thing to do would be to simply copy our 100 A design, and attach the two outputs together, as in Figure 6.16. You shouldn't be surprised that this works perfectly because the inductors are identical. The reason, of course, is that the simulator has two *identical* copies, and so the inductances are identical, the capacitors' ESRs are identical, the trip-points are identical and so on. In fact, there would be something seriously wrong if the currents *were* different![7]

If we change some of the parameters,[8] however, we find that there is not just a mismatch in the inductor currents, but one converter ends up putting out the entire 200 A and the other zero. If you vary things even more, one of them may source *more* than 200 A and the other *sink* current from the first!

Clearly, this is not acceptable. The issue turns out to be that each converter is sensing at the output of the system, so each is directly changed by what the other is doing. The slightest imbalance between the slices is amplified, resulting in complete instability.

The solution, it turns out, is to buffer the sense-point of the slices from each other. We do this, as shown in Figure 6.17, by adding in a small inductor in each slice after its sense-point (1 μH with 125 $\mu\Omega$ winding resistance, a multi-source part) and only after these recombine the two outputs into the actual 200 A output.

6 We avoid calling this "multi-phase" as the word phase will have a different meaning when we turn to AC supplies.

7 Actually, if they were different, it might be the resolution and accuracy of the simulation, which, over enough time steps, integrates the error into something noticeable.

8 Note that we're only talking about parameter tolerance; this does *not* take into account aging effects, as we may assume that these are equal between two identical converters.

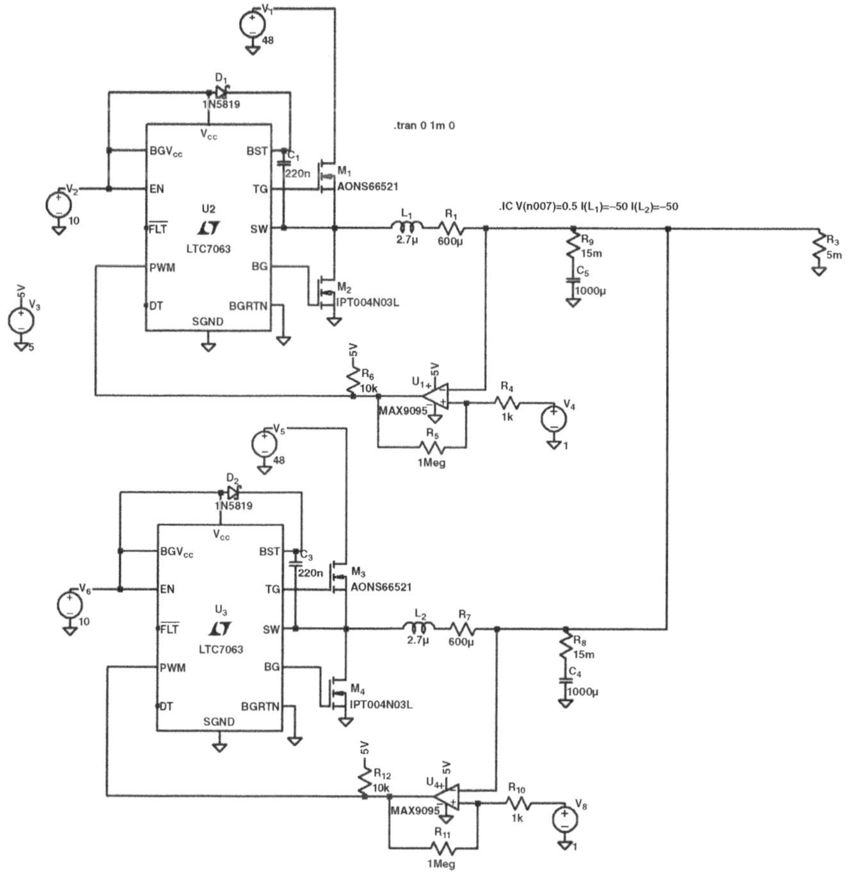

Figure 6.16 Model of conceptual two-slice bang–bang buck at 200 A.

We've also added in one more 1 mF aluminum-polymer capacitor so that the load doesn't see an inductive impedance.

The results seem quite good. If the bottom input voltage drops from 48 to 42 V (this number was picked to speed up the simulation), the voltage ripple at the output is 1 mV at a frequency of 164 kHz. The current on the top converter is 116 A, and on the bottom it is 84 A.[9]

---

9 To get these results requires waiting for the simulation to reach steady-state, which means extending the run-time to 20 msec (sic), which is over 3000 switching cycles, very stiff! To make this feasible, we have LTspice record only the final 10 msec. However, this still generated a file of >1.3 GB, and the run-time on a high-end workstation was eight minutes! You were warned.

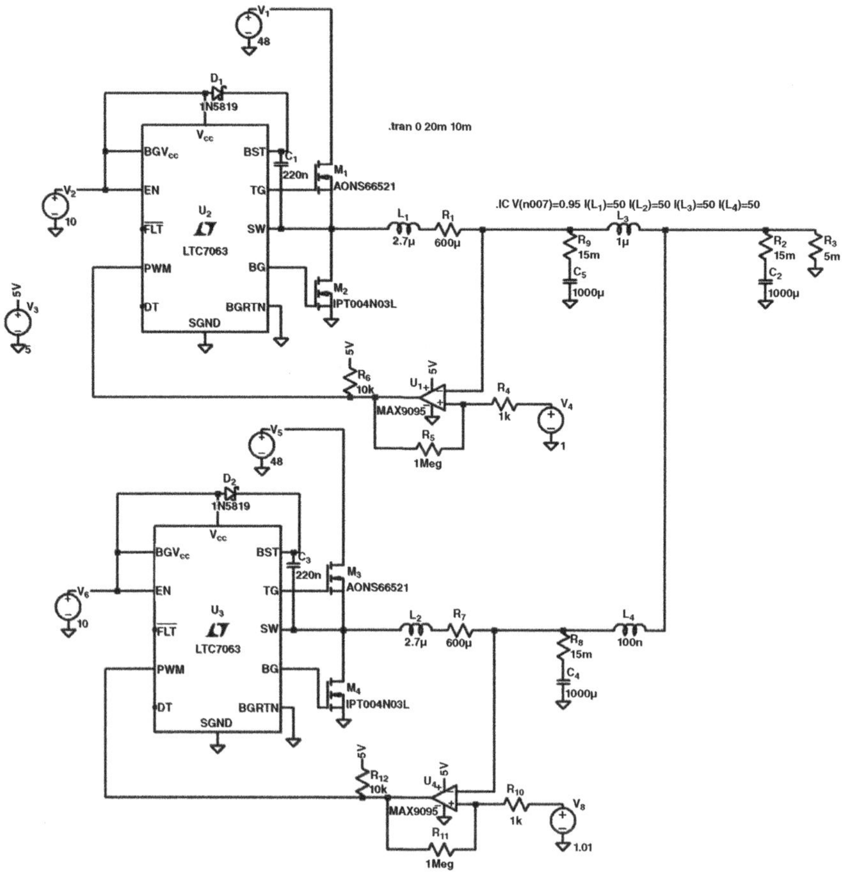

Figure 6.17 Two-slice current-balanced bang–bang 200 A at 1 V.

We also tried raising the bottom reference voltage from 1.1000 to 1.1001 V. After 20 msec, the average output voltage was 1.1004 V, with ripple voltage of only 10 μVpp. The upper inductor current was 104.9 A, while the lower was 95.9 A.

To make sure of this result, we also tried raising the bottom reference voltage by 1% rather than just 0.01%, from 1.00 to 1.01 V. The average output voltage after 20 msec was 1.008 V, with ripple voltage of 1.2 mVpp. The upper inductor current was 71 A, while the lower was 130 A. Probably the right solution will be to have a single voltage reference, which is then passed between slices using instrumentation amplifiers to remove ground voltages.

So this system can be used to make a two-slice bang–bang converter, and presumably can be extended to multi-slice. Of course, this certainly needs to be prototyped for verification.

6.8 Example: Very Low Current

For our final example, we want to look at the advantage of using a bang–bang converter for very low power, as bang–bangs have a minimal number of active components. In particular, we are going to design a boost converter that converts two AA alkaline cells (in series) to 5 V, at a current of 1 mA, for a power output of 5 mW.

Clearly, for this very low power SMPS, we want to minimize control components' current drain, in order to maximize efficiency. The bang–bang boost converter's only ICs are the comparator and the reference voltage. Of course, there is also the transition losses of the transistor, which we can adjust by controlling the switching frequency, the $R_{DS,on}$ loss of the transistor, for which we can select a suitable MOSFET, and the V_f of the diode. There are also resistors, but these can be made very large value.

6.8.1 Robust Design

Let's start with a specification as shown in Table 6.11. The battery's useful range is expected to be from 1.55 V/cell when new down to 1.00 V/cell when almost fully discharged. We have set a target efficiency of 70% because that is typical of custom ICs for this type of power level, but we will make an effort to do better than this. The target efficiency is set at the voltage at which the battery spends most of its time 2.6 V (1.3 V/cell). We haven't set a switching frequency, as we're going to decide this based on a trade-off between the inductor size and the MOSFET switching losses.

Table 6.11 Specification and measurements of two AA cells to 5 V at 1 mA, bang–bang boost converter.

Parameter	Specification	Design
Minimum input voltage	2.00 V	
Maximum input voltage	3.10 V	
Minimum output voltage	4.75 V	
Maximum output voltage	5.25 V	
Minimum output current	0 A	
Maximum output current	1 mA	
V_{ripple}, max	500 mVpp	
50% step max. undershoot	50 mV	
50% step max. overshoot	50 mV	
Efficiency, 1 mA, 2.6 Vin	70%	

As with the bang–bang boost in an earlier example, we start by estimating the values of the power components. Using the same equations, we have the inductance as

$$L = \frac{V_{\mathrm{in}}^2 \mathrm{DC}^2}{2fP} \tag{6.33}$$

with a peak current of

$$I_{\mathrm{pk}} = \frac{2P}{V_{\mathrm{in}}\mathrm{DC}} \tag{6.34}$$

V_{in} ranges from 2.0 to 3.1 V, and since we want to operate in DCM, we pick the lowest value for V_{in}.[10] We can select a maximum duty cycle of 40%, and the only thing remaining to calculate the inductor value is the frequency; let's try 100 kHz as a first guess. We get

$$L = \frac{2\,V^2 0.4^2}{2 \times 100\,\mathrm{kHz}\,5\,\mathrm{mW}} = 640\,\mu H \tag{6.35}$$

with a peak current of

$$I_{\mathrm{pk}} = \frac{2 \times 5\,\mathrm{mW}}{2\,\mathrm{V} \times 0.4} = 12.5\,\mathrm{mA} \tag{6.36}$$

The peak current is very low, so the winding resistance of the inductor and the $R_{\mathrm{DS,on}}$ losses of the MOSFET won't matter. We might consider reducing the inductance by increasing the frequency, but this might necessitate a faster and thus higher supply current comparator, so we stick with 100 kHz. We end up with a 680 μH inductor, and in a reasonably small-sized surface mount part, we end up with a typical 17 Ω.

For the output capacitor, we have

$$C = \frac{I_{\mathrm{pk}}\mathrm{DC}}{2fV_{\mathrm{pp}}} = \frac{12.5\,\mathrm{mA} \times 0.4}{2 \times 100\,\mathrm{kHz} \times 50\,\mathrm{mV}} = 500\,\mathrm{nF} \tag{6.37}$$

and 470 nF is readily available as a ceramic.

As for the transistor, we don't, as already mentioned, care much about the $R_{\mathrm{DS,on}}$, although the gate charge should be relatively low. Of importance, though, is that the gate has to be driven by only 2 V, since that's the minimum input.[11] So we

10 Higher V_{in} will cause the current to reach its peak sooner, so inductor current will certainly be discontinuous if it is at lower V_{in}.

11 We can't connect the input and output together, as then the minimum operating voltage would be a diode drop below 2 V, and the unit would never start.

need a MOSFET with a gate threshold voltage of not more than, say 1.5 V. Since the power is low, we look for a MOSFET in a SOT-23; there are a number of these available from multiple vendors, so we pick something like the BSS138, with typical parameters of 2.5 Ω and 1.7 nC gate charge.

Finally, for the diode, the 4148 type has a typical V_f of 400 mV at 12.5 mA during the 60% of the time it's conducting.

6.8.2 Modeling

So the circuit we're considering has a model as shown in Figure 6.18. Since we're so concerned about even minuscule amounts of power loss, this time we're careful that we model everything and that *everything* is powered from the input—even the comparator and the reference voltage.

There are a few things we've done to translate our parameters into a schematic.

- We've increased the resistors on the timer to reduce power consumption, while decreasing the capacitor to maintain the 100 kHz.
- We've added a 3.3 Ω gate resistor to suppress high-frequency ringing.
- The LT431 model in LTspice uses 1 mA, whereas we want to use the type of 431 that regulates down to 100 μA. The LT1004-1.2 can be used as low as 10 μA, but we bias it so that it uses the 100 μA we will actually have.

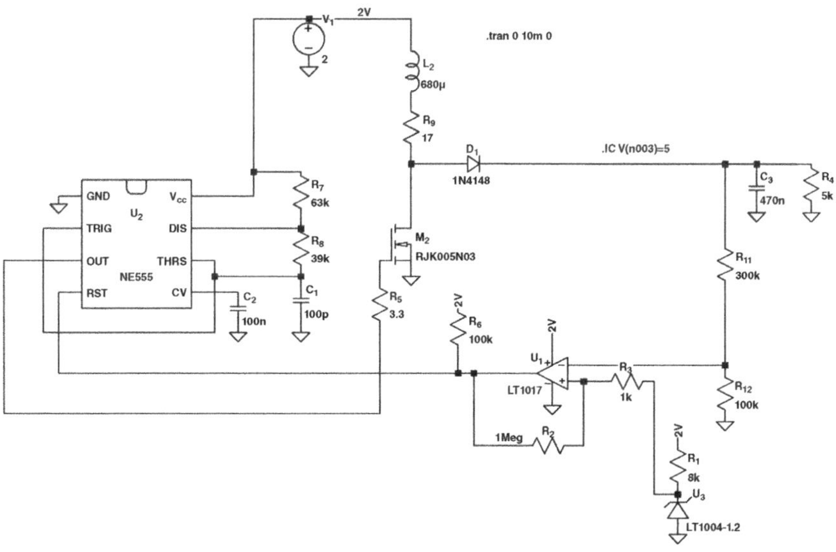

Figure 6.18 2 AA cells to 5 V at 1 mA bang–bang initial schematic.

- There aren't many low-gate threshold MOSFET models in LTspice, we found the RJK005N03 in a SOT-23, which, looking at its datasheet, has a $V_{gs,th}$ of 1.5 V at 1 mA, an $R_{DS,on}$ of 400 mΩ and a gate charge of 2 nC. With a minimum transconductance of 500 mS, at 2 V it should be capable of conducting 250 mA.

Running this model[12] we assemble a Table 6.12 of power dissipations at 2 V input. Some of the power dissipations will, of course, be higher at 3.1 V input.

We systematically investigate the components dissipating the highest power, as with 5000 μW output power, these losses would cause the efficiency to be terrible.

To start with the highest power component first, measuring the V_{CC} current to the NE555 model shows 1 mA. However, our plan is to use the (multi-sourced) CMOS version of the 555, which has a guaranteed maximum current of 200 μA at 2 V. To fix the model, we will power the NE555 model from a separate source (which follows the voltage of the input voltage), while adding in a 200 μA load to the input power.

Next highest power is the 4148. We try a higher current Schottky, the 1N5817, which reduces power in the rectifier from to 337 μW.

The inductor resistance also seems to be too high. We choose a different type of multi-source surface-mount 680 μH inductor, somewhat larger, with typical 1.3 Ω.

Table 6.12 Power dissipation in initial very-low current schematic.

Component	Pd (μW)
555	1938
R7	55
R8	2
Inductor resistance	1205
Transistor	77
4148	1530
R11	47
R12	16
R1	73
LT1004	117
LT1017	252
R6	31
Total	5343

12 You need to add an extra zero to Reltol to get accurate enough measurements.

Our improved schematic shown in Figure 6.19 then has output power at 2 V of 5.180 mW and input power of 6.569 mW, for an efficiency of 79%, quite good for this low a power level. At 3.1 V input, the numbers are 5.179 mW / 7.583 mW = 68%. Efficiency at the specification point is 5.044 mW / 6.533 mW = 77%.

We now start filling in the rest of the specification Table 6.13. The first thing we find is that at 3.1 V input, the output voltage ripple is enormous. We trace this back to the response-time of the controller, which allows the peak current in the inductor to get quite high before turning off the transistor. As a solution, we increase the output capacitor by ten to 4.7 μF, still a common ceramic.

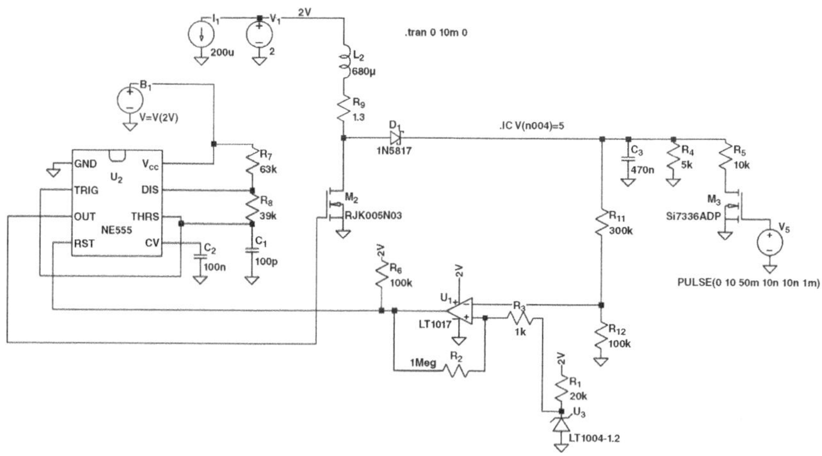

Figure 6.19 2AA cells to 5 V at 1 mA bang–bang improved schematic.

Table 6.13 Specification and measurements of two AA cells to 5 V at 1 mA bang–bang boost converter.

Parameter	Specification	Design
Minimum input voltage	2.00 V	2.00 V
Maximum input voltage	3.10 V	3.10 V
Minimum output voltage	4.75 V	4.98 V
Maximum output voltage	5.25 V	5.29 V
Minimum output current	0 A	0 A
Maximum output current	1 mA	1 mA
V_{ripple}, max	500 mVpp	330 mVpp
50% step max. undershoot	50 mV	0 mV
50% step max. overshoot	50 mV	0 mV
Efficiency, 1 mA, 2.6 mV	70%	77%

Due to the bang–bang control and the larger output capacitor, there is no over- or undershoot during load steps, though of course the ripple changes.

6.8.3 Battery Life

Although not directly important to the construction of out SMPS, we will take this opportunity to comment on how long this power supply will last before the battery is empty. Alkaline cells, as we've suggested, run from about 1.55 V down to 1.00 V. How long this takes is rated in mA-hrs for a given cell size. For the AA cells, a typical number is anywhere from 1700 to 2800 mA h. Supposing that we have one of the better quality ones at 2800 mA h and we are discharging it at the average voltage of 1.3 V. We will be operating at an average current draw of 2.51 mA, and so we will be able to operate the 5 V for something like 2800 mA h/2.51 mA = 1116 h, which is seven weeks.

This lifetime seems pretty good, depending on what the particular application is. However, we observe that at no load, which we model as a 5 MΩ, we are drawing 458 μA, which at 3.1 V is 1.419 mW, 22% of the input power at full load. Unless the load needs to be on continuously, this current draw, when unneeded, will seriously compromise the lifetime of the battery.

We look at the idle powers in Table 6.14. We observe that more than half the power drawn in no load is due to the ICs, particularly the timer and the comparator. What is needed could be as simple as an idle shutdown: A MOSFET to

Table 6.14 Power dissipation at no load for very low current bang–bang.

Component	Pd (μW)
555	520
R7	107
R8	37
Inductor resistance	5
Transistor	3
5807	87
R11	47
R12	16
R1	93
LT1004	83
LT1017	286
R6	65
Total	1349

disconnect these items from the supply voltage when the output load asserts that it's not operating. The reader may care to design a robust circuit to do this.

6.9 Notes for the Hobbyist

Unlike LDOs, there are usually not readily available SMPS for most applications. This is why power supply engineering remains a profession: each one seems to have different requirements.

The bang–bang converters designed in this chapter can be a very good starting point for the hobbyist. For one thing, they are easy to understand: if the output voltage is too low, the transistor turns on. If the output voltage is too high, the transistor turns off.

Bang–bang converters are also easy to simulate. There's nothing complicated going on, really, so they simulate easily and quickly. And finally, bang–bang converters are pretty easy to get going on the bench. Ripple too much? There are only a few components to mess with, the inductor, the capacitor, and the hysteresis resistor. So you can get what you want by tinkering (although again, the author strongly recommends that you do the calculations and do the simulations before you build anything. The calculations are just high school algebra, and the simulation takes just a few minutes to set up).

The one thing that can be a little tricky is the comparator selection. If it has too much offset, or is too slow, that might not be apparent on the bench, though it should have been clear in the simulation when you measured the efficiency. An easy way to avoid the issue is to buy a (relatively) high-performance comparator, rather than using the LM339 you have in your junk box. Since you're only building one circuit anyway, the extra $US0.50 shouldn't concern you.

7

PWM Converters

7.1 Introduction to PWMs

7.1.1 Background

Pulse-width modulation (PWM) is the standard method used in most power supplies to control the power stage. As such, we want to look at how to make such a converter using multi-sourced parts.

This is actually more challenging than you might at first suppose. Over the last few decades, PWM ICs have been developed in great numbers ... there are at least several thousand types. There is voltage- and current-mode control, hard-switching, resonant and quasi-resonant, buck, boost and buck-boost, and on and on in a variety of packages and a large variety of options. And supporting this huge diversity is a large number of companies, each producing its own unique design. The unfortunate result, as we have all found out, is that there are almost no second sources for PWM control ICs. If a manufacturer can't deliver parts, or goes out of business, all of the SMPS have to be, at the least, redesigned and re-laid out, re-tested, re-certified, etc.

In this chapter, we will look at ways to design a PWM SMPS robustly, using multi-source parts. This will limit the designs to "old-fashioned" parts, but as we will be showing in the examples, that doesn't mean that the performance will be old-fashioned: our examples range from 10 W to 22 kW, and with a switching frequency from 100 kHz to 2 MHz, comparable to some of the most modern designs.

7.1.2 PWM Structure

Unlike bang–bang controllers in Chapter 6, and unlike PFC controllers in Chapter 8, PWM controllers switch at a constant frequency. Every so many microseconds they turn on their main switch transistor. Then, after a period set by the difference between the output voltage and the set-point voltage, they turn the transistor off until the next cycle starts.

The ratio of the on-time to the period of the oscillator is known as the duty cycle (DC). When the DC is increased, the amount of energy transferred to the output is increased, and thus the output voltage increases, and *mutatis mutandis* for a decrease in the DC.

To understand better the operation of a PWM control, consider Figure 7.1. We're showing a buck power stage, though, of course, PWM can be applied to any type of converter. In particular, we're showing an n-channel MOSFET, but this is just for ease of conception; a p-channel will be the same except that the gate drive will be inverted.

The output voltage is divided down and subtracted from the reference voltage. This is done by an op-amp (the error amplifier) with a shaped frequency response. In this conceptual drawing this is just a capacitor, but below we will examine the criteria for, and design of, a specific type of compensation (as the frequency response is known) for the particular power components used and specifications required. With a pole at the origin (the capacitive feedback), the error amplifier will output a voltage to force the (divided down) output voltage to be equal to the reference voltage.

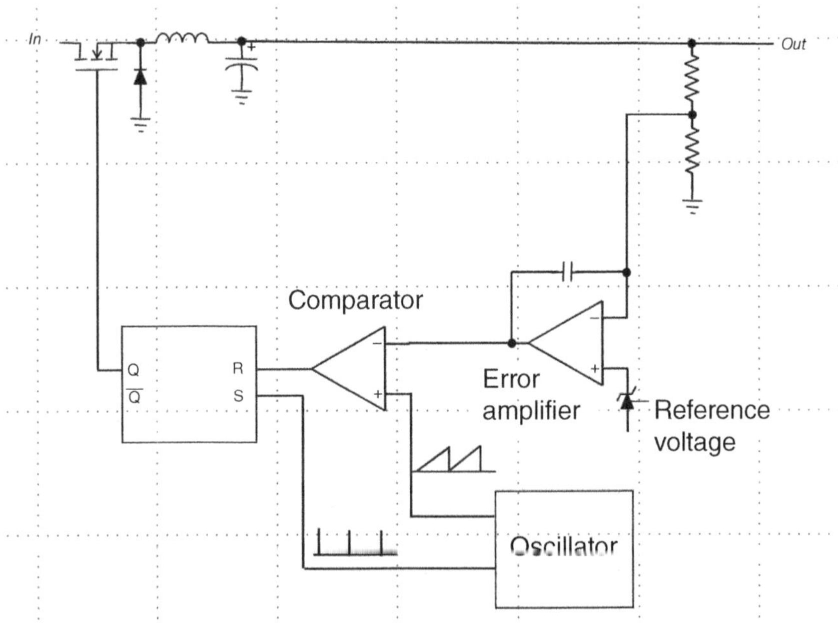

Figure 7.1 PWM concept.

The output of the error amp is compared with a sawtooth waveform, referred to as a ramp. The ramp is generated by an oscillator running at the intended switching frequency. When the ramp just starts, the output of the error amplifier will (typically) be greater than the ramps voltage, and so the output of the comparator will be low. Then, as the ramp rises, it will cross the error amplifier's voltage and go high for the rest of the cycle until the ramp starts over again.

The oscillator also produces a pulse train at the switching frequency, consisting of very short pulses. Each such pulse sets the output of a Set-Reset (SR) flip-flop, the reset being done when the comparator goes high. The output of the flip-flop is what drives the MOSFET.

The end result of this sequence is that once every cycle, at a fixed frequency, the MOSFET is turned on. When the ramp reaches a level equal to that of the output of the error amplifier, the MOSFET is turned off. If the time to reach this level is longer, the DC is longer, and so this is, in fact, pulse-width modulation at a constant frequency.

7.2 Robust PWMs

As noted in the beginning of this chapter, there are a very limited number of PWM ICs that are multi-sourced. Our portfolio will be using the following three methods for PWM control:

1. We will demonstrate that the 555 timer's pulse width can be adjusted while switching at a constant frequency by modulating its control pin. This is not as limiting as it may *a priori* seem, as there are multi-sourced versions of the 555 that will run at 1 MHz or more.
2. The 3845 is a multi-sourced PWM IC that has been used for many years for offline conversion not requiring PFC. It has all the control parts built in to it, but still requires quite a number of external components.
3. We will also build our own PWM controller using a 555 to generate the oscillator and ramp, an op-amp and reference voltage for the error amplifier, a discrete comparator, and an SR flip-flop. While not as elegant as the other designs, it will enable us to reach very high frequencies and allow us to construct our own robustly designed multi-slice converter.

To complete the list, we will *not* be using

4. The two-555 method, or 556, has a serious issue that makes it unsuitable for reliable operation as a PWM controller. We will be specific about this problem in Section 7.3.

7.3 Components of PWMs

7.3.1 Clock

Since we need a clock for a PWM, if we don't have a control IC with one built in, we will have to build our own. Naturally, the 555 comes to mind. The only real constraint is that we would like a narrow pulse. This requires two things:

1. The frequency is set by the two resistors and the capacitor. A constraint is that the current drawn should be (at least relatively) low so that the sum of the two resistances should be high. And finally, to make the pulse narrow, the ratio of the first resistance to the second should be large. The narrowness of the pulse sets the maximum DC of the converter, as during the pulse the switch cannot be turned on (depending on the polarity).
2. The 555 has an output whose DC has a minimum of 50%. To get the right polarity, we'll invert the output to get the narrow pulse, either using a (fast) comparator or else a logic gate inverter.

A model that generates the set pulse for the flip-flop is shown in Figure 7.2 (without the inverter). It has an output frequency of 100 kHz, with an (inverted) pulse width of 585 nsec. The DC is thus limited to 6%. Of course, this could be further reduced by further reduction of the ratio of R_1 to R_2; and increasing their sum while

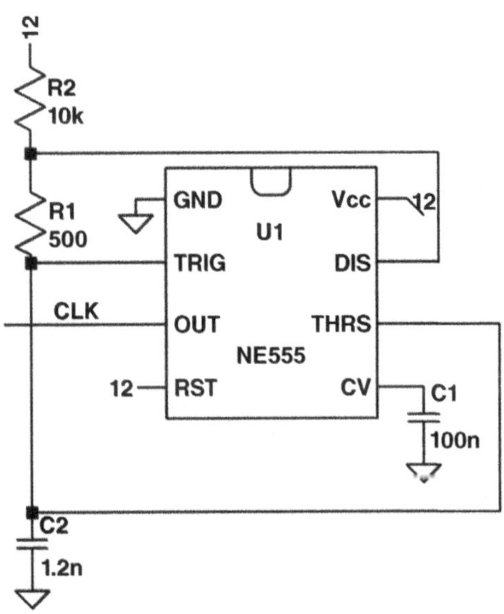

Figure 7.2 Generating a clock signal.

decreasing C_2 would reduce current consumption. The circuit could naturally be run on some voltage other than 12 V.

7.3.2 Ramp

A ramp is needed in a PWM to set the DC, and the ramp must be synchronous with the clock. The error amplifier is compared to the ramp, and when the two are equal, the switching transistor is turned off, setting the DC.

Now it is traditional to have the ramp be linear. Linearity makes the analysis of the transfer function of the SMPS (relatively) easy. However, it is not *technically* necessary for the ramp to be linear. The only *real* requirement is that it be monotonic, that is, always increasing; it should never be constant or decreasing. The error amplifier will simply adjust its output voltage to generate whatever DC is necessary to get the correct output voltage.

Now in PWM ICs, the ramp actually is linear. It is generated by a current source flowing (usually) into an external capacitor. And this, of course, is designed in such a way as to be synchronous with the clock in the IC.

But since only monotonicity, not linearity, is required of the ramp, this is not necessary. In fact, it is rather inconvenient for our design to require linearity, as (low-current) current sources aren't multi-source available.

One way to generate a ramp would be to use an RC. It's exponential, but you could use just the initial part and then amplify that. But then you would need some method of (rapidly) discharging the capacitor. Another method would be to use another 555 and integrate its output. But using an additional IC seems unattractive.

In fact, the easiest method is to simply use the clock output we've already generated and form a ramp from that. That way, the clock and the ramp are synchronized, and no extra IC is required.

As shown in Figure 7.3, the clock output charges a capacitor through a resistor, and discharges it rapidly almost to ground through a Schottky. Peak ramp amplitude goes to 9 V, which is three-fourth of the 12 V signal from the error amplifier. The ramp thus generated is not too badly nonlinear, and is certainly monotonic.

7.3.3 Error Amplifier

A key component of many SMPS is the error amplifier. This is the op-amp that subtracts the feedback from the reference and thus generates a signal determining the DC in such a way as to adjust the output voltage appropriately.

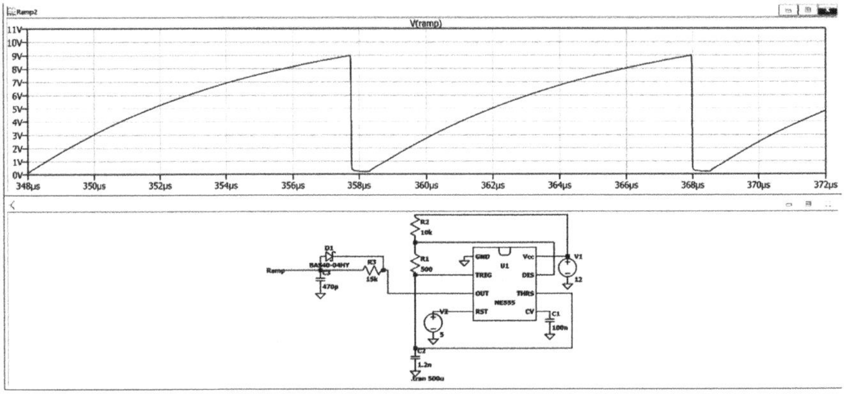

Figure 7.3 Generating a ramp from the clock.

An error amplifier needs some gain, and this gain is almost always frequency dependent in order to assure the small-signal stability of the power supply. The way to do this is by using Venable's K method.[1] Here's a synopsis.

Let's suppose you've measured the open-loop response of a power supply.[2] The two numbers that you want to find from the Bode plot are the gain, G, and the phase, P, at the bandwidth, f, that you want to achieve. You also need to select the phase margin, M, which should typically be 45°.

From this information we calculate the boost, which is defined to be

$$\text{Boost} = M - P - 90° \tag{7.1}$$

and the required gain, G, which is the negative of the measured gain (because this will add up to zero dB, which is to say, the bandwidth).

If the boost is 0° or less, you can use a Type I compensation, which is just the feedback capacitor that we default to, as shown in Figure 7.4. The value of the capacitor should be

$$C_1 = \frac{1}{2\pi f G R_1} \tag{7.2}$$

Here, R_1 and R_{Bias} are the resistor divider that you're using to scale the output voltage down to the reference voltage. Remember to convert G into a number, don't leave it in dB!

1 [19]. It is also shown in [9].
2 How to do this using a closed-loop measurement, which is vastly easier on the bench, in shown in the examples.

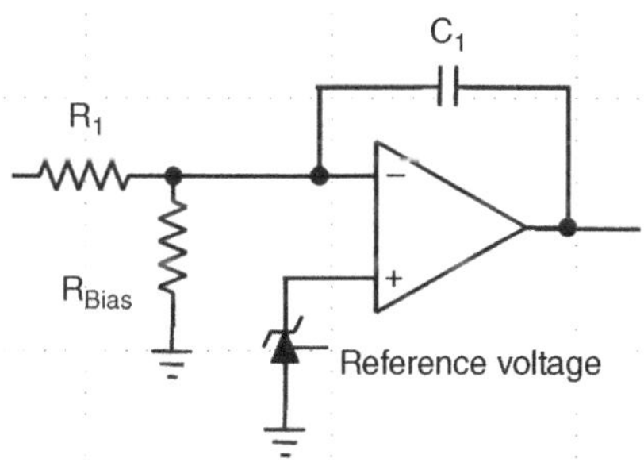

Figure 7.4 Type I compensation.

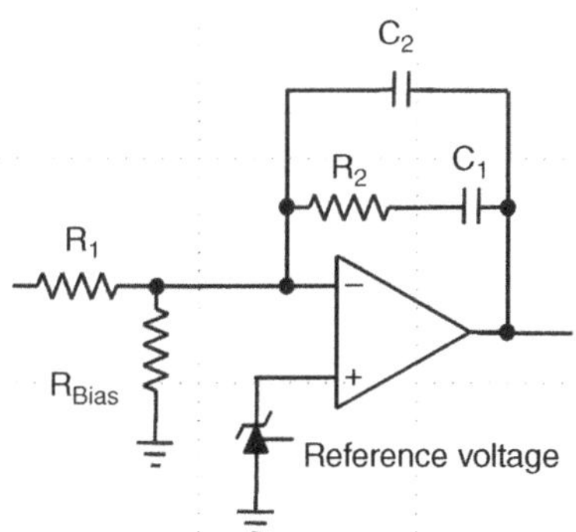

Figure 7.5 Type II compensation.

If the boost required is greater than 0° but less than 75°, you can use a Type II compensation, shown in Figure 7.5. The equations for this are

$$K = \tan\left(\frac{B}{2} + 45^\circ\right) \tag{7.3}$$

$$C_2 = \frac{1}{2\pi f G K R_1} \tag{7.4}$$

$$C_1 = C_2(K^2 - 1) \tag{7.5}$$

$$R_2 = \frac{K}{2\pi f C_1} \tag{7.6}$$

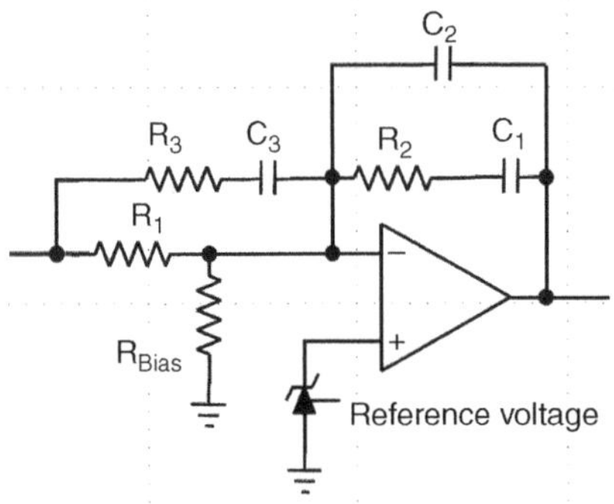

Figure 7.6 Type III compensation.

Finally, if the required boost is greater than 75° but less than about 160°, you can use a Type III compensation, as shown in Figure 7.6. The equations for this are

$$K = \left[\tan\left(\frac{B}{2} + 45°\right)\right]^2 \tag{7.7}$$

$$C_2 = \frac{1}{2\pi f G R_1} \tag{7.8}$$

$$C_1 = C_2(K - 1) \tag{7.9}$$

$$R_2 = \frac{\sqrt{K}}{2\pi f C_1} \tag{7.10}$$

$$R_3 = \frac{R_1}{K - 1} \tag{7.11}$$

$$C_3 = \frac{1}{2\pi f \sqrt{K} R_3} \tag{7.12}$$

If more than 160° of boost are required, chances are reasonable that you've measured or calculated something incorrectly; or else that you're trying to give a bandwidth that isn't possible for the converter, such as greater than one-tenth of the switching frequency.

As a practical matter, the exact values of the resistors and capacitors don't matter, since everything is in dB anyway. For example, if you calculate 942 pF, 1 nF is just fine; 1 nF/942 pF = 1.06 = 1.2 dB.

7.3.4 The Dual 555

In this subsection, we want to show why another historical PWM scheme, using two 555s, is unsuitable for use.

Exar[3] published a scheme wherein a first 555 is used in astable mode to generate a fixed clock frequency, which would be the PWM switching frequency. This clock then triggers a second 555 acting as a monostable oscillator, that is, a one-shot. Each time the first oscillator puts out a falling edge, the second oscillator begins its single pulse. The length of time of the single pulse is set both by the second 555's RC and also by a modulation input applied to the control pin, which sets when the threshold is reached for the RC to turn off the output pulse.

A model showing the concept is shown in Figure 7.7, using two 555s; of course, the same can be done with a single 556, which is two 555s in one package, and which is also multi-source available. On the right side of the figure, a 555 is being operated as an oscillator running at about 100 kHz and with 50% DC. This acts as the trigger for the second 555 on the left side. This second one is configured as a one-shot set for about a 4 μsec pulse. The actual pulse length is controlled by the voltage being applied to the control pin of the second 555. The output of the second 555 is inverted by a small-signal transistor with a pull-up, as we've done previously, to provide a controlled DC, which can vary from 0 to 50%, as the 555 output is 50–100%.

The result of this control is shown in Figure 7.7. We sweep the control input from 0 to 5 V, and the DC measurements are shown in Table 7.1. Immediately obvious is that the control range is very narrow: the DC only starts decreasing from 50% at 3.0 V, and by 4.15 V it's very close to 0%. Worse, above 4.15 V the DC jumps back up to 25% and the period doubles to 20.5 μsec.

The narrow range of control, from 3 to 4.2 V isn't really a problem. The control voltage will be coming from an error amplifier configured to increase the control

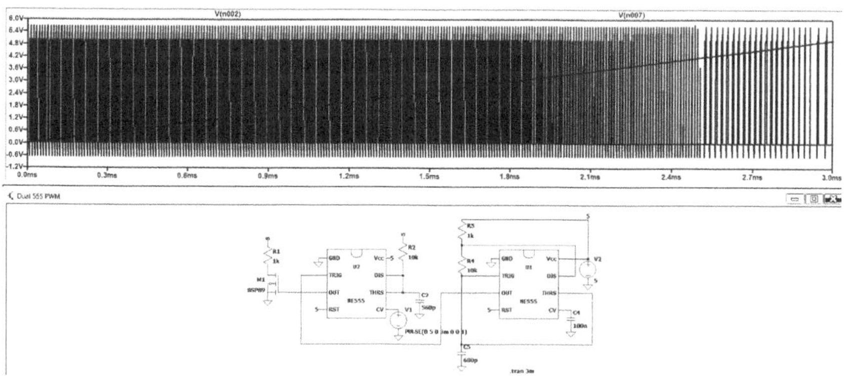

Figure 7.7 Dual 555 PWM - Not for use!

---

3 Exar's "Timer Data Book" of 1981 is the earliest reference the author could find detailing this method. [11]

Table 7.1 Dual 555 timer PWM response.

Control (V)	On-time (μsec)	Period (μsec)	DC (%)
0.0	5.24	10.25	51
0.5	5.19	10.27	51
1.0	5.12	10.27	50
1.5	5.17	10.24	50
2.0	5.23	10.28	51
2.5	5.18	10.26	50
3.0	4.93	10.34	48
3.5	3.37	10.32	33
4.0	1.19	10.38	11

voltage if the output voltage is too high, and so it will eventually reach the active control region. But the discrete jump in the DC and frequency above 4.15 V is completely bad—since it's in the opposite direction, the error amplifier won't know how to recover once it's entered this region.

What we *could* do would be to ensure that the control voltage never goes this high. For example, we could divide down the output of the error amplifier with a resistor divider so that at 5 V out it has a maximum of 4.0 V. This could work in theory; in practice, you would need a wide margin to be very sure to stay away from that jump up in DC. Naturally, this will present a limit to the minimum attainable DC, perhaps 20% or more. And there's then no way to get to very low DCs, which is necessary as the load becomes light.

We thus feel that this venerable method is unsuitable for modern PWM power supplies.

7.4 Example: Buck 10 W

In this first example, we're going to design and model a 10 W SMPS, 12 input to 5 V output at 2 A. We'll be using the method talked about in detail in the previous sections, using a 555 to generate the clock and ramp, an op-amp and comparator, and an inverter and an SR flip-flop to do the PWM logic. As this is our first example in this chapter, everything will be done simply, and in particular, we'll be running this low-power converter at a switching frequency of only 100 kHz. We will, nonetheless, be measuring the load-step response and bandwidth of the converter, this latter requiring building a state-space average model and then compensating the error amplifier.

Table 7.2 12 to 5 V at 2 A, 100 kHz specification.

Parameter	Specification	Measured
V_{in}, min	11.4 V	
V_{in}, max	12.6 V	
V_{out}, min	4.75 V	
V_{out}, max	5.25 V	
$V_{ripple,pp}$	10 mV	
Bandwidth, min	1 kHz	
Phase margin, min	30°	
50% step overshoot	100 mV	
50% step undershoot	100 mV	
Startup overshoot	250 mV	
Efficiency	90%	

7.4.1 Specifications

As always, our first step is to write a specification, Table 7.2. We have targeted a bandwidth of 1 kHz, which is 1% of the switching frequency, and have specified the response to load step from half to maximum load current. We've also specified the overshoot permitted on startup and have a reasonable efficiency target for a simple buck.

7.4.2 Robust Design

Our selection of a (p-channel) MOSFET and a diode is straightforward at this low current and low voltage. The two parts of the power stage that need to be selected, then, are the inductor and the capacitor.

We can select the capacitor by thinking about the converter's bandwidth and what the step response has to be. Since the bandwidth is going to be 1 kHz, the response time is going to be something like 500 μsec (one half of the time to settle). During that time, the capacitor has to provide the output current.

Thus, doing the math

$$C \times V = C \times 250\,\text{mV} = I \times t = 1A \times 500\,\mu\text{sec} \tag{7.13}$$

so C must be no less than 2 mF, possibly more.[4]

4 Is this an outrageously large capacitor? Not for a switching frequency of 100 kHz in voltage-mode. This capacitor is multisource available, for example, in a 12.5 mm diameter by 21.5 mm height package with a lifetime of 8000 h at 105 °C for $0.27 at 1000 pieces.

Since we're doing voltage mode control, we will have a resonance between the inductor and capacitor, and this resonant frequency needs to occur at a frequency rather higher than the bandwidth. We'll see why this is as soon as we do a preliminary Bode plot. The resonance occurs at

$$f_{\text{resonant}} = \frac{1}{2\pi\sqrt{LC}} \tag{7.14}$$

Since we want a bandwidth of 1 kHz, the resonant frequency ought to be at least greater than that and we have

$$1\,\text{kHz} \leq \frac{1}{2\pi\sqrt{L \times 2\,mF}} \tag{7.15}$$

so that $L \leq 12.7\ \mu$H. We pick a typical multi-sourced part with 10 μH inductance with a winding resistance of 5 mΩ, and we'll get an idea how much that should be changed by looking at the Bode plot and the step response.

7.4.3 First Model

Putting these values in and selecting the other components, we are starting with a model shown in Figure 7.8. The MOSFET, as we've seen in other designs, doesn't matter too much; we have one with a 13.5 mΩ RDS,on and 11 nF of Qg.

The diode is the most important for the efficiency. We've chosen a 2 A one whose datasheet shows a typical V_f of 370 mV at 2 A at 25 °C. (Voltage rating of 30 V is enough to meet our 50% derating at 12 V input.) If that's not good enough to meet the efficiency specification, we can later go back and choose a bigger one, or

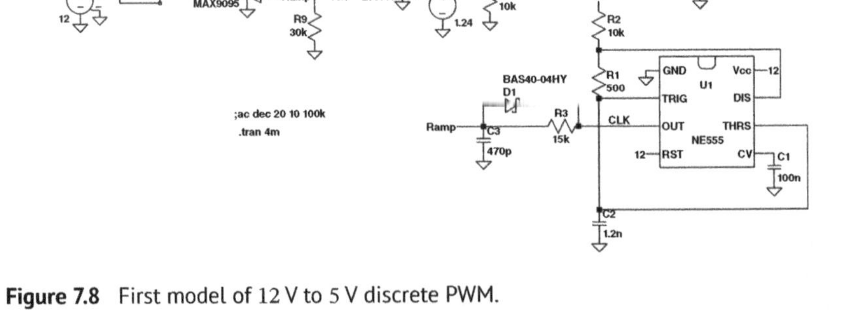

Figure 7.8 First model of 12 V to 5 V discrete PWM.

estimate how much lower it will be by letting it heat up. (Typical V_f for this diode drops to 340 mV at 75 °C.)

We've put in a large capacitor for the error amplifier compensation; we'll see below that this is part of measuring the bandwidth and phase margin.

We've divided down the output of the error amplifier so that it doesn't go higher than the ramp (three-fourth of the rail). The LTspice comparator only runs on 5 V, so we presumably can zener down to five from the input to get that. (You can't have it run from the output because, otherwise, how would it start up?) The output of the flip-flop is one volt, so we multiply it back up to 12 V. We use the $\overline{Q}$ output to get the polarity right, since we're using a p-channel transistor. And finally, there's a small gate resistor to suppress high-frequency ringing by the MOSFET.

A few more operational details. We've added in a step load that takes the load from half to full. We've also added in an AC (only, no DC) source of magnitude 1 mV. It takes whatever the output voltage is and adds to it 1 mV of AC voltage to the feedback resistor divider. The frequency of the AC being added is what will be swept by the AC analysis in the state-space average model. It has no effect during transient analysis.

Accordingly, we've also added in an AC sweep command. As shown, the command, instead of being started with a period as is the .tran command, is started with a semicolon. The command with the period is the one that is executed. This way, you can switch between transient and AC sweep without having to reenter the information. You just put the period on the one you want to run and change the period to a semicolon on the analysis you don't want.

7.4.4 State-space Average

We already suspect that there's going to be an issue with the LC resonance, because we've started with an inductor value that puts that resonance at the desired bandwidth. So even before we try to compensate the error amplifier, we ought to take a look at the Bode plot of the initial design and see if there's going to be a problem. We need to have a (Bode) plot of the phase and gain as a function of frequency of this system.

Now, this can't be done with the current model. The trouble is that there is switching going on. This happens at 100 kHz and has edges that are very much faster than that. If we were to try to measure the transfer function at even a single frequency, say 100 Hz, we would have to switch for one thousand cycles before we could extract the response. And of course, we need to find the response at very many frequencies to construct a plot.

However, by token of the same frequency ratio, there is no particular reason that we need to know what is going on at the switching frequency (and much less so at the edge frequency) in order to understand what is happening at frequencies much lower than these. And in fact, that is what we are going to do by constructing a state-space average of the switching portion of the converter.

A state-space average, as the name implies, averages the performance of (only) the digital portions of the circuit over one switching cycle. Thus, the transistor and diode are replaced with an average power model, while the inductor and capacitor remain as discrete components. The error amplifier also remains discrete, while the ramp, comparator, and flip-flop are also averaged.

To understand the process of state-space averaging in detail, the discrete converter we are designing in this section is particularly well-suited, as all the components are apparent, not buried inside an IC.

To start with, then, consider the action of the transistor and diode in the present buck converter. Let's call State I the state where the transistor is on and the diode is off. During State I, a voltage of V_{in} is applied to the input of the inductor (the left-hand side of the inductor in the schematic). State I lasts for a percentage of time equal to DC.

In the other condition, State II, the transistor is off, and the diode is on, and thus a voltage of $-V_{\text{f}}$ is applied to the input of the inductor (which may, if the input voltage is high enough, be considered to be 0 V). State II lasts for a percentage of time equal to 1 – DC.

If there were a condition in which the inductor current went to zero, and both the transistor and the diode were off, then that would be State III. We'll meet such a case in the next example.

So the model of the transistor plus the diode, averaged over a complete switching cycle is

$$V_{\text{out}} = \text{DC} \times V_{\text{in}} + (1 - \text{DC}) \times (-V_{\text{f}}) \tag{7.16}$$

or approximately $V_{\text{out}} = \text{DC} \times V_{\text{in}}$. We can thus model these transistor and diode power components by a voltage source applied to the input of the inductor, the voltage source being controlled by the DC.

But now, where does the DC come from? The DC is set by the time when the output of the error amplifier is equal to the ramp voltage, which is a function of time at the switching frequency. In equations,

$$\text{DC} = \frac{V_{\text{EA}}}{\text{Ramp Amplitude}} \tag{7.17}$$

where we are ignoring the short dead time when the ramp is zero during which the SR flip-flop is initially set.

So we have now modeled all the components operating at the switching frequency by continuous (non-switching) functions. This is the meaning of state-space averaging. We end up replacing the high-frequency components with a voltage-controlled voltage source, of magnitude V_{out}, which depends on V_{in}, and

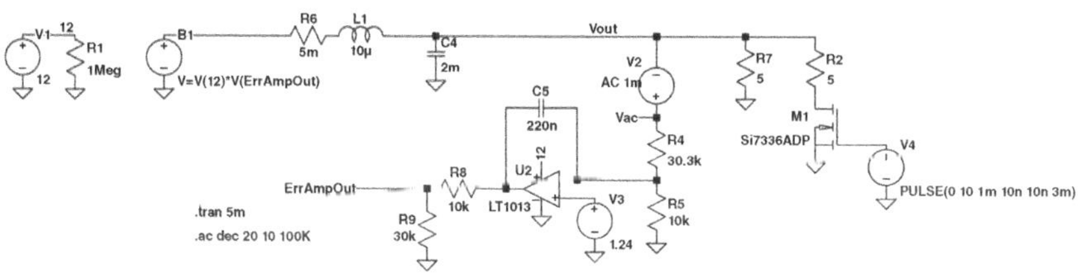

Figure 7.9 State-space average model of 12 to 5 V at 2 A discrete PWM.

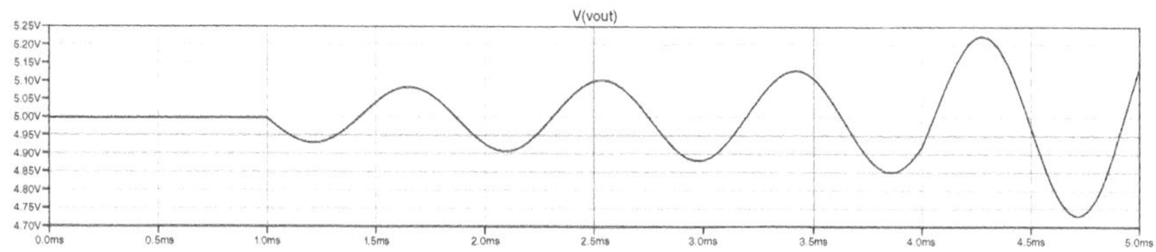

Figure 7.10 Step response of initial schematic of 12 to 5 V at 2 A discrete PWM.

is controlled by DC, which in turn depends on the error amplifier output. We thus replace all of the switching-frequency components in our model with a **B**-source to get our state-space average model in Figure 7.9.

Here we see that much of the circuitry, including the clock and ramp generator, has been absorbed into the transfer function of the **B**-source, which is multiplying the input voltage by the error amplifier voltage to produce the voltage that is being applied to the input to the inductor. This exactly corresponds to how the switching works in the real circuit, with everything else still in place. In fact, this model looks very much like the original model with the left-hand side and the 555 removed.

Let's start by running this model in transient mode. As shown in Figure 7.10, the output voltage at the initial operating point can be measured to be 4.997 V, so that the model works correctly. But when the load step gets turned on, not too surprisingly, the model breaks into oscillation. The oscillation frequency can be measured to be 1.27 kHz and so is clearly due to the LC resonance. It's interesting to note that when the load step turns off, this oscillation doesn't stop.

To understand all the issues, let's also look at the Bode plot (we revert to full load). Looking at Figure 7.11, we immediately see the issue. There is a fairly sharp peak at 1.1 kHz, where the LC resonance is. It looks like it goes up a good 10 dB from where it would otherwise be. You can't just ignore it and try to compensate at a frequency above this resonance, because there is a 180° phase shift (L and C each contribute 90°), and so you can't add enough phase boost in the compensation.[5]

5 And incidentally, this is why current-mode control is favored for this sort of design, because it effectively removes one of the poles, and thus make the phase boost for higher bandwidth achievable.

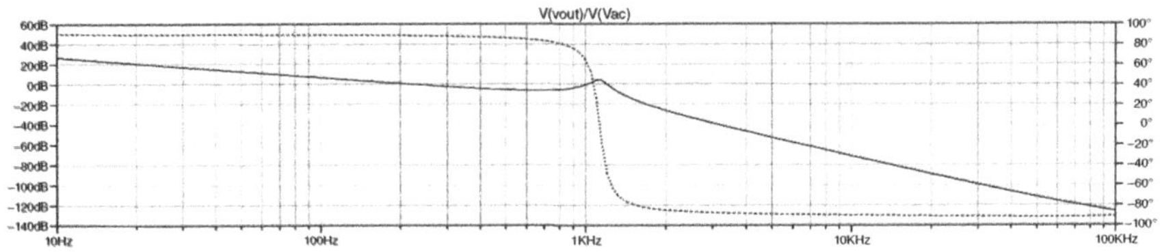

Figure 7.11 Initial schematic Bode plot of 12 to 5 V at 2 A discrete PWM.

You also can't just compensate it at a slightly lower frequency than the resonance because of the peaking. You might cross zero dB at a lower frequency, but then the Bode plot would still go back above zero dB at this peak, and then the power supply would still oscillate at that frequency.

There are at least two things we could do. For one, we could make the inductance smaller, which would move the resonance frequency up. For a second, we could increase the damping of the RLC tank, which effectively amounts to increasing the winding resistance of the inductor (the ESR of the capacitor is likely to be small, and increasing it will worsen the step response).

How do we know that the RLC is underdamped? At 1 kHz the inductor impedance is

$$Z_L = 2\pi f L = 2\pi \times 1\,\text{kHz} \times 10\,\mu\text{H} = 60\,\text{m}\Omega \tag{7.18}$$

which is an order of magnitude higher than the winding resistance.

We decide to do both, decreasing the inductance to 3.3 μH and increasing the winding resistance to 50 mΩ (such parts are multi-sources available). Doing this latter isn't a problem, it will dissipate only something on the order of

$$P_L = 2A^2 \times 50\,\text{m}\Omega = 200\,\text{mW} \tag{7.19}$$

which out of the 10 W output is only 2%, so that's acceptable.

7.4.5 Stable Model

We thus change the model as shown in Figure 7.12. Now we're ready to redo the Bode plot, shown in Figure 7.13. The crossover frequency is now only 206 Hz. At our desired bandwidth of 1 kHz the phase margin is fine at 49°, but the gain is low at −13.4 dB.

What we need to do now is to back out the effect of that 220 nF capacitor in order to find the open-loop response, at which point we can then apply Venable's method (discussed above) to compute a compensation that gives us the desired performance.

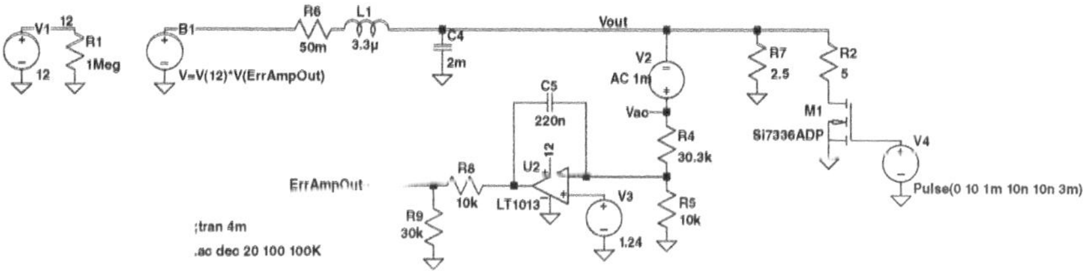

Figure 7.12 State-space average model with resonance suppressed of 12 to 5 V at 2 A discrete PWM.

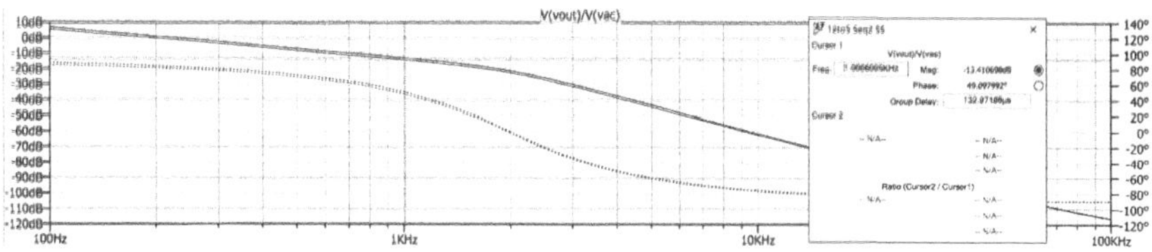

Figure 7.13 Bode plot with resonance suppressed of 12 to 5 V at 2 A discrete PWM.

The 220 nF works with the 30.3 kΩ to form a pole at[6] which shows up in the Bode plot as an increase in the slope.

$$f = \frac{1}{2\pi RC} = \frac{1}{2\pi \times 30.3\,\text{k}\Omega \times 220\,\text{nF}} = 24\,\text{Hz} \tag{7.20}$$

Again, we don't actually care what the frequency response is, we want to *design* the error amplifier's compensation to get that 1 kHz bandwidth. Since we don't need any phase boost, Venable's method tells us that a Type I compensation is fine—and that's what we already have! So we want a gain of 13.4 dB = 4.68 beyond what we already have, which is to say, we want to reduce the compensation capacitor from 220 nF to 220 / 4.68 = 47 nF.

One further consideration is the gain at 1 kHz. The capacitor at that frequency has an impedance of

$$Z = \frac{1}{2\pi fC} = \frac{1}{2\pi \times 1\,\text{kHz} \times 47\,\text{nF}} = 3.4\,\text{k}\Omega \tag{7.21}$$

which is only a tenth of the feed resistor. To improve accuracy at millisecond time scales, we're going to add a minimum gain to the error amplifier, which is to say, put a 1 MΩ resistor in series with it, which gives a minimum gain of 30. This results in the final Bode plot shown in Figure 7.14. The bandwidth is now 1.0 kHz with a phase margin of 49°.

Changing the error amplifier compensation and reverting to the switching model, the step response is shown in Figure 7.15. The unit starts up smoothly from

6 Remember the 2π, since this is frequency, not angular frequency.

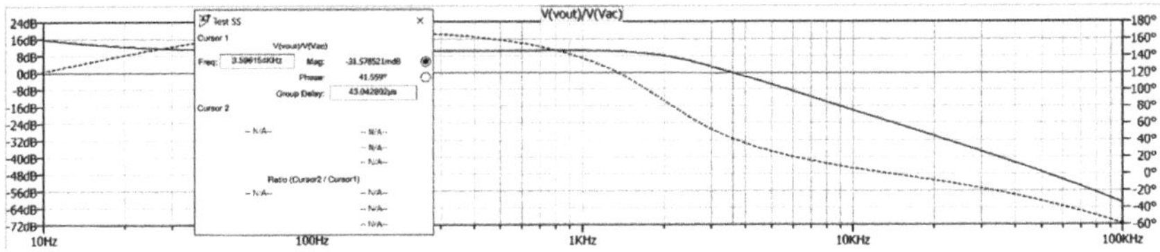

Figure 7.14 Final Bode plot of 12 V to 5 V at 2 A discrete PWM.

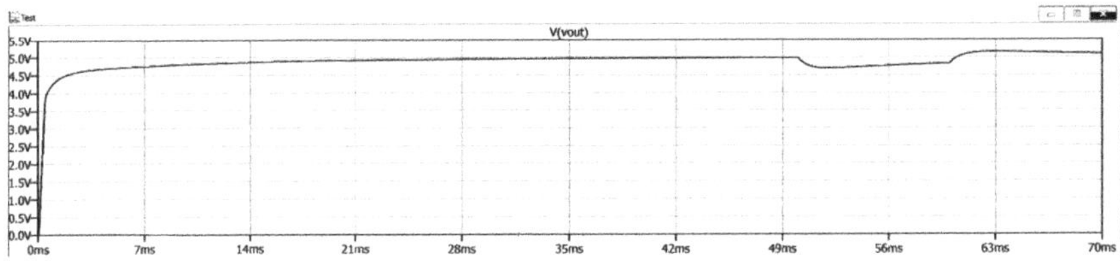

Figure 7.15 Final startup and load step response of 12 to 5 V at 2 A discrete PWM.

zero volts and has no overshoot. The output reaches to within 1% of final voltage (4.95 V in 25 msec. The load step causes a drop of 300. Load step recovery, of course, depends on the load, since this buck is not synchronous. The output ripple is of course not visible, it's only 3 mVpp.

7.5 Example: Boost 100 W

For this, our second example, we want to demonstrate the use of the multi-sourced 3843 family of ICs. These have all the parts of the previous design built in to them, and so will be simpler component-wise. However, there are some issues to be watched out for when using the 3843 family, and we will be sure to point those out. And more significantly, and unlike our other examples in this book, the 3843 family is current-mode, necessitating a state-space average model that is the most complicated to understand.

This design will be a boost rather than a buck, going from 12 V to 48 V at 2 A, which is 96 W, an order of magnitude more power than in the previous example. We will also be operating at the same switching frequency, 100 kHz. Since we're building a flyback, we want to limit the DC to 50%, and since we have 12 V as the input, we don't want to need a higher voltage to run the IC. We thus select the 3845 for this design.

There's another reason to limit the DC. It's known that current-mode converters operating at more than 50% DC have a peculiar type of instability: they have subharmonic oscillations, which is to say they alternate every other switching cycle between a long pulse and a short pulse. A cure is known, which is to add in ramp compensation, but to make things easy we're going to keep the DC below 50%.

It's also possible to operate in continuous-conduction mode (CCM) rather than discontinuous-conduction mode (DCM), the former meaning that the inductor current doesn't return to zero every cycle. While there's no particular reason to operate in CCM at this low power level, another reason not to operate in CCM is that as the output power decreases, there is eventually a changeover from CCM to DCM no matter what values the components have. So avoiding this shift, although it's not a great big deal, seems desirable.

7.5.1 Specifications

As previously, we start with the specifications for this design (Table 7.3). The only significantly different part from the specification we've had so far is that we've reduced the expected efficiency; we anticipate that to keep the inductor a reasonable size, which is probably going to end up with a more significant winding resistance loss. Of course, this will be at least in part compensated by the V_f of the rectifier diode being a less significant percentage of the output voltage.

Table 7.3 12 V to 48 V at 2 A, 300 kHz specification.

Parameter	Specification	Measured
V_{in}, min	11.4 V	
V_{in}, max	12.6 V	
V_{out}, min	45.6 V	
V_{out}, max	50.4 V	
$V_{ripple,pp}$	1 V	
Bandwidth, min	1 kHz	
Phase margin, min	30°	
50% step overshoot	1 V	
50% step undershoot	1 V	
Startup overshoot	2 V	
Efficiency	85%	

7.5.2 Design

As a starting point, let's use the design from the datasheet, shown in Figure 7.16.[7] The datasheet says that this is a 25 W design.

- First off, of course, we've decided we're going to use the 3845.
- We aren't coming in off the AC line, so the bridge (and the 1 W resistor R_1 being used as a fuse[8]) aren't needed.
- Since the input voltage is 12 V and we're using the 3845, special power circuitry isn't required. We can just power everything from the input. The tertiary winding and its associated rectification and filtering can be removed, as can the startup 2 W power resistor R_2.
- The tertiary winding is also acting as an output voltage sense, and we can replace this with a direct feedback.
- Modern MOSFETs are avalanche-rated, and so we won't bother with the (two!) snubber circuits.
- In fact, since we're not coming in off the AC line, the input and output can be on the same ground. Thus, no transformer is needed at all, just an inductor in the drain of the switching transistor.

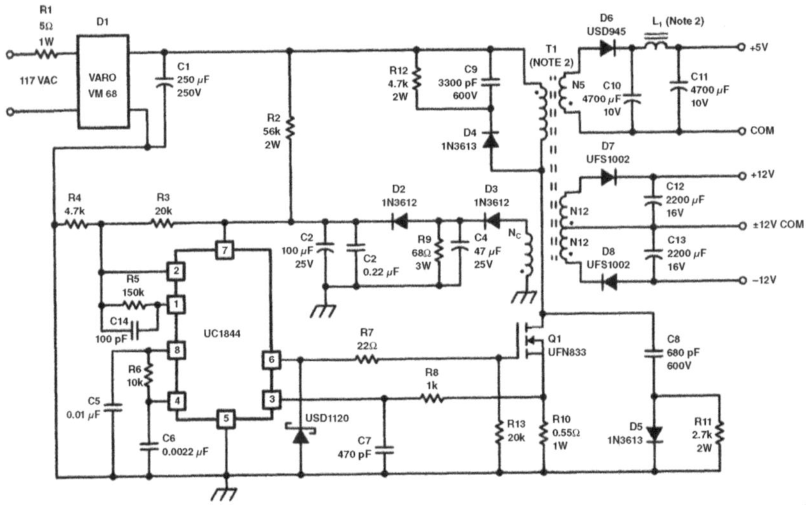

Figure 7.16 12 to 48 V at 1 A starting point.

7 By the way, although we will not be pursuing it in this book, this schematic gives a pretty good idea of how to go about using the 384x family of ICs to produce regulated DC voltages offline when PFC isn't needed.

8 This is no longer allowed by UL.

- We're going to beef up the design a bit (in particular the MOSFET and the 48 V output diode) since we're running 100 W all on a single output.
- The 5 V output has an extra LC filter on it. This is perhaps used to get the 30 mVpp ripple on the 5 V they show, but probably won't be necessary for our design. In any case, you'd have to be quite careful with an additional resonant tank after the 5 V output. Does it ring during a load step?

We still need to select the inductor and output capacitor. We again use the equations in Chapter 6 (Bang–Bang Controller), assuming a maximum DC of 40%. The required peak current is

$$I_{pk} = \frac{2P}{V_{in}DC} = \frac{2 \times 96\,\text{W}}{12\,\text{V} \times 0.4} = 48\,\text{A} \tag{7.22}$$

and the inductance is

$$L \leq \frac{V_{in}^2 DC^2}{2fP} = \frac{12\,\text{V}^2 \times 0.4^2}{2 \times 100\,\text{kHz} \times 96\,\text{W}} = 1.2\,\mu\text{H} \tag{7.23}$$

The RMS current is

$$I_{RMS} = 48\,A\sqrt{\frac{0.4}{3}} = 17.5\,\text{A} \tag{7.24}$$

Observe that this is rather higher than the average current (96 W / 12 V) = 8 A, because of the sawtooth current waveform as well as because it's being operated in DCM. This inductance will turn out to be smaller than needed, particularly as the timing capacitor we end up with will give us a frequency of 78 kHz. We end up with a 2.7 μH inductor, which is multi-source available in a surface-mount package at 1 mΩ (or less).

For the capacitor,

$$C \geq \frac{Q}{V_{pp}} = \frac{I_{pk}DC}{2fV_{pp}} = \frac{48\,\text{A} \times 0.4}{2 * 100\,\text{kHz} \times 1\text{V}} = 96\,\mu\text{F} \tag{7.25}$$

We select half of this, readily available from multiple vendors, as four 10 μF, 63 V ceramic capacitors.

7.5.3 Modeling

Digesting all this material results in a preliminary design model shown in Figure 7.17. A number of comments are in order.

- The 3845 is modeled by the LT1245 in LTspice, which is the equivalent.
- For the transistor, we have picked a device with only 3.8 mΩ RDS,on but with 28.5 nC of gate charge; in other words, a hefty device. Whether that is necessary can be decided once we measure the efficiency.

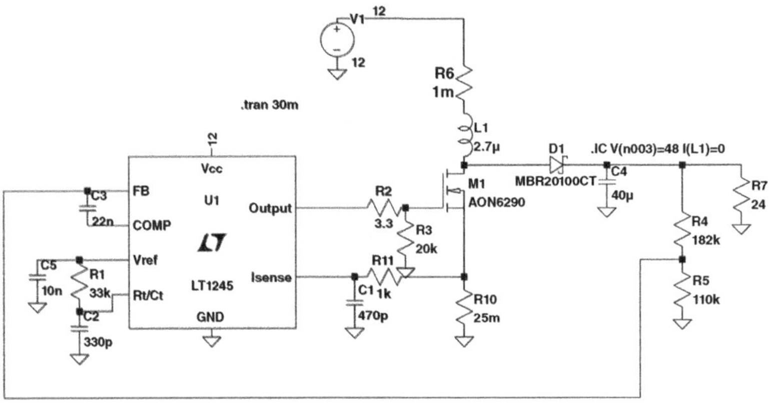

Figure 7.17 Model for 3845 boost.

- The diode is a 10 A, 100 V Schottky in a TO-220 package. At the average current during conduction (8 A), it has a typical V_f of 830 mV. Whether a smaller device would be adequate can be decided, again, after we look at the efficiency.
- Per the original schematic, there is a 20 kΩ pull-down on the transistor gate. This is to ensure that the transistor isn't accidentally left on during power-down of the IC, as was discovered the hard way by early users.
- Since the peak current is 48 A and the 3845 has a maximum current-sense voltage of 1 V typical, we have chosen a 25 mΩ current sense resistor. In actual practice, this is too small (and the minimum guaranteed value is actually 900 mV), but this is good enough for what we're doing.
- Finally, for the compensation, we have put in a 22 nF Type I. How did we get to that value? In the lab, we would put in a 1 μF. But that gives a time to settle of a substantial fraction of a second, which is impossible for LTspice running at 100 kHz. So the author reduced the capacitor one step at a time until it was just marginally stable. Startup response (with the output voltage almost right) can be seen in Figure 7.18. Of course, it will be much improved once we've properly compensated the design, though not entirely, since the 1245 model takes 300 μsec to start oscillating.

The next step will be to decide on the compensation needed to meet the specifications. Since the design is marginally stable, we can get some results already. Looking at the simulation during the last millisecond of the simulation, we see that

1. The switching frequency is, as mentioned, 78 kHz.
2. The output voltage after settling in is very close to 48.0 V.

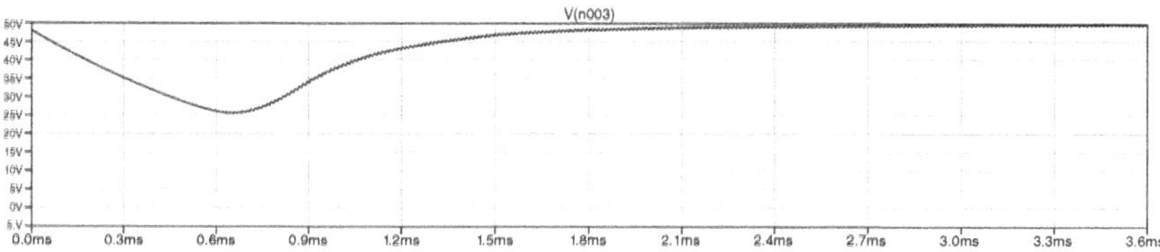

Figure 7.18 100 W, V_{out} with 22 nF as a guess for compensation.

3. Peak-to-peak ripple is only 540 mV (this is at full load).
4. Output power is 95.98 W and input power is 101.52 W, for an efficiency of 94.5%. So doubtless, the transistor and diode can be downgraded unless the higher efficiency is desired.

7.5.4 State-space Average

To answer questions about stability, we need to compensate the converter, and for this we need to build a state-space average model.

This model is much more complicated to derive.[9] The converter is in current mode rather than voltage mode and so there are two loops, one nested inside the other. There is an inner current loop hidden inside the IC and controlling the inductor current, and then, in addition, there is an outer voltage loop controlling the inner loop and the output voltage (this outer loop is also hidden inside the IC).[10]

In the previous state-space average model, we ended up replacing the switching components with a DC-controlled voltage source. What the current-mode control does is to control the current through the inductor, and so we need to replace the voltage source with a DC-controlled current source.

What we need, then, is to derive the average current going into the diode. The peak current through the inductor is

$$I_{pk} = \frac{V_{in}}{L} t_{on} = \frac{V_{in}}{L} \frac{DC}{f} \tag{7.26}$$

And the average current through the inductor, since this is a sawtooth, is

$$I_{avg} = \frac{I_{pk}}{2} DC = \frac{V_{in} DC^2}{2fL} \tag{7.27}$$

9 If you want to skip the derivation and just copy the result, that's fine.

10 Although this is the only example we give in this book of current-mode control, since it is not often necessary, be aware that you can fairly straightforwardly add it to voltage-mode controllers such as the 555 or discrete examples.

Now we know the current as a function of time, since it starts at zero each switching cycle,

$$I(t) = \frac{V_{\text{in}}}{L} t \tag{7.28}$$

and the length of time the transistor is on is set by the time when the current sensed by the sense resistor is equal to that of the error amplifier:

$$V_{\text{EA}} = I(t) \times R_S = \frac{V_{\text{in}}}{L} t_{\text{on}} \times R_S \tag{7.29}$$

so that

$$\text{DC} = t_{\text{on}} f = \frac{V_{\text{EA}} f L}{V_{\text{in}} R_S} \tag{7.30}$$

We can now substitute this in to find the average current. We have

$$I_{\text{avg}} = \frac{V_{\text{in}} \text{DC}^2}{2fL} = \frac{V_{\text{in}}}{2fL} \left(\frac{V_{\text{EA}} f L}{V_{\text{in}} R_S} \right)^2 = \frac{V_{\text{EA}}^2 f L}{2 V_{\text{in}} R_S^2} \tag{7.31}$$

Now we just need to find the error amplifier output. Unlike most of the other examples in this book, this error amplifier *is inside the IC*. In order to build a model, we need to extract it.

Looking at Figure 7.19, which is from the 3843 datasheet, we see the structure. Note particularly the divide by 3 at the output of the error amplifier; that's how they get down to the 1 V level for the current ramp. The two diodes also need to be included; model measurement (as well as common sense) shows that the forward voltage of 4148 diodes is pretty close to what's inside the IC.

We thus build the state-space average model shown in Figure 7.20, with the error amplifier extracted from the IC and built as in the datasheet (observe that the COMP pin output of the LT1245 is not the error amplifier output. The output is rather at the end of the diode and divider chain). The error amplifier in the 3845 has a GBW of 1 MHz, which is similar to that of the LT1013, which we're using in the model.

From this we can see where all the switching pieces have gone. The **B** current source contains the inductor (2.7 μH), the switching frequency of the 3845 (78 kHz), and the current sense resistor (25 mΩ). The inductor no longer shows up in the circuitry like the capacitor does; this is what is meant when it is said that the current-mode control "gets rid of" the inductor, now it's merely a parameter.

Again, if this derivation seemed complicated[11] the reader can simply copy and modify this state-space average model for other design parameters.

---

11 And to be honest, re-deriving this from scratch after all these years made the author's head hurt for a bit.

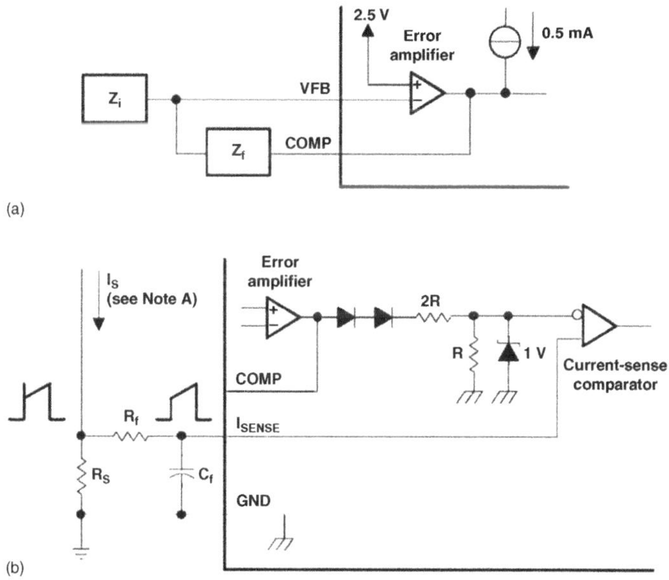

Figure 7.19 Internal error amplifier on the 3845 (a) error-amplifier configuration: error amplifier can source or sink up to 0.5 mA. (b) current-sense circuit: peak current (I_S) is determined by the formula $I_{S(max)} = 1\ V/R_S$; a small RC filter formed by resistor R_f and capacitor C_f may be required to suppress switch transients.

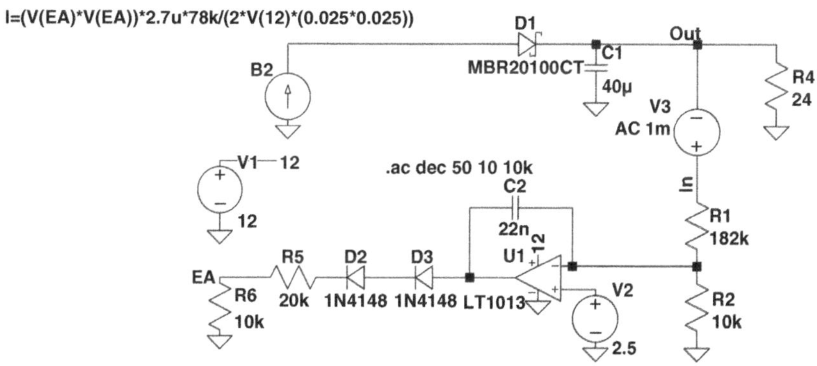

Figure 7.20 State-space average model of 3845 converter.

We verify that the output is 48 V. We can now run the AC sweep, shown in Figure 7.21. Observe that there is now only a single pole rolling off the transfer function, as the inductor has been effectively removed. The cross-over frequency is 709 Hz with a phase margin of only 13°, explaining why there is some

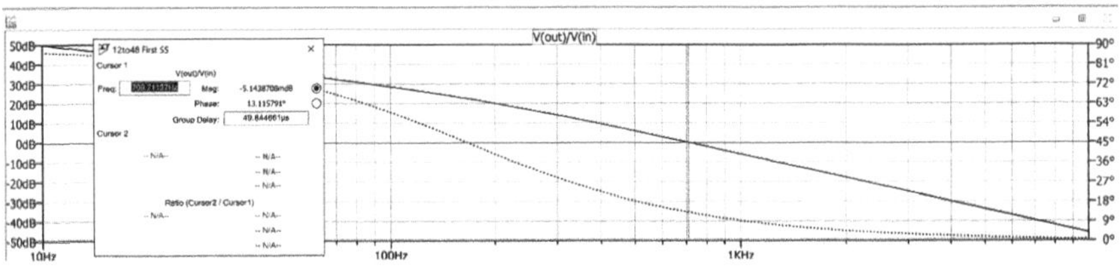

Figure 7.21 Bode plot of 3845 converter, showing effect of current-mode control.

oscillation in the output. At our desired bandwidth of 1 kHz, the gain is −6 dB and the phase is 9°, and at 1 kHz what we have now is a gain of

$$Z = \frac{\chi_C}{182\,\text{k}\Omega} = \frac{1}{2\pi \times 1\,\text{kHz} \times 22\,\text{nF} \times 182\,\text{k}\Omega} = 0.04 \tag{7.32}$$

so we need a gain of 6 dB more, that is, twice this, 0.08, and 45 − 9 = 36° of boost. This calls for a Type II amplifier. Doing the calculations,

$$K = \tan\left(\frac{36}{2} + 45°\right) = 1.96 \tag{7.33}$$

$$C_2 = \frac{1}{2\pi f G K R_1} = \frac{1}{2\pi * 1\,\text{kHz} * 0.08 * 1.96 * 182\,\text{k}\Omega} = 5.6\,\text{nF} \tag{7.34}$$

$$C_1 = C_2(K^2 - 1) = 15.9\,\text{nF} \tag{7.35}$$

$$R_2 = \frac{K}{2\pi f C_1} = 19.6\,\text{k}\Omega \tag{7.36}$$

and inserting this into the model, we indeed get a bandwidth which we measure to be exactly 1 kHz with a phase margin of 45.3°.

7.5.5 Step Response

For large-signal stability, we measure the response to a step in the load current. As shown in Figure 7.22, the load step is turned on at 5 msec, and the output is again a stable 48 V about 1.5 msec after that. At 10 msec the load step is turned off, and the output is again a stable 48 V after about 2 msec (the boost converter can't sink current). Maximum over and undershoot are each 4 V, which could be improved by increasing the output capacitance and/or increasing the bandwidth. The 45° of phase margin can be seen in the very slight undershoot during recovery from unloading, and could be further improved (if it matters) with a little extra phase margin.

Before moving on, a comment about current-mode control. Current-mode was originally invented for just the reason we have seen in this example: the presence of an LC resonance makes high bandwidth very difficult. And when the switching frequency was only 100 kHz, the inductor and capacitor would necessarily be

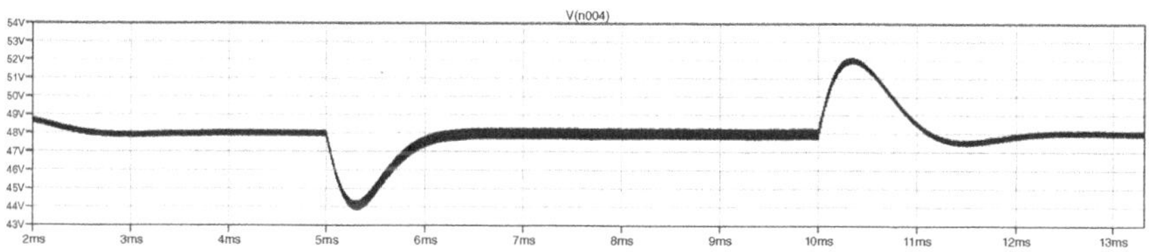

Figure 7.22 12 V to 48 V converter step response.

large, prohibiting high bandwidth. However, nowadays, it is straightforward to get switching frequencies in excess of 1 MHz, as we do with robust components in some of the examples in this book. This means the inductor and the capacitor can be greatly reduced in value, in turn causing the resonant frequency to become very high. For modern converters, then, voltage mode works fine for most applications, and indeed it is the author's opinion that current mode is somewhat obsolete. For this reason, this present section contains the only current-mode example in the book, although many of the other designs could be built with current-mode using the techniques described here.

7.6 Example: Buck Multi-slice 22 kW

In this section, we're going to become considerably more aggressive with high-performance, robust power supplies. We're going to increase the power level by more than two orders of magnitude (22 kW). And we're also going to increase the (effective) switching frequency by a factor of 20, to 2 MHz.

In this example, we're going to produce 400 VDC from a three-phase power source, nominally 480 VAC (this design is not bidirectional, though it could made so). We're using Delta not Y three-phase, though it wouldn't take much to convert it. Output power is 22 kW. To deal with this high power, we're going to show how to divide the power into four (interleaved, that is, phase-shifted) slices (each slice being a power stage; we reserve the word "phase" for the AC line), each carrying 5.5 kW and operating at 500 kHz. This will produce a ripple voltage at 2 MHz. This will be the converter at the highest power at the highest frequency in this book.

This converter could be used for charging the battery in an EV, as it is the right power level and voltages; however, the control scheme would need to be more complicated. This would involve several op-amps, and their design is better left for a less full section than this. In fact, we charge a battery in the chapter on isolated supplies, to which the interested reader is referred. In this chapter we are just doing a DC/DC conversion. Our focus in this example will be on the high power, and how to obtain the interleaving. Of course, in today's world, power supplies like this are done with fancy high-pin-count microcontrollers; we show how to do it using only a handful of 555s, a few op-amps and a couple of logic gates, and four half-bridge gate drivers, one for each slice.

7.6.1 Specifications

We need some specifications, shown in Table 7.4. The input is intended to be three-phase Delta, and so each individual phase is 480 root 2. We then have ±10% on each side. We'll show this when we get to the modeling section.

Efficiency target is 85%. This isn't spectacular, since the currents are relatively low, but we're operating each slice at 500 kHz in order to minimize size, and this frequency will affect switching losses.

7.6.2 Design

For this design, we're going back to the discrete implementation of a buck control we used in the first example in this chapter. There are two main differences.

1. Since this is synchronous, we need to drive not only the "buck" transistor but also the "rectifier" transistor. To do this, we will be using a generic synchronous gate driver.
2. We want to have four interleaved slices, so the main clock will be running at 2 MHz, and then we will divide this down into four interleaved clocks, each at 500 kHz, using flip-flops. Each clock will then in turn generate its own ramp for its own slice.

Table 7.4 480 VAC to 400 V at 22 kW, 2 MHz specification.

Parameter	Specification	Measured
V_{in}, min	527 V	
V_{in}, max	743 V	
V_{out}, min	300 V	
V_{out}, max	432 V	
$I_{ripple, pp}$	5%	
Bandwidth, min	1 kHz	
Phase margin, min	30°	
Efficiency	85%	
Cost target	US$25	

7.6.3 Modeling

Due to the large size of this converter design, we will break up the explanations for building the model into parts. We first look at modeling three-phase (Y) power, as there seems to be a lot of discussion on the web about this. Next we look at the power stage of a single synchronous slice, including its gate drive, which has model limitations to be overcome. Ultimately, the slice will be reduplicated to make four slices. We have additional parts looking at interleaved clock generation, and for the control circuitry, both for the individual slices and for all of them functioning as a single power supply.

With this in place, we start by looking at how to drive the buck. It's synchronous, so we want to use a multi-source half-bridge driver to avoid designing analog for the non-overlap time to avoid shoot-through. This turns out to be tricky, though, since, although there are suitable multi-source drivers available, LTspice doesn't have them in its library. And the ones it *does* have don't work outside their specified voltage range, unlike other parts in their library.

To fix this, we have to do something tricky. We build a "little power supply" that runs on low voltage, and then use the signals to and from it to run the main power slice. You can think of the "little power supply" as being part of an extended model of a high-voltage synchronous gate driver. Again, the only reason for this is the lack of a suitable model in LTspice.

We then proceed to build out the four slices. We generate out-of-phase clocks and ramps using some digital models. They are all derived from a master clock to ensure they are all running at exactly the same frequency.

Next we look at the control of these slices, using standard error amplifier design. This is now a working system, but we still have to address current-sharing between the slices. We opt for limiting the maximum current output by each slice, and we design a current sensor for each. And then we design the maximum current circuitry, which is done by limiting the DC of each slice individually.

The final item needed is to generate the 10 V housekeeping supply used to run all this circuitry; we put in a small bang–bang buck.

Although the 400 VDC output could well be an EV battery being charged from a three-phase outlet, the additional circuitry required to ensure the appropriate I–V control of the battery current is not unique to robust design, and has been omitted in the interest of space.

This is by far the most complex converter in the book, and a block diagram is included to help guide the reader through the design process, Figure 7.23. The reader should not feel that this design is overly complicated; for those who have designed power supplies at this power level before, the size and complexity should feel familiar, and the number of components, and of distinct components, in the BOM is very much on the low side compared with the norm.

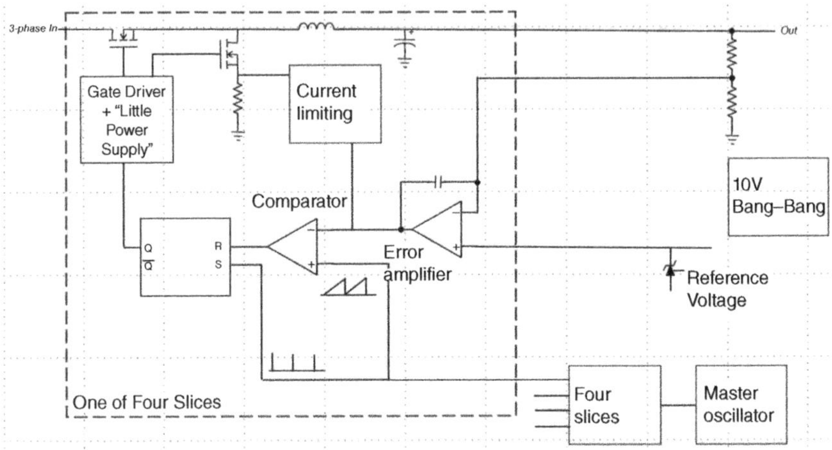

Figure 7.23 Block diagram of multi-slice buck.

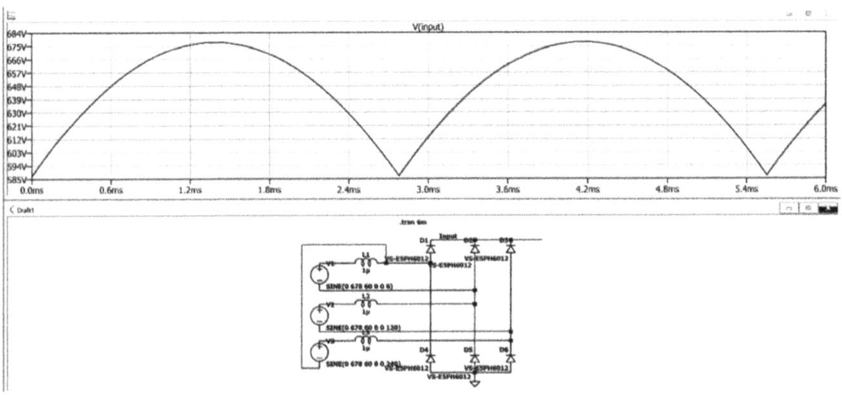

Figure 7.24 Three phase delta input with rectifier.

7.6.3.1 Three-phase Power

The first thing we make is a model of three-phase rectified power. We need this, rather than just testing at various DC levels, because we need the converter to have enough bandwidth to follow the input variations: rectified 400 VAC is at 2×60 Hz $= 120$ Hz, and then there are three out-of-phase sources, so the input ripple voltage is at 3×120 Hz $= 360$ Hz. If we made the bandwidth too low, the output would have ripple at this frequency. Figure 7.24 shows the LTspice circuit (Delta not Y) and the resulting line voltage after the three-phase bridge.

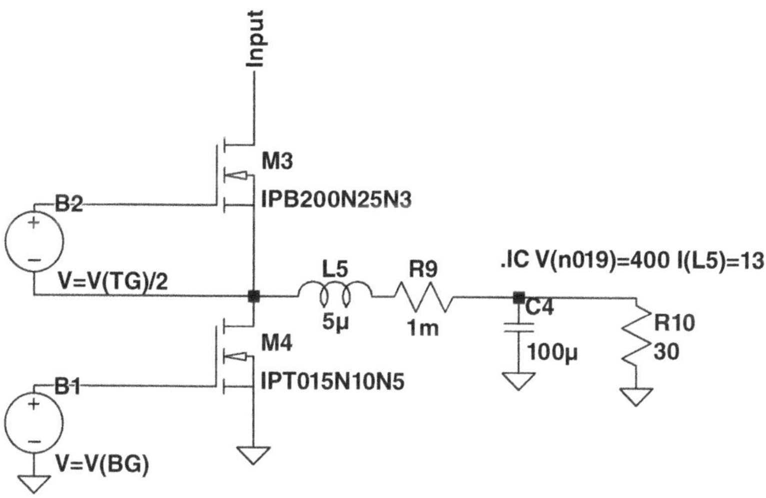

Figure 7.25 5.5 kW slice power stage.

Observe that each phase is of peak amplitude $480\sqrt{2} = 678$ V. The diodes are rated at 60 A, and their reverse voltage is 1200 V. This voltage rating is not necessary for the simulation, but in the real design we're going to want some margin over the 678 V to be able to protect them during lightning. We've added in 1 µH of inductance in each line, necessary because otherwise LTspice throws a fault (voltage sources in parallel); in any case, this is probably lower than the real line impedance. The simulation shows where the minimum input voltage in the specification comes from (remember the 10% tolerance).

7.6.3.2 Synchronous Buck Power

As the first step in building the LTspice model, we create the main power stage Figure 7.25, using the gate drive signals from the "little power supply" as ideal floating sources to drive the MOSFETs; it will be shown how these are to be generated in the next section. These are used because they reproduce the rise and fall times, as well as the dead times, of a high-voltage gate driver, while getting around the voltage limitation of the half-bridge drivers that LTspice supports. The inductance and the capacitance come from our standard buck design practice, and we use the load resistor as the appropriate one-quarter of the total eventual load.

It is this power stage that we will be studying.[12] In the ultimate design, there will be four of these slices in parallel, running at identical frequency but phased 90° apart from each other, to minimize the output ripple.

12 Where, you may well ask, is the power factor correction circuit? Most three-phase power supplies don't do PFC, and so we won't here either. It is possible to build a three-phase PFC, but takes quite a bit of circuitry.

7.6.3.3 Synchronous Buck Driver

The next part is the most complicated modeling in LTspice we'll come across in this book. We need to model a synchronous buck, running from the 500–700 V input and producing 400 V output. What we'd *like* is to use a synchronous half-bridge driver, so we don't have to do an analog design of the non-overlap gate drives to prevent shoot-through.[13]

Here's where the main problem comes in. There are multi-source 1200 V half-bridge drivers available. However, they aren't in LTspice. We try using the LTC7060 (or the LTC7063), which are rated at 100 or 150 V, respectively. But these must be more modern models, for, unlike other LTspice models that never break at any voltage, these ICs don't work above 100 or 150 V, respectively.

So we have to get clever. *Bearing in mind that this is just a trick due to LTspice limitations and not components that we will actually use*, we build a little half-bridge power stage running from the same 10 V that powers an LTC7060 half-bridge driver; see Figure 7.26. The MOSFET control signals from this will then produce the appropriate non-shoot-through gate drives to the top and bottom MOSFETs in the main power stage; we will (ultimately) control this IC from the error amplifier and then copy its gate drive outputs to run the actual power stage. You could think of this as being a complicated way of using the error amplifier to generate the gate drives in the actual power stage.

In this "little power supply" schematic, we're applying a 20% DC, 500 kHz signal to the input as a test, in order to view the response. In particular, note that

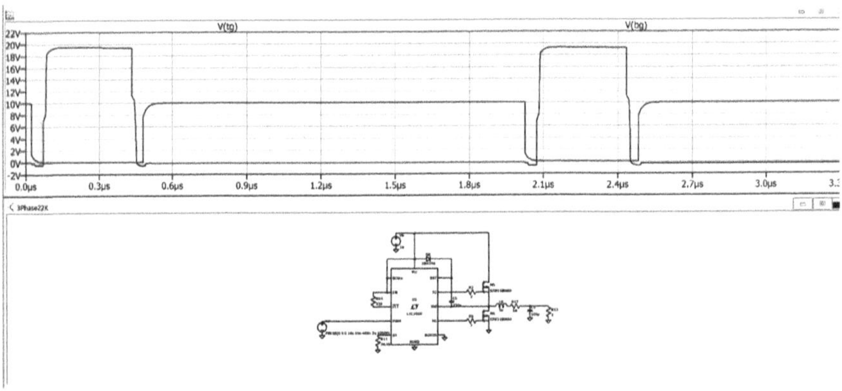

Figure 7.26 Gate drive circuitry for 5.5 kW slice.

13 If both MOSFETs were to be on at the same time, even momentarily, they would form a short to ground.

1. The top gate drive signal is at 20 V, because the input voltage and the drive voltage are both 10 V. We'll divide this by two when we use it in the main power converter.
2. The dead time, that is the non-overlap time needed to ensure no shoot-through, is a clean 49 nsec. We can adjust this by adjusting the resistor on the DT pin when we measure the response of the main power converter, either up if there is still shoot-through, or down to improve efficiency.

Again, you can think of this entire circuit as just being used to create a high-voltage half-bridge gate driver in LTspice; there is no intention of actually building this circuit. And in any case, the main power stage is driven by ideal replicas of the gate drives in this circuit, which of course aren't real.

7.6.3.4 4-slice Clock and Ramp

Now we need to design the clocking circuitry for four such slices running interleaved, that is, all at exactly the same switching frequency, but running 90° out of phase from each other. The identical frequencies ensure that there won't be beat frequencies between the slices. The 90° phase separation makes the effective frequency four times higher than the actual switching frequency,[14] so that the output ripple voltage is also at four times the individual slices' switching frequency, and thus at one quarter of the amplitude.

We start by creating a 2 MHz "master clock," as shown in Figure 7.27. It doesn't particularly matter what the DC is. We can use one of the multi-source high frequency 555s in the design; in the LTspice model the NE555 isn't frequency limited.

Next, we want to divide this down into four phases. To do this, we use the traditional digital method using D-flip-flops (these parts are, of course, multi-source); see Figure 7.28. First, we have a single D-flip-flop that divides the master clock signal down by a factor of two, producing a square wave.[15] We use the two outputs of the flip-flop to each run two additional D-flip-flops, whose outputs are, as desired, all at exactly one quarter of the master clock frequency, and phased 90° apart.

Finally, we use each of these signals to generate ramp and clock signals for the control circuitry, one for each slice. As shown in Figure 7.29, for one of the four, we first have to multiply up the 1 V output of the D-flip-flops' output to 10 V to make it compatible with the rest of the circuitry. In reality, of course, this would be unnecessary as the logic family can be selected to work on 5 V or even 10 V.

---

14 If they were running in phase, it would be no different than simply having a single large converter at that frequency. This is why we bother with the four slices at all.

15 Observe that the digital parts in LTspice don't have supply voltages, they run on 1 V invisibly. The input can be 0 V and anything higher than half a volt; the outputs are 0 V and 1 V. The lower right corner is an (almost invisible) connection to ground.

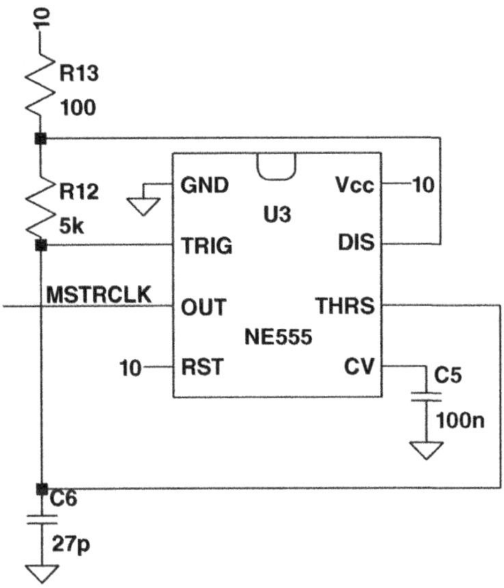

Figure 7.27 2 MHz master clock.

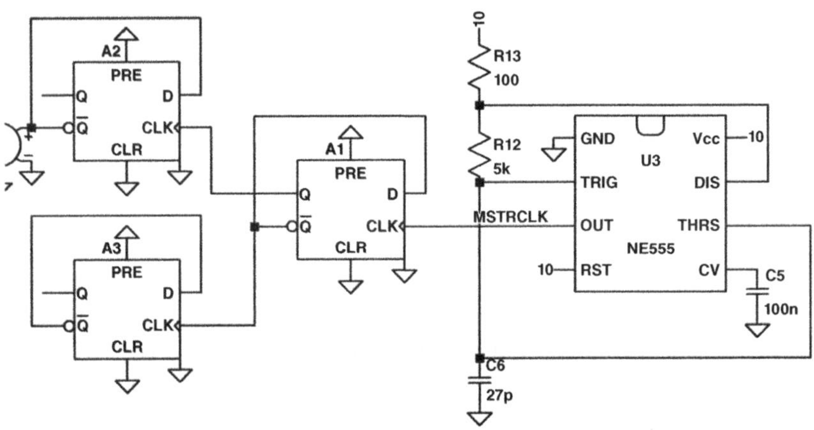

Figure 7.28 Four-phase clock.

Since we want a short pulse for the clock and a long pulse for the ramp and have a square wave coming in, we AC couple the square wave so that only its edges produce a signal, and use this as a trigger for a monostable 555 to generate the slice's clock. We produce a ramp for the slice the same way as previously. Results are shown in Figure 7.30. The individual slice clock frequency is 491 kHz; the pulse-width is 233 nsec = 11%; and the ramp amplitude is 7 V. We will divide down the

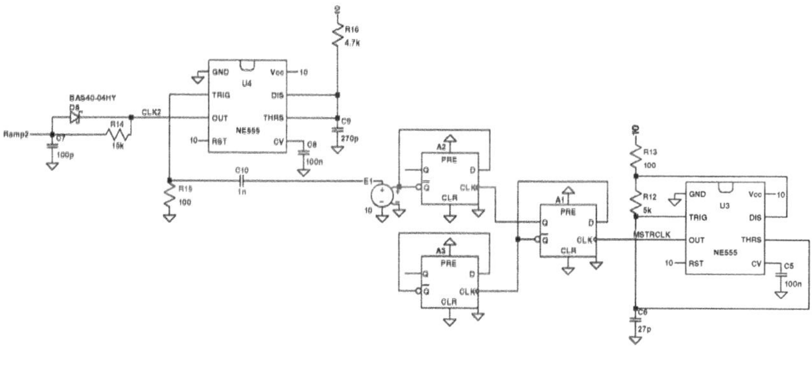

Figure 7.29 Ramp and clock generation. Single phase shown.

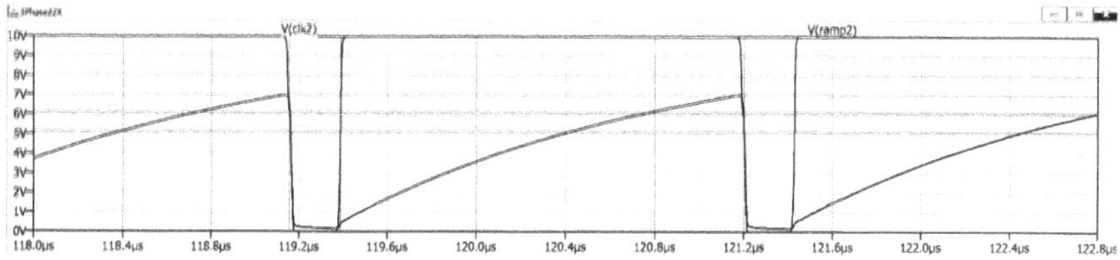

Figure 7.30 Clock and ramp signals.

error amplifier output to ensure we stay below this level (otherwise, we would be allowing it to go to 100% DC).

Two observations on this circuit, the first of which is technical. For some reason, the circuitry in the inside of the 555 and the logic gates doesn't start up immediately; it takes about 100 μsec for normal operation to begin. Since the bandwidth of this converter will be small, that won't be an issue; otherwise, we could just hold the power stages off for a startup delay.

Second, there may be some question about the need to have two 556s, a 555 and a quad D-flip-flop, which is four ICs, to generate the four clocks. While there are two four-slice buck controllers from different vendors listed on distributors' websites, one is in an SO-24 and is marked "obsolete," while the other is in a QFN-48 and costs US$4.83 at 1000 pieces from distributor. So this robust design actually seems pretty good.[16]

7.6.3.5 Control Circuitry

We need to control all four slices, and so we need to give a little thought about the best way to do this.

16 If the reader is trying to run the model at this point, the author found it necessary to check the box "Skip initial operating point solution" in the transient command for the simulation to run.

1. One possibility would be to have each slice have its independently controller. Each already has its own ramp, so it could independently measure the output voltage and have its own error amplifier. However, then the error of each section would add up independently.
2. Various pieces mentioned could be made one-for-all. The extreme case would be to have a single voltage measurement, a single error amplifier, and a single voltage reference for all the slices together. The problem with this is that the slices will (typically) be on different PCBs, or at least on different sectors of a single large PCB. Transmitting these various analog voltages around will incur noise pickup.

In the end, we choose the second option because it uses fewer ICs. We will design the current monitoring circuitry so as to obviate the problem of different slices having different operating points due to noise pickup.

Since we are just building a DC/DC converter, all we need to control is the output voltage. Since it's a buck, we can build an error amplifier for it the same as previously. And since the bandwidth we've targeted is only 1 kHz, it seems natural to just use a Type I compensation.

So, now we just have a single voltage divider (in reality with three resistors in series because of the 200 V rating of a 1206 resistor), a single 431 reference and a single error amplifier compensated by a capacitor (which we choose to give a time constant of a couple times the 3 msec repetition rate of the rectified three-phase AC), all of which is shown in Figure 7.31. We will determine the actual value for the compensation capacitor when we've built a state-space average model, in Section 7.6.4.

We now add in the rest of the control circuitry, the comparator, the D-flip-flop and the "little power supply." A complete slice is shown in Figure 7.32. The entire circuit will consist of four of these.

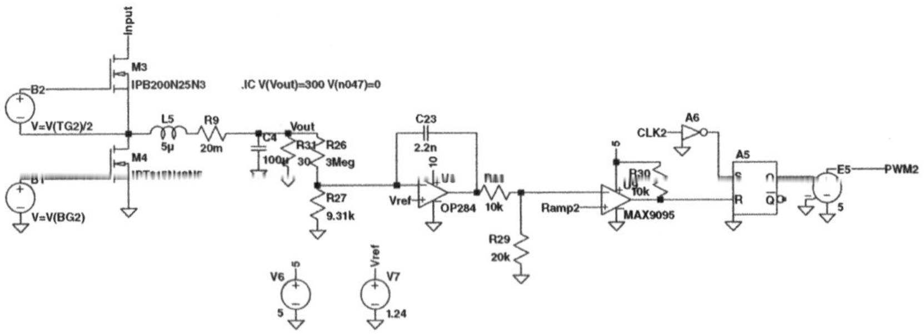

Figure 7.31 Compensation network for 5.5 kW slice.

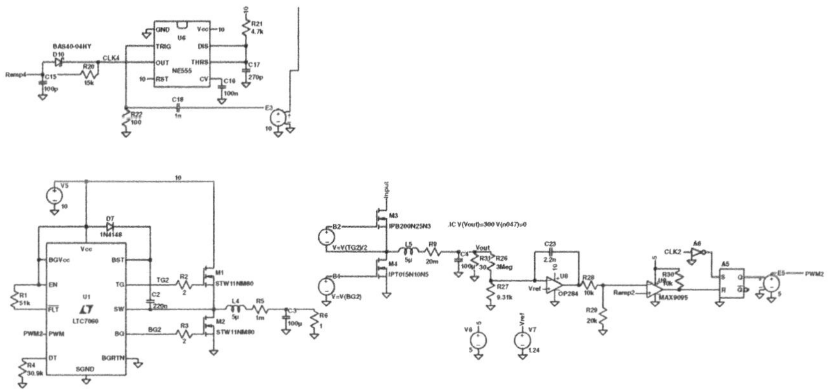

Figure 7.32 One 5.5 kW slice complete schematic.

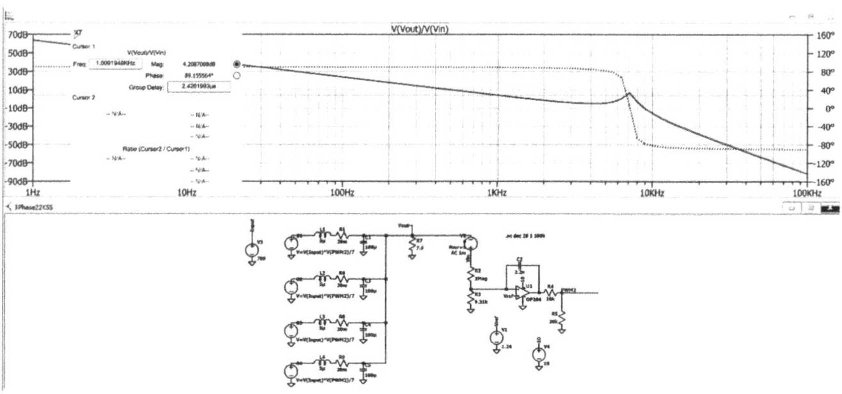

Figure 7.33 22 kW four-slice state-space average model.

7.6.4 State-space Average

Although the value of the compensation capacitor used in the schematic results in stable operation, we still need to determine the appropriate compensation to get the desired bandwidth and phase margin.

To determine the compensation, we build a state-space average model. This is particularly important in this converter, which has so many components that LTspice and the computer struggle to simulate it. A state-space average model is straightforward for a single slice and is done the same way as for the buck converter earlier in this chapter. While we could amalgamate all four slices into a single one with appropriate scaling of components, we simply put four of the models in parallel. The state-space average ignores the interleaving, as it occurs at the switching frequency. The model is shown in Figure 7.33.

As may be seen, the gain at 1 kHz is 4.2 dB with 89° of margin, showing that our guess for the Type I capacitor was pretty darn good. The resonant tank is up at 7 kHz, and thus does not interfere with a simple Type I compensation. Thus, compensation is simply reducing the gain of the loop by 4.2 dB = 1.62, and so we increase the 2.2 nF to 3.9 nF.

7.6.5 More Modeling

7.6.5.1 Current Monitoring

One thing we haven't yet dealt with is ensuring that one of the four slices doesn't deliver all the current while the other three are idle. Obviously, this won't work thermally. Of course, this is not apparent in our simulation because all of the components in each slice are exactly identical!

There are at least two choices for what to do.

1. One choice would be to measure the current in each slice, and then add them up and force them all to be one-fourth of the total. This is in fact frequently done, but requires inter-slice communication of analog signals, which is subject to noise. An option to get around the noise would be to communicate this information digitally, which is the most frequent, but then you end up with A/Ds and DACs and microcontrollers, which robust designs should avoid when possible.
2. A better option is not to force the individual slices to have the same current, but to only limit each one to its maximum allowable current. If only three-fourth of the maximum current is demanded by the load, one of the slices (in the extreme case) will deliver zero current while the other three will each deliver their maximum current. While this might *a priori* seem wasteful, each slice is after all *designed* to deliver maximum current, so they're already thermally prepared for this.

In any case, we will below address how to do this with the second scheme; the immediate preliminary requirement is that we find a suitable method of measuring the current at all.

Measuring the current takes a little thought. We can't measure the load current, because all four slices feed in to it; we want the current in each slice individually. Measuring the current at (say) the output of the inductor with a resistor is problematic, because the output, and therefore the resistor, is sitting up at 400 V. Building an instrumentation amplifier to get rid of that voltage and leave (for example) only 100 mV requires very large common-mode rejection, and will result in high-precision resistors and an expensive op-amp or three.

The preferred method is rather to measure the current in the source of the rectifier MOSFET. This is ground-referenced, and while physical construction may be tricky, at least the only common-mode signal is that due to ground currents in the copper.

The current in the rectifier is, of course, modulated by the on/off nature of the rectifier, at 500 kHz. However, for current balance, we don't need this information. Since the purpose of current balancing is to make sure none of the slices are overloaded, it can be done at a small fraction of the switching frequency of the slice.

In particular, we can simply average the rectifier's sensed current. It's true that the sensed voltage is (somewhat) dependent on the DC, but we know that this should be close to 400 V / 635 V = 63%. So we can just set the limit accounting for this factor, and include a bit for the tolerance of the DC.

Now that we have a plan, we have to think how to *actually* measure the current in a slice. Each slice is carrying a maximum of about 5.5 kW / 400 V = 13.75 A. If that generates 1 V, we would be dissipating 14 W in each of the slices' resistors! We need to generate much less voltage for the signal, and then amplify it. The 358 type op-amps have a maximum offset voltage of 7 mV (at room temperature). If we want to have, say, 1% accuracy, that would require 700 mV of input signal from the current sense resistor, which we've already decided is too much power.

We turn instead to the OP284, which is available in the LTspice library. It has a maximum offset voltage of 125 μV at room temperature, so 1% accuracy requires 12.5 mV signal, which will be 172 mW at 13.75 A, and so can be a 500 mW resistor. Similar op-amps are available from multiple vendors.[17]

We thus need a current sense resistor of 12.5 mV / 13.75 A = 925 μΩ; we can select a 1 mΩ resistor in 1%, which are common and inexpensive multi-source parts in (for example) 1206.

7.6.5.2 Current Balancing

Now we need to do the current limiting. Recall that we're not trying to force the currents in the four slices to be equal; we're just trying to ensure that no one slice takes much more than one quarter of the maximum total load current.

What we're going to do is quite simple. We want the maximum slice current to be 13.75 A, and we have a 13.75 mV signal representing that. If we add some margin, we can say we have 15 mV. If we multiply this by 84, we can get 1.24 V to compare with the (already in use) 1.24 V reference voltage. This doesn't have to be fast, and so we'll filter it to get rid of the switching frequency. In fact, we'll make it quite slow, as thermal time constants on a slice are going to be very slow.

The circuit we end up with is shown in Figure 7.34. The circuit control has been greatly simplified to make it possible to simulate. The 2 msec simulation we ran

17 Caution. As with all op-amps, looking at the *listed* V_{offset} on the distributor's page isn't good enough as the *typical* is listed; you have to actually open the datasheet to see the maximum value, which sometimes is an order of magnitude higher than the typical.

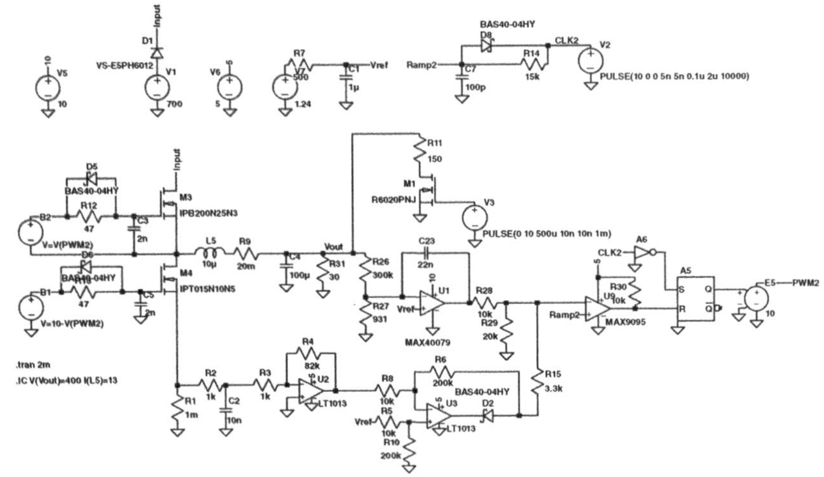

Figure 7.34 One 5.5 kW slice (simplified) with current balancing.

took some 10 minutes on a super-workstation and generated a 1.3 GB file. In particular, we got rid of the "little power supply" and replaced it with running the MOSFETs directly with the PWM signal, adding in a crude turn-on delay using an RC.

A single-op-amp amplifier is used to amplify the measured current.[18] Remember that the lower MOSFET is acting as a rectifier (instead of a diode), and so the current comes up from ground, and thus the signal is *negative*. The amplifier is thus an inverter.[19] We filter the signal with a 10 µsec RC to remove gate charge and switching frequency signals, though this could be made much longer.

The amplifier's output is subtracted from (not compared to) the reference voltage using a single-op-amp instrumentation amplifier, which can now be an inexpensive 358 since the signal levels are large.

The gain of this instrumentation amplifier is small, only 20. That way the difference between the measured current and the maximum allowed current can control the output without hitting rails. The amplifier is in a diode-sink configuration. As long as the current is less than the maximum, the output of this amplifier is high, and is thus cutoff by the diode. When the output of the current is higher than

18 In real life, you may need to make the amplifier an instrumentation amplifier due to ground offsets. High currents cause voltage drops on traces. So what's wanted is a Kelvin connection to the current sense resistor, that is, traces that come immediately from the component, not distantly connected, plus some way to remove ground offset noise.

19 Yes, you can have a negative signal inverted by an op-amp operating with only a positive rail. The non-inverting input is at ground, so the op-amp works to make its other terminal ground as well, causing its output to be positive.

the maximum, the amplifier starts decreasing its output until it gets low enough to turn on the diode, and so it sinks some current from the signal that set the DC.

Thus, this overcurrent system acts to cut back on the DC smoothly, instead of just turning the slice off. Since every slice has this circuit, each will only take at most approximately the maximum allowed current per slice, and the others will pick up the rest. If there really *is* an output overcurrent all four slices will go up to the maximum and stop.

Observe that we don't have to include this circuit in the state-space average model because it's not supposed to be active during steady-state operation.

The result of this circuit is best understood by looking at the output voltage. We are operating at maximum load, and then at time 500 μsec we turn on an extra 2.7 A load.

As shown in Figure 7.35, the voltage has almost settled in by the time the extra load turns on (the simulation started saving data at time 500 μsec to limit file size). The voltage abruptly drops and does not recover, indicating that there is insufficient current capability to run the load. After the extra load is turned off 1 msec later, the output voltage begins to recover.

7.6.5.3 Example: Internal Voltage Generation

We have one loose end in this design, and we've done so many things already that it would be easy to forget it: generation of the internal 12 V and 5 V to power all the ICs. Since our input is 635 V, even 1 mA would be quite a bit of power; the many ICs we're using are bound to be more than that, so an LDO is contraindicated.

The one tricky thing is that the input voltage is so high. We need a high-voltage gate driver to run the buck MOSFET. But actually, we already have such a thing: we're already using it for the main power slices! We can use the same gate driver, with much smaller components for the smaller currents, to generate 12 V from the input. Since the on-time would be very small (12 / 635 V = 1.9% times 2 μsec = 38 nsec), if we tried to do a PWM at the 500 kHz clock frequency, it will be much more convenient to do a bang–bang converter.

To start the whole thing up (after all, the 12 V has to come from *somewhere* before this converter is running!), we can have a startup circuit that only powers

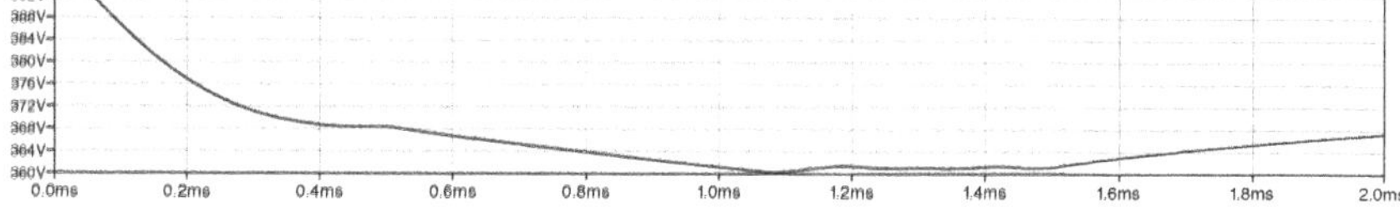

Figure 7.35 One 5.5 kW slice response to current overload.

this converter.[20] A high-voltage BJT with a 10 V zener on its base will work; then once this supply is running, it can be fed back in to the BJT output to reverse-bias it off. See Figure 7.36 for an example of what we're discussing.

The base of the BJT is fed by four 500 kΩ resistors in series (for the voltage rating). We use this value instead of our usual 1 MegΩ because high-voltage BJTs have relatively low gain. We generate the 5 V with a resistor and a zener. We use a small high-voltage MOSFET (though LTspice doesn't have any suitably high-voltage MOSFETs, the lower-voltage ones don't break) and a diode. The buck MOSFET is run with the charge pump in a synchronous driver.

The circuit works as expected. The output varies from 13 to 12 V with a period of 1.5 msec. Load is set at 5 mA. After startup, the BJT is back-biased and dissipates only 37 mW.

7.6.6 Results

The results are pretty spectacular. Figure 7.37 shows that the four currents match very closely, and that the effective frequency is four times that of a single slice, 2 MHz. Figure 7.38 shows that the peak-to-peak voltage ripple on the output is an astonishing 1 mV! Of course, this doesn't include the effect of the ESL of the capacitors. Clearly, we could reduce the output capacitance by a large amount, though since presumably this would involve re-doing the error amplifier compensation we leave this for the reader.

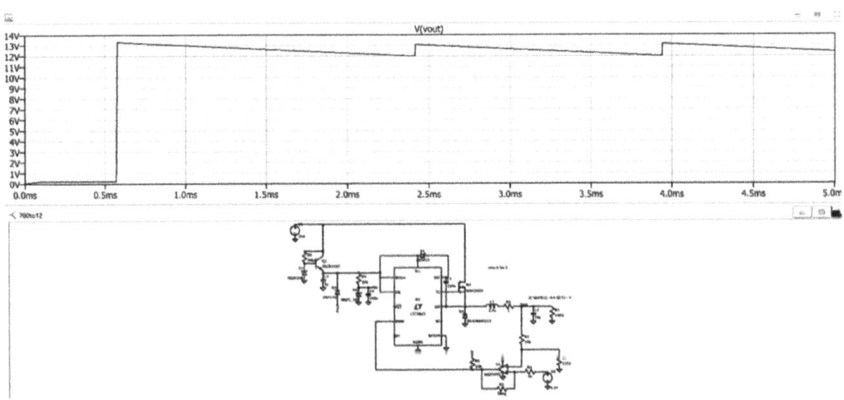

Figure 7.36 700 V to 12 V internal power source with startup.

20 This is known as a bootstrap.

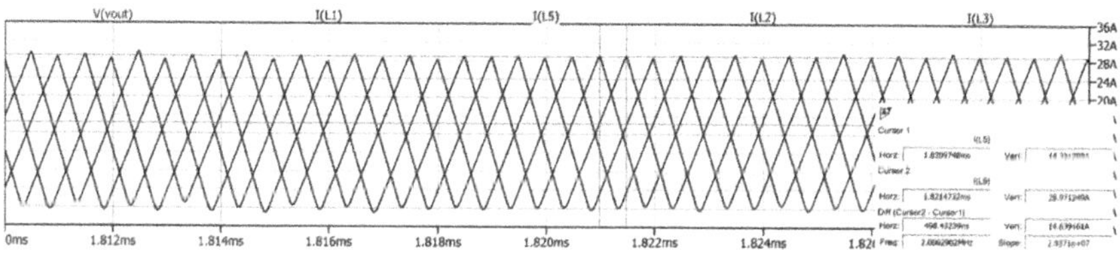

Figure 7.37 Currents in the four inductors, 22 kW.

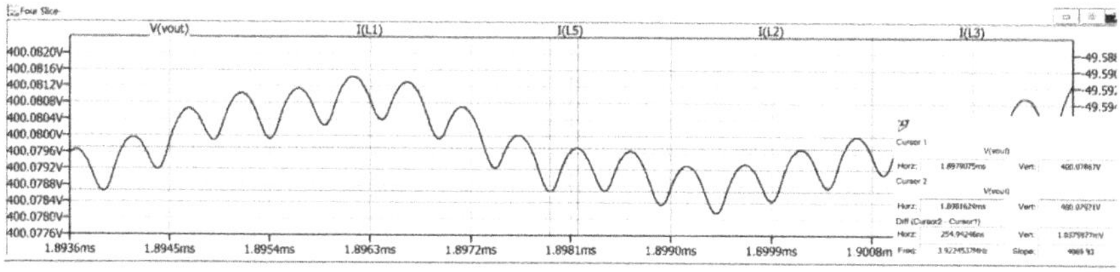

Figure 7.38 Output voltage with all four slices, 22 kW.

Table 7.5 Major component cost of three-phase four-slice 22 kW converter.

Part	Quantity Used	Price Per ($)	Total ($)
1000 V half-bridge gate driver	4	1.39	5.54
1000 V 350 mΩ MOSFETs	16	1.98	31.68
5 μH 13 A inductor	4	0.28	1.12
Op-amps	16	0.08	0.64
SR flip-flop	4	0.09	0.36
D flip-flop	2	0.09	0.18
Inverter logic gate	4	0.09	0.36
High-speed 555	5	0.32	1.60
TOTAL			41.80

7.6.7 Cost

As our last step in this lengthy example, we want to examine the issue of the BOM cost of the circuitry we've designed. As stated at the beginning, we estimate the cost by looking for the 1000 piece pricing at a distributor, or the price of 1000-pieces times the number of each device actually used in the total circuit.

This includes the current-balancing circuitry and all four slices. We only consider the major components, not the resistors or ceramic capacitors; and since all the parts are multi-source, we generally pick the cheapest one the distributor offers. Looking at Table 7.5, we see that the **total BOM** for the *complete converter* is in the ballpark of $40; the high-end FPGA typically used for this sort of design could by itself alone cost this much!

7.7 Notes for the Hobbyist

Bang–bang converters are really quite nice for many applications. They turn on as soon as the voltage goes too low, and turn off as soon as the voltage goes too high. So when should you consider a PWM SMPS?

It's the "as soon as" part that can be an issue. If you have a load step, the bang–bang may not respond as quickly as you need. There can be a propagation delay through the comparator, which causes the bang–bang to over or undershoot its target voltage. Of course, you can then get a faster comparator, but these are somewhat harder to choose and find, and can be more expensive. The other fix can be a bigger capacitor on the output, but "bigger" may be a *lot* bigger, and start getting into more serious money.

A PWM may be a better choice. Although you typically can't get more bandwidth from a PWM than one-tenth of the switching frequency, this latter can be more than a megahertz, as we showed in the examples in this chapter. A hundred kilohertz of bandwidth is good enough for almost all applications.

Of course, there is another choice, a bang–bang SMPS followed by a high-speed LDO. We show how to do this in the final chapter of this book. The downside is that now you have two converters in series, which means two circuits to debug, and of course, two efficiencies in series as well.

8

PFC Converters

8.1 Basic Idea of PFC

A power factor correction (PFC) circuit is a special switch-mode power supply (SMPS) designed to draw a sinusoidal current from an AC source. Power factor (PF) itself is a measure of how close the line current is to being sinusoidal. It typically, but not necessarily, produces a high-voltage DC output, which may then be used as the input to a DC/DC converter second stage to get to the voltage that's actually desired.

8.2 Example: Passive PFC

Although at the start of this chapter we said that a PFC circuit is designed to draw a sinusoidal current from the AC, this does not necessarily require a SMPS. What's actually required by governments is only that the power factor be greater than 0.9; however, that may be achieved.

If you have a limited range of operating conditions, it may be possible to construct a PFC circuit with nothing more than two capacitors, three diodes, and a resistor; no IC is needed at all! A prototypical circuit is shown in Figure 8.1.

The circuit is designed for a constant-power load of 200 mA at 120 VAC, which is 24 W. The two capacitors are aluminum electrolytic because each sees half of the (DC) line voltage (83 V), which would be both expensive and not readily available in ceramic. The diodes are just standard recovery type, the same as are used in the bridge. The resistor's power dissipation depends on the load, and we will evaluate it during modeling.

The circuit is simulated as shown in Figure 8.2. The load is shown as a constant current, which is a good model of a load that varies not at all or very slowly with

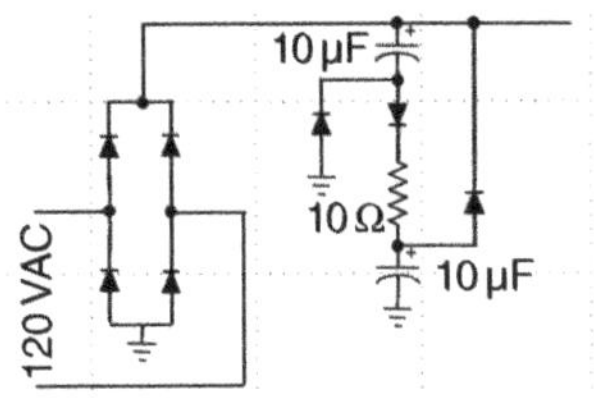

Figure 8.1 Passive PFC circuit.

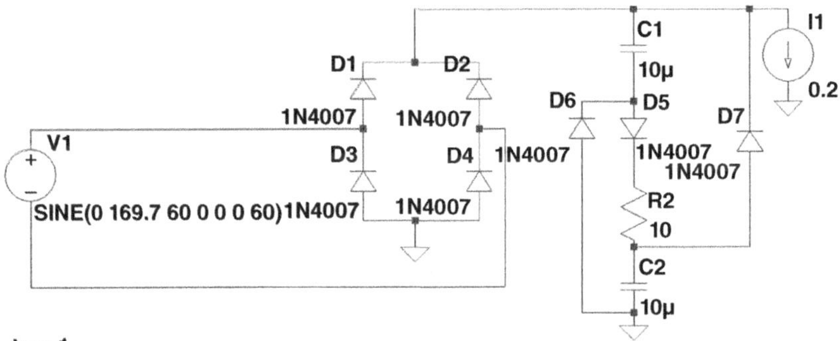

Figure 8.2 Passive PFC model.

time, such as a light-emitting diode (LED) lamp. With this circuit, we measure a power factor[1] of 0.936, which is good and meets specifications.

In Table 8.1, we vary the parameters to show the sensitivity of the circuit.

The results show this circuit to be surprisingly robust. The only situation in which the PF drops below 0.9 is when the circuit is used for a universal input; with some tinkering, it might be possible to raise this to 0.900 without dropping the 120 VAC result below that threshold. It is straightforward to design the circuit to get a good PF if it is to be operated *only* at 230 VAC.

The limitations of this circuit, however, are also clear. First off, two aluminum electrolytic capacitors are required, and they naturally have to be selected to be rated for long life. As the power level goes up, so does the required capacitance, as well as the power dissipation in the resistor. Furthermore, the ripple current in

1 Power Factor is measured at the input source, as $\frac{\text{Power}}{V_{\text{RMS}} \times I_{\text{RMS}}}$, all three of which are available in LTspice: left-clicking on the waveform name will show a table of the average (for the power) and the RMS (for the voltage or the current). The simulation is run for 1 s (60 cycles) so that the initial conditions don't significantly impact the measurements.

Table 8.1 Passive PFC parameter variation effects.

Variation	PFC	Power in R2 (mW)
None	0.936	42
V_{in} = 132 VAC	0.935	45
V_{in} = 230 VAC, Load = 0.1 A	0.881	31
C1 = C2 = 12 μF	0.935	47
Load = 0.1 A	0.921	18
Load = 0.4 A	0.933	94

the capacitors goes up, and at some point, the combination of all of these requirements becomes too expensive and physically large; building a PFC SMPS becomes cheaper and smaller.

If this is to be used for higher voltage AC input, the capacitance voltage rating goes up as well; at 230 VAC, the capacitors will need to be rated 200 VDC.[2] Of course, a SMPS PFC converter also requires a high-voltage output capacitor, usually rated 400 VDC or so for the 230 VAC line. So when possible, the passive PFC circuit is a good choice.

Thus, there are a number of relatively constant-power power supplies where the passive PFC circuit works well, but at some point, you will want to have an active circuit to perform the PFC. As a reference point, the author tried to build a passive PFC at a 1,000 W load; there simply weren't any capacitors available that could do this for under US$10. And the power in the resistor went over 10 W, although this is only a 1% loss, which is likely to still be better than a PFC SMPS.

8.3 Constant On-time Controller

While not capable of producing a perfect power factor, a constant on-time controller can produce PFC at considerably higher power than the passive PFC circuit we looked at in Section 8.2.

Looking at Figure 8.3, we can see the basic concept of the constant on-time PFC. The rectified line is a slowly time-varying positive voltage. For any particular moment when the rectified line is V, when the transistor turns on, the current (we're in either discontinuous conduction mode or boundary conduction mode, so the current starts from zero by assumption) ramps up to $\frac{V_{DC} \times \text{ton}}{L}$. Ton is by

2 RMS line voltage times 10% times the square root of two divided in half, rounded up to the nearest common voltage value.

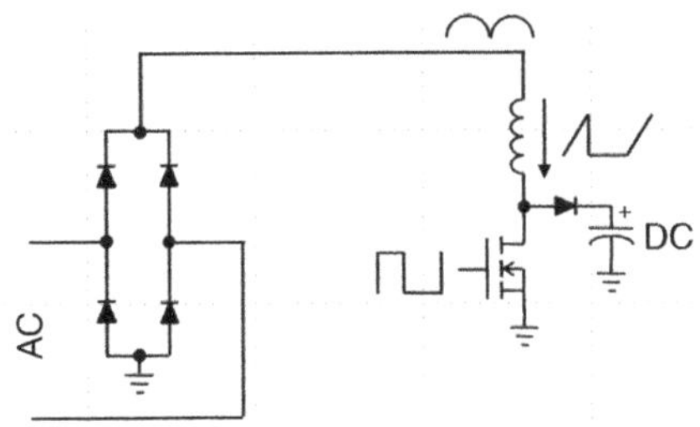

Figure 8.3 Constant on-time controller concept.

definition a constant, so the peak current the inductor reaches is linearly dependent on V. Thus, as the input voltage slowly varies sinusoidally, so does the peak current drawn each switching cycle. And this is exactly what's needed to make a good power factor!

Now we've determined how the peak current changes with V_{in}, but we haven't yet said how to assign just what ton ought to be. The output of the PFC is supposed to be (at least approximately) a DC voltage somewhat higher than the highest input voltage (as otherwise the boost can't work).

However, the energy delivered per cycle by the inductor is determined by the peak current. If (on average) there is too much power, the output voltage on the capacitor will rise; conversely, if on average there is too little peak current, the output voltage will drop.

So, it's clear from this what's needed: we need to (slowly) control the on-time so that the output voltage is on average what we want. To put some numbers to this, the frequency of the AC line is 50–60 Hz; the frequency of the switch is something like 70–130 kHz, so that the on-time has to be less than 7–14 μsec; and to regulate the output voltage so slowly that it doesn't respond to the 50–60 Hz input voltage variation, the loop bandwidth will be 10–20 Hz.

What we've decided we need for power factor correction is a constant on-time. You'll observe that we didn't say anything about the off-time. This varies with how much power is needed, since the on-time plus the off-time is the period, one over the frequency. And thus, in a constant on-time controller, the frequency varies.

So now we can imagine how to make a constant on-time: we can use a 555 timer. The circuit for this is very simple and is indeed shown in the datasheet as the astable vibrator, redrawn here as Figure 8.4.

This will produce a constant on-time when attached to the boost circuit; now we need to be able to adjust the power level, which is done by *modulating* the on-time. The following idea and its implementation seem never to have appeared in the literature heretofore.

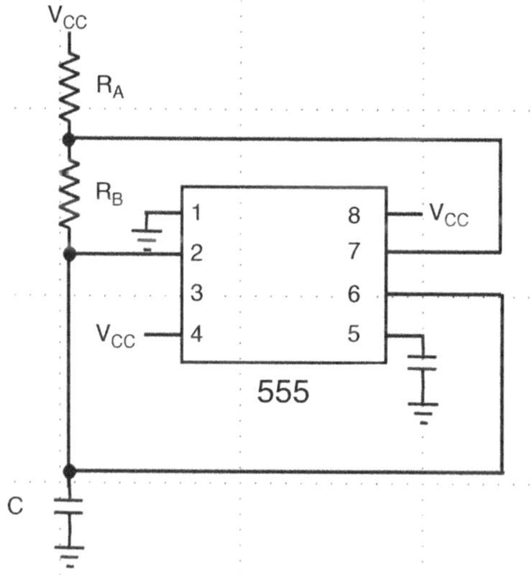

Figure 8.4 555 astable oscillator.

The 555 works by alternately charging and discharging its timing capacitor, which is to say putting current into it and taking current out of it, using the two resistors. Its transitions between the two states are set to occur when the capacitor's voltage is at 1/3 and 2/3 of the 555 supply voltage.

If we were to *change* the amount of current, we could affect the amount of time in the two states. While it's possible to build a robust discrete current source or sink, something similar can be accomplished with a voltage applied through a resistor. And so in Figure 8.5, we do just that. We have an op-amp that is going to end up being the error amplifier of a power supply, feeding a resistor into the junction of R_A and R_B.

To see what this accomplishes, let's build a simulation model of it, shown in Figure 8.6. We've (for the moment) modeled the output of the op-amp as a voltage source, labeled "Control," which we sweep from 1 to 5 V. The output of the 555 is inverted by the transistor, producing the node labeled "Output." We invert it because the 555 has a minimum duty cycle of 50%, and we rather want a *maximum* duty cycle of 50%.

The output vs. the control is shown in Figure 8.7. Below about 1.7 V, there is no output. At about 1.8 V, there is an output with a low duty cycle. As the control voltage increases, the duty cycle increases as well. The measured parameters of the output vs. the control are listed in Table 8.2.

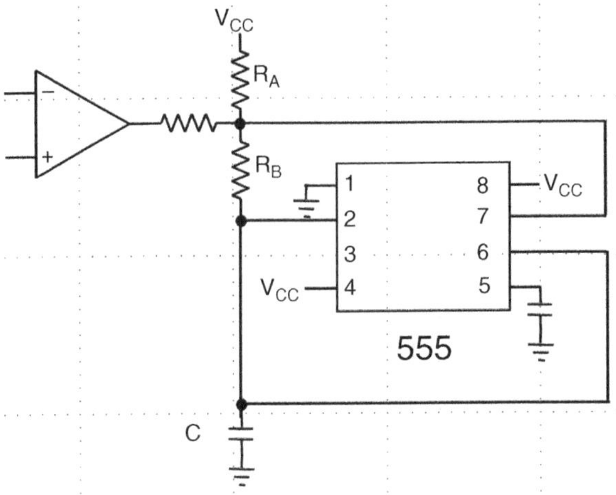

Figure 8.5 Variable astable oscillator.

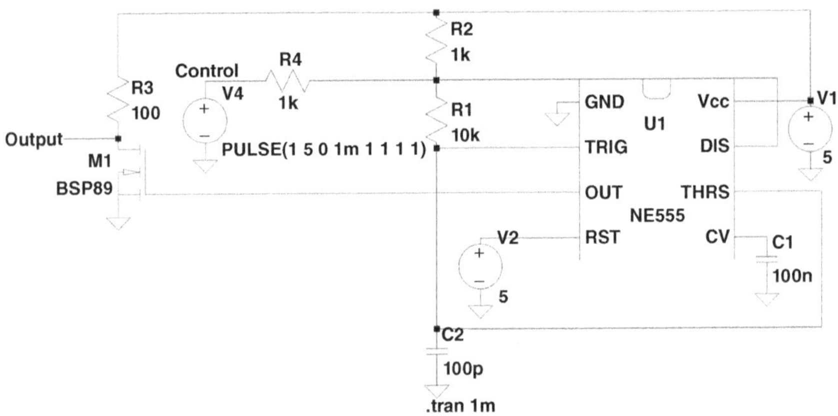

Figure 8.6 Variable 555 astable model.

Looking at this table, it is immediately apparent that the on-time is constant (within the hand-measurement accuracy). It's a higher frequency than we wanted for our PFC, but that's just a matter of selecting the appropriate value of the capacitor. The duty cycle ranges monotonically from 0% to almost 50%. We can thus use this novel control scheme to build our constant on-time PFC.

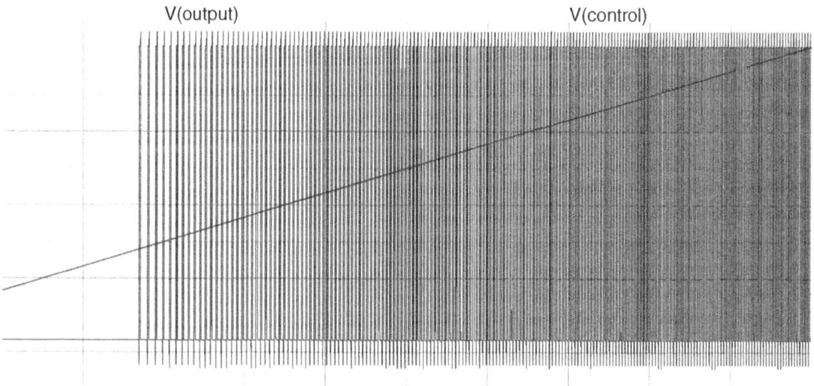

Figure 8.7 555 output with controlled on-time.

Table 8.2 Performance of controlled 555.

Control (V)	On (μsec)	Off (μsec)	Frequency (kHz)	DC (%)
1.50	0			0
1.67	0.85	4.62	183	16
2.00	0.84	2.65	287	24
2.50	0.83	1.81	379	31
3.00	0.85	1.44	437	37
3.50	0.85	1.21	485	41
4.00	0.87	1.06	518	45
4.50	0.86	0.93	559	48

8.4 Example: 120 VAC, 100 W

Now that we have a method to create a robust constant on-time controller, we will construct our first design: a 120 VAC input PFC rated for 100 W. We will be designing not just the power stage and the control system, but also a startup circuit and output overvoltage protection. We won't be designing the input electromagnetic interference (EMI) filter, as how to do this is well-covered elsewhere.[3]

3 See [9].

8.4.1 Robust Design

The PFC power stage is normally just a non-isolated boost converter, with isolation, if necessary for the system, being done in a second-stage converter; refer back to Figure 8.3.

This first example is 120 VAC input, so adding in 10% for line variation and multiplying by $\sqrt{2}$, we find the peak voltage is 187 V. Therefore, we'll set the output voltage to be 200 V and thus need our output capacitor, diode, and MOSFET to be at least 250 V.

In order to avoid confusion, let's clarify that there are three different timescales in this design. The first is in seconds or more. It is at this timescale that we have a power output of 100 W, and so the output current at 200 V will be 500 mA. We select a 1 A Schottky (the author's experience is that high-voltage Schottkies work better than ultrafasts in this application). Following the recommendations in Chapter 3, we look for 50% voltage derating, thus for a 400 V, 1 A multi-sourced Schottky. Lots of manufacturers make a 650 V, 2 A silicon carbide Schottky, so we can pick that.

A second timescale is at the line frequency, approximately 8.3 msec for half a line cycle. At this timescale, the voltage varies from 0 to 168 V, and the current, since we are power factor correcting, also goes up by a factor of square root of two. At 120 V, 100 W is 833 mA, so at line peak it is 1.178 A, giving us a peak power during the line cycle of 1.178 A × 168 V = 198 W. This is double the average because both the voltage and the current are root two higher at the peak, and therefore, the power is double.[4]

For selecting the other power components, we need to know the equations governing the switch-mode operation of the converter. This occurs at the third timescale, which is the switching frequency, about 130 kHz, or about 7.7 μsec. We are planning to operate in discontinuous-conduction mode. ("Discontinuous" means that the inductor current returns to zero every cycle.[5]) Starting with the equation for an inductor, for every switching cycle we have

$$\frac{V}{L} = \frac{dI}{dt} \tag{8.1}$$

4 It's not *exactly* double because of rounding errors.

5 Modern PFC control ICs frequently use Boundary-Mode Conduction or Critical Conduction Mode, which are two names for the same thing: the inductor current just barely reaches zero every cycle. We can accomplish this by choosing an inductance close to the value we calculate, rather than substantially above or below that value. Just by the way, the author prefers the term "boundary-mode conduction" (BCM) because "CCM" normally means continuous conduction mode and "critical conduction mode" would confusingly have the same acronym.

The current starts at zero when the switch turns on, so the peak current is

$$I_{\text{pk}} = \frac{V \times \text{ton}}{L} = \frac{V \times \text{DC} \times T}{L} = \frac{V \times \text{DC}}{f \times L} \tag{8.2}$$

with the period T being one over the frequency, and DC being the duty cycle.

We also need the equation governing how much power is transferred by the switching each cycle. Since the energy stored in the inductor depends on the current,

$$E = \frac{1}{2} L I_{\text{pk}}^2 \tag{8.3}$$

the power transferred depends on how often this is done,

$$P = \frac{1}{2} L I_{\text{pk}}^2 f \tag{8.4}$$

We can put these two equations together to get an expression for the power,

$$P = \frac{V^2 \text{DC}^2}{2Lf}. \tag{8.5}$$

As just indicated, it is common with PFC converters for 120 VAC line to make the peak-power frequency somewhere around 130 kHz. The reason for this is that government-conducted-EMI regulations start at 150 kHz, and by having the peak switching frequency stay below this, we can minimize the size of the EMI filter required (the first in-band frequency will thus be the second harmonic at 260 kHz). Let's suppose that in our 555 circuit we have a maximum duty cycle of 40%. Using a 390 pF timing cap, this will give a maximum frequency of 130 kHz (actual simulation at line peak suggests using 240 pF). So, the maximum on-time is 40% / 130 kHz = 3.1 μsec.

We can put these numbers into our equations, 200 W maximum, 168 V maximum, 40% duty cycle maximum, 130 kHz maximum:

$$200\ \text{W} = \frac{168\ V^2 \times 0.4^2}{2 \times L \times 130\ \text{kHz}} \tag{8.6}$$

and so the inductance should be 87 μH (or less; if it's less, the on-time will be lesser to compensate, but it can't be more than this, because we're already at maximum on-time to remain in DCM). Using this, we find the peak inductor current to be

$$I_{\text{pk}} = \sqrt{\frac{2P}{Lf}} = \sqrt{\frac{2 \times 200\ \text{W}}{87\ \mu\text{H} \times 130\ \text{kHz}}} = 5.95\ \text{A}. \tag{8.7}$$

Does this seem high to you? Remember that this is the *peak* current during a single switching cycle at line peak. Average current during a switching cycle at line peak is

$$I_{\text{avg}} = \frac{I_{\text{pk}}}{2}\text{DC} = \frac{5.95\ \text{A}}{2} \times 40\% = 1.19\ \text{A} \tag{8.8}$$

remembering that this is a sawtooth. 1.19 A × 168 V = 200 W, as it's supposed to. RMS current is

$$I_{\text{RMS}} = I_{\text{pk}}\sqrt{\frac{\text{DC}}{3}} = 5.95\,\text{A}\sqrt{\frac{0.4}{3}} = 2.17\,\text{A} \tag{8.9}$$

We select a surface-mount inductor of ≤68 μH with a saturation current a bit more than 5.95 A (current rating, which determines how hot the inductor gets, only needs to be 2.17 A), and we find that there are a number of such parts, with about 100 mΩ winding resistance that easily meet all the requirements and are multi-sourced.

The transistor needs to be 250 V VDS or more and carry 2 A average; these are very common, and we can simply pick one with suitably low Q_g.

Finally, we need the output capacitor, which will also be 250 V, a long-life aluminum electrolytic. It has to bridge the output power during the few milliseconds while the rectified AC voltage is low. We can estimate 100 W/200 V = 500 mA, so we conservatively choose something that will bridge for a (half) line cycle, something like

$$C = \frac{Q}{V_{\text{pp}}} = \frac{It}{V_{\text{pp}}} = \frac{500\,\text{mA} \times 8.3\,\text{msec}}{20\,V_{\text{pp}}} = 207.5\,\mu\text{F} \tag{8.10}$$

We pick a 220 μF.

8.4.2 Control Circuitry

So far, we have the power stage and the 555 as shown in Figure 8.8. Now, we need to hook up the error amplifier. We want the output voltage to be 200 V, so we need to divide this down to the reference voltage, for which we can use a 431. Following the voltage ratings in Chapter 3, we need two 0805 resistors on top, which we'll pick to be 1 MΩ each. The bottom resistor should then be 12.477 kΩ, and we can pick a 12.5 kΩ, all at 1%.

As for the compensation, this is actually pretty straightforward. We know that the bandwidth of the PFC should be very low, low enough that the 60 Hz line cycle doesn't affect it much. So say we make it 10 Hz, which is 100 msec. Since the feedback impedance is 2 MΩ and we want a 10 Hz pole, 100 msec/2 MΩ = 47 nF can be our Type I capacitor on the feedback of the error amplifier.

We haven't yet talked about connecting the 555 to the MOSFET. We know it needs to be inverted; for the moment, since the frequency is (relatively) low, we will try just an inverter and a pull-up resistor as we showed in an earlier section.

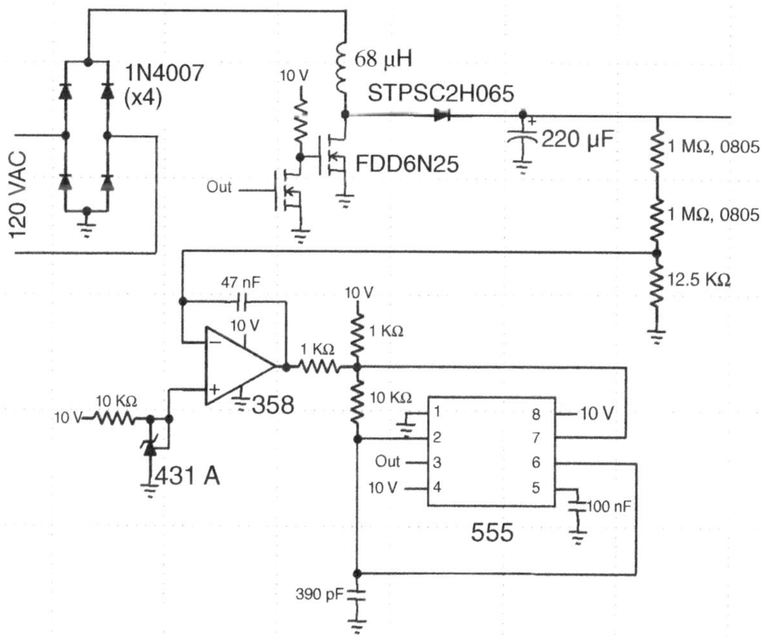

Figure 8.8 100 W PFC with controlled on-time, first cut.

8.4.3 Modeling

Time to build a model, shown in Figure 8.9. We can, for the most part, use the model we built in Section 8.4.2. We make the usual changes: the 431 circuit is replaced by a voltage primitive, and the inductor has its winding resistance placed externally to its inductance.

There are a few significant changes from the schematic. We've dropped the voltage running the 555 and driving the MOSFET gate from 10 V to 5 V, because we're using just a pull-up resistor to turn the MOSFET on, so this reduces the power dissipation in the pull-up by three quarters.

We've added in a 100 nF capacitor at the input to the inductor, for local power bypass, that is, so the high-frequency current being pulled by the inductor doesn't upset the rest of the power supply circuit. We can't make it very big, because otherwise that would ruin the power factor. A general rule of thumb is to make this about 100 nF per 100 W of offline power.

As usual, we've added in an initial condition that the output voltage starts at the expected voltage, though this isn't necessary.

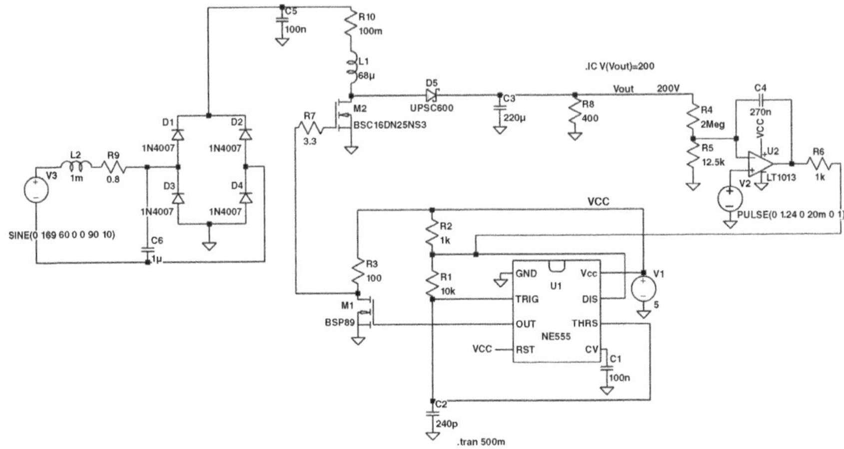

Figure 8.9 Model of 100 W PFC with controlled on-time.

Most significantly, we've added in an EMI filter at the input. A little thought shows why. The inductor current is going to be coming from the line. If there is no impedance from the line to the inductor, this sawtooth current will show up directly on the AC. This will upset the PF measurement, not to mention ruin the EMI. The filter, as is common, consists of a 1 mH multi-sourced inductor with a typical 800 mΩ winding resistance and a 1 μF capacitor line to line. This latter can probably be reduced once we've measured the PF and EMI.[6]

To get good measurements of the PF, we are going to have to simulate for multiple line cycles. We've set it up for 500 msec, which is 30 cycles. This means the simulation is very stiff, so it's going to take a lot of time and generate a very large file.

Running this first for just a few cycles, we observe (after the startup transient dies off) that the response of the circuit is still too fast. We make the final change to our schematic of increasing the compensation capacitor from 47 to 270 nF.

Finally, we run the model. The output voltage varies between 195 and 207 V, which is just fine for the input to a following power stage. A picture of the input voltage and current is shown in Figure 8.10. The current is more or less sinusoidal and clearly closely in phase with the line voltage.

We compute the power factor

$$PF = \frac{\text{Power}}{I_{RMS}V_{RMS}} = \frac{111.03\ \text{W}}{976.69\ \text{mA} \times 121.20\ \text{V}} = 0.938 \tag{8.11}$$

6 Observe that the line-to-line capacitor is *not* the same as the post-bridge capacitor. It has to be rated for the AC line, not DC.

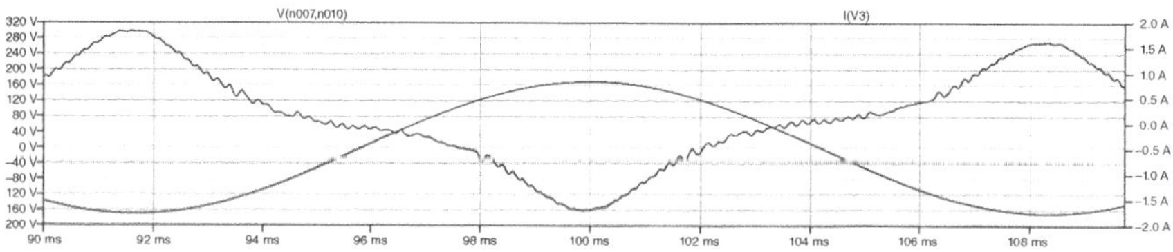

Figure 8.10 Input current and voltage on 100 W PFC with controlled on-time.

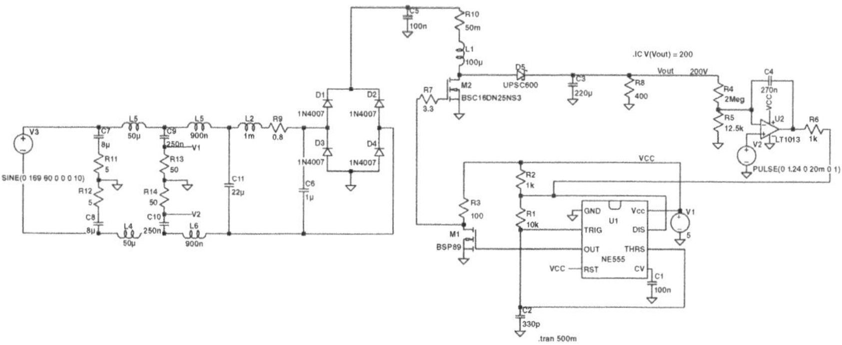

Figure 8.11 100 W PFC with controlled on-time with LISN.

showing that we have, indeed, corrected the power factor in a 100 W converter using only an op-amp and a 555 timer!

8.4.4 Electro-magnetic Interference

The other thing we need to look at is the EMI, as we always must for offline converters. We use the model of a Line Impedance Stabilization Network (LISN) from Chapter 4 as shown in Figure 8.11, Earth ground connections are taken as ground, since our input is two-wire, per [17].

We measure the voltage from V1 to V2 and do a Fast Fourier Transform (FFT) on it. This is tricky; follow these instructions exactly. You click on the "View" tab, and select "FFT." You then click on *both* V(v1) and V(v2). You also select "Use current zoom Extent" (sic) to only grab the part of the waveform that is after the startup transient. Pressing "OK" pops up a new window that asks you to "Select Waveforms to Plot." Click on the *first* one, V(v1) (only), and then Alt-double click to edit it to V(v1,v2). Finally, clicking "OK" pops up the FFT you actually need.

Again multiplying the value by one million to get the result in dBμV, we get Figure 8.12. We see that the highest noise above 150 kHz occurs at 191 kHz and is 33 dBμV, clearly passing. Indeed, this is quite substantially below the

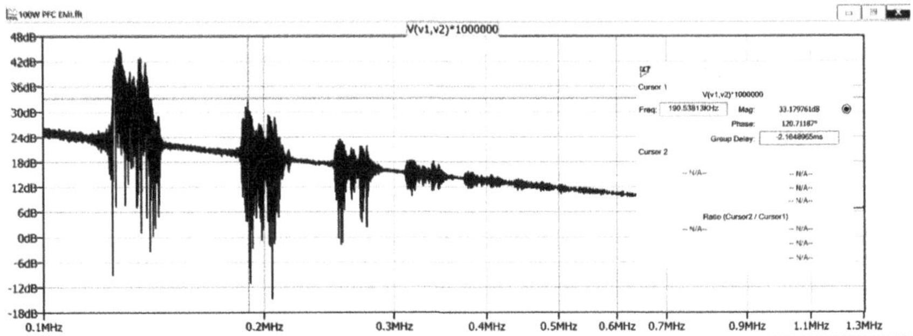

Figure 8.12 100 W PFC with controlled on-time EMI spectrum.

requirement, and we should think about reducing the 1 μF AC line capacitor, as it is large and expensive. This is left for the reader.

8.4.5 Startup Circuit

This model is great; it passes both PF and EMI requirements, but a moment's thought shows that something is missing. Where did that 5 V internal supply come from?

What we're going to do is to use a linear regulator to generate 5 V. Since this is a boost converter, as soon as input power is on, the output capacitor will charge up to 169 VDC. As soon as the converter is running, it will boost up to 200 VDC. Since we're going to linearly regulate this down, it'll be less lossy to regulate from the rectified input at 120 V_{RMS} instead.

Our model shows that the current drawn on the 5 V is 45 mA average, with all but 4.4 mA average going to the gate drive. Replacing the pull-up resistor with an actual gate drive reduces this to 6.9 mA, for a total of 11.3 mA. At 120 V average, this produces a power loss of 1.4 W. While this isn't wonderful (it's a 1% efficiency hit), the other choice is to build a little switching regulator, which seems excessive for this converter. We could also look for a MOSFET with less gate charge, as the model is presently using one with 8 nC, which is only moderate.

The circuit shown in Figure 8.13 regulates from the rectified AC line down to 5 V. As soon as there's more than 6 V on the input line, the zener on the base of the bipolar junction transistor (BJT) turns on, and the emitter follows this voltage (minus V_{BE}), giving about 5 V. The resistor can be a large value, as the gain of the BJT is large. The BJT itself must be chosen to take 250 V and be in a package capable of handling the power dissipation. The capacitor is selected large enough to ride through the time during which the AC line is <6 V, which of course isn't for very long.

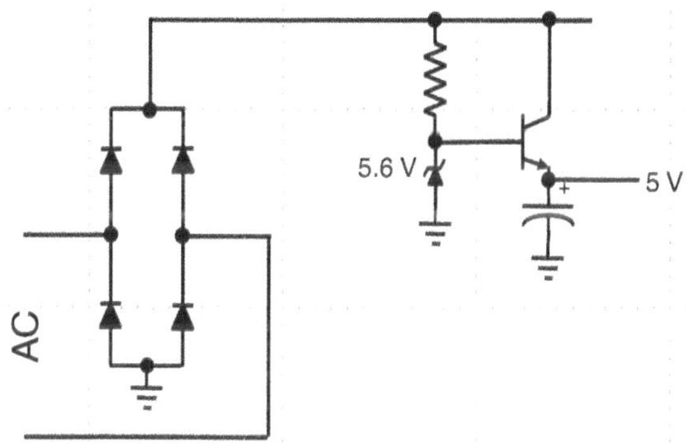

Figure 8.13 Offline 5 V supply.

8.4.6 Output Overvoltage Protection

One more thing is missing from this PFC circuit. PFC ICs typically provide some protection if the output voltage goes too high. This is particularly desirable, as the control bandwidth of the converter is very low, so that at no load the output voltage may quite substantially overshoot before the 555 is turned off.

As shown in Figure 8.14, we put a comparator on the output voltage (using the same divider we already use for the feedback voltage). We set another 431 to be a higher voltage (with a setting divider whose impedance is high compared with that of the resistor feeding the 431). You could, of course, use only a single 431. If the output voltage goes more than 10% high (220 V), the comparator immediately goes low, turning off the 555 within one switching cycle. In normal operation, the comparator output is open collector, and the RESET pin is pulled high (active) by the pull-up resistor.

In case you were wondering, PFCs don't need a soft start. Since their bandwidth is less than that of the line frequency, it takes multiple line cycles before they reach steady-state output voltage anyway. And since it's a boost, the output voltage rises up to the line peak almost instantly as soon as input power is present.

8.5 Example: 230 VAC, 100 W

We're going to take the previous design and do a quick modification for it to run on 230 VAC. Looking at Figure 8.15, we have

- Increased the boost inductor to 100 μH;
- Increased the output capacitor to 1,600 μF;
- Changed the resistor divider to get 375 V output (with three series 1 MΩ resistors for voltage rating);

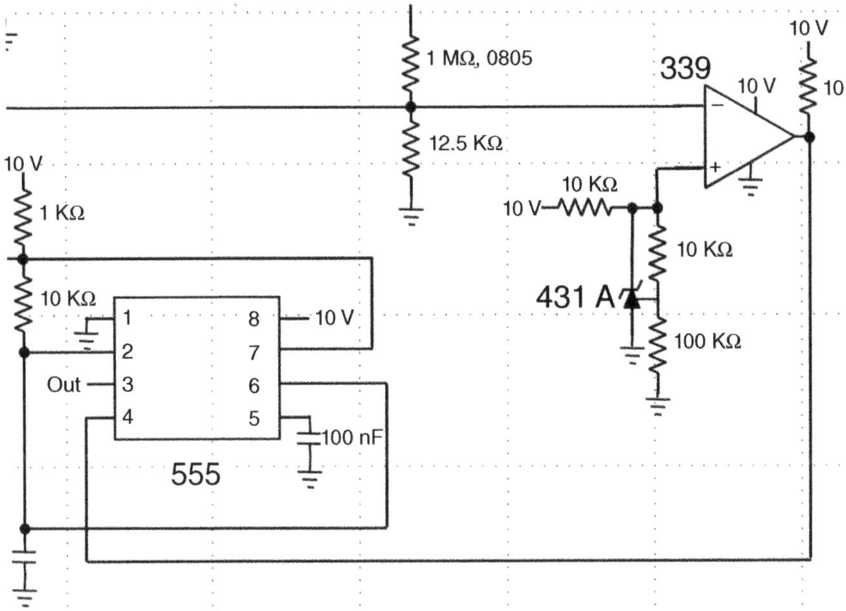

Figure 8.14 Over-voltage for a 555 PFC.

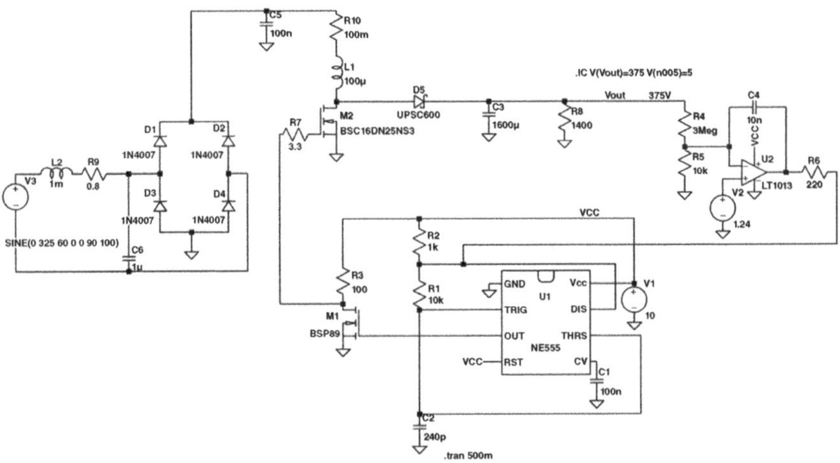

Figure 8.15 Model of controlled on-time 100 W PFC with 230 VAC input.

- Changed the load resistor to maintain 100 W;
- Decreased the compensation capacitor to 10 nF; and
- Reduced the power factor feed resistor to 220 Ω because of the increased dynamic range needed to do power factor correction at a higher line voltage.

Naturally, we would also increase the voltage of the MOSFET and the capacitor. The results are that $V_{in} = 230.37$ V, $I_{in} = 0.50315$ A, $P_{in} = 104.24$ W, and $P_{out} = 98.197$ W. This yields efficiency = 94.2% and $PF = 0.900$. Peak-to-peak ripple voltage on the output is 623 mV, although we didn't include the ESR of the output capacitor, which will be the dominant effect.

Can this be extended to a universal input? This will depend on the dynamic range of the 555 and is left for the reader to explore.

8.6 The 9910 as a PFC

For the final multisource PFC controller, we're going to consider the 9910. It was invented by Supertex as an LED driver and was so popular that there are now several manufacturers of it. We will be using it to do general-purpose PFC, not specifically for driving LEDs.

The 9910 is fundamentally a boost converter that runs current-mode by sensing the MOSFET source current. As originally designed, as shown in Figure 8.16 from the datasheet, the LED current goes through the inductor and the MOSFET, and thus controlling the MOSFET current (in switch mode) controls the LED current. The current sense signal is a ramp (because of the inductor), and the MOSFET is turned off when the signal reaches 250 mV.

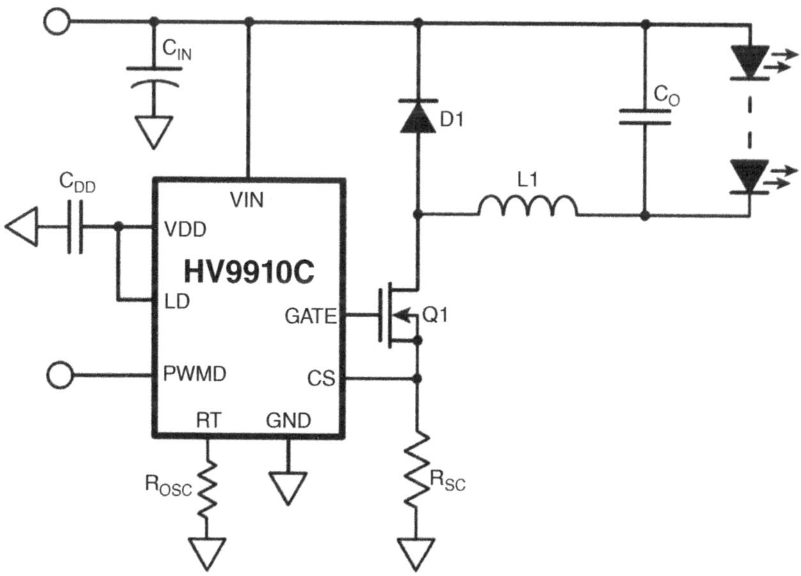

Figure 8.16 9910 basic circuit.

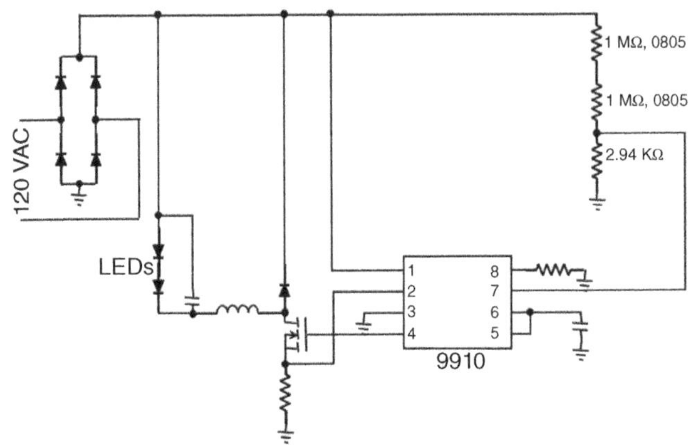

Figure 8.17 9910 LED driver with PFC.

Now the 9910 has not just one but *two* dimming methods: a digital control on the PWM dimming (PWMD) pin, and an analog control on the linear dimming (LD) pin. The digital control is just an on/off, which is intended to run at some kilohertz. The analog control pin adjusts the current at which the MOSFET is turned off: when the voltage on this pin is 250 mV (or more), the maximum current allowed is set by the current-sense resistor. But as the voltage on the LD pin is decreased below 250 mV, the turn-off current is proportionally reduced.

A number of years ago, the author realized that this provides a method to use the 9910 as a PFC, since the LD pin acts as a multiplier. All it takes is to make the voltage on the LD pin proportional to the input voltage. Since then, the author's invention has been widely used.

A PFC LED driver conceptually using the author's idea is shown in Figure 8.17. It is just like the circuit shown on the 9910 datasheet but additionally has a resistor divider from the rectified line that has a peak of 250 mV at $120\text{ V} \times \sqrt{2} = 169\text{ V}$. This signal is fed into the LD pin, controlling the peak current. Since the input is (rectified) sinusoidal, so is the control signal and thus the peak current, and thence the average current, which is what is required for a good power factor.

8.7 Example: PFC 277 VAC, 5,000 W

For our final example in this chapter, we build a high-power PFC, 5,000 W. This is for LED stadium lighting and runs from 277 VAC,[7] which is phase-to-phase in a three-phase 480 VAC system. The LED floodlight itself is a large panel containing

7 This voltage is commonly used for lighting, for example, in office buildings.

20 strings of 12 LEDs each, 240 LEDs total. Each LED[8] is being run at about 20 W, and at 150 Lm/W, this is 3,000 Lm/LED, for a total of about 750,000 Lm. This is a good application for LEDs, as the light it is replacing (Metal Halide) uses 20 kW, and thus there is a 75% reduction in power usage.[9]

8.7.1 Robust Design

The input voltage is so high that we can still use a simple boost (recall that power is proportional to the *square* of the input voltage). Peak input voltage is $277\,\text{VAC} + 10\% \times \sqrt{2} = 430\,\text{V}$, and so we will run the output at 450 VDC. And for this example, we will be using the 9910.

Let's select components. At 5,000 W output at 430 V, the output current is only 11 A, and so we can even still use a diode as the rectifier rather than needing synchronous rectification. In particular, there are multiple vendors with ultra-fast diodes rated at 1,200 V, and since peak current will be much higher than average, we choose an IDWD30E120D7, which has an average current rating of 52 A, a V_f of 3.75 V at 80 A, and runs US$1.96 at 1,000 pieces. With a 450 V output, the diode losses, even with this high a V_f will be less than 1%.

The MOSFET needs to be 600 V or more. There is a wide selection of such parts available; we can pick one with the appropriate $R_{DS,on}$ after we simulate (or calculate) RMS current.

For the other power components, we use the same equations as before. Frequency is set as usual at 130 kHz, maximum duty cycle is 40%, and peak power is double the average at 10,000 W, and so

$$10{,}000\,\text{W} = \frac{430\,V^2 \times 0.4^2}{2 \times L \times 130\,\text{kHz}} \tag{8.12}$$

and so the inductance should be less than 11.4 μH with the peak inductor current of

$$I_{pk} = \sqrt{\frac{2 \times 5{,}000\,\text{W}}{11.4\,\mu\text{H} \times 130\,\text{kHz}}} = 82\,\text{A} \tag{8.13}$$

and RMS of

$$I_{RMS} = I_{pk}\sqrt{\frac{DC}{3}} = 82\,\text{A}\sqrt{\frac{0.4}{3}} = 30\,\text{A} \tag{8.14}$$

8 For background on LEDs, see the author's work [2].

9 If you're curious about thermal management, it turns out that something like 20% of the input power to the LEDs is converted to light, so thermal load is more like 4,000 W.

We select an inductor of 7.5 μH at a saturation current of 118 A and a current rating of 31 A, such as the IHDM1107BBEV7R5M20, which has a winding resistance of 2.6 mΩ. There are a number of such parts from various vendors; this one costs US$2.84 at 1,000 pieces.

Finally, we need the output capacitor, which will also be 630 V, a long-life aluminum electrolytic. It has to bridge the output power during the few milliseconds while the rectified AC voltage is low. We can estimate 5,000 W/450 V = 11 A, so we conservatively choose something that will bridge for a (half) line cycle, something like

$$C = \frac{11\,\text{A} \times 8.3\,\text{msec}}{50\,\text{V}_{\text{pp}}} = 1{,}800\,\mu\text{F} \tag{8.15}$$

We pick two parallel 1,000 μF aluminum electrolytic capacitors rated at 550 V each with a lifetime of 18,000 hrs at 85 °C and a ripple current rating of 7.59 A at 10 kHz. These run US$31.40 each at 1,000 pieces, but after all, it's a 5,000 W supply; no sense in skimping on the pennies for a shorter-life part.

8.7.2 Modeling

Control of this PFC is going to be done with a 9910, for which we've already built a model in Chapter 4. Looking at Figure 8.18, we've selected a reasonable model of a diode (rated 600 V at 15 A), and we've put in a pair of MOSFETs of 165 mΩ each (the model's voltage is only 250 V. We've numerous times mentioned the deficiency of adequate MOSFET models in LTspice, but the models don't break).

The resistor dividers have three 1 MΩ on top for voltage ratings, and we've thrown in a 5 V supply, which could be the 9910 internal 7.5 V regulator if the current drawn is less than 5 mA.[10] Otherwise, you can make a little bang–bang to generate it. Finally, we have a 2.2 mΩ current sense resistor, and we've selected the error amplifier's compensation to make it slow.

Running the simulation (again, the simulation is slow, and the file is big), we see in Figure 8.19 that the output voltage varies between 449 and 468 V. Power factor is 5,352 W/(20.295 A_{RMS} × 281.74 V_{RMS}) = 0.936. Efficiency is 5,271 W/5,352 W = 98.5%, both very respectable numbers;[11] the latter implies that the total losses in the PFC are only 81 W (not counting the 5 V supply).

We mentioned above that we have two capacitors, each rated at 7.59 A RMS current. We measure ripple current into the capacitor at 20 A_{RMS}, and so we should look at the possibility of either getting a higher-rated pair or going to three of them.

10 The 9910 regulator generically works up to 450 V, but you need an external resistor to drop voltage for power dissipation.
11 High-power PFC converters are often in the 98–99% efficiency range.

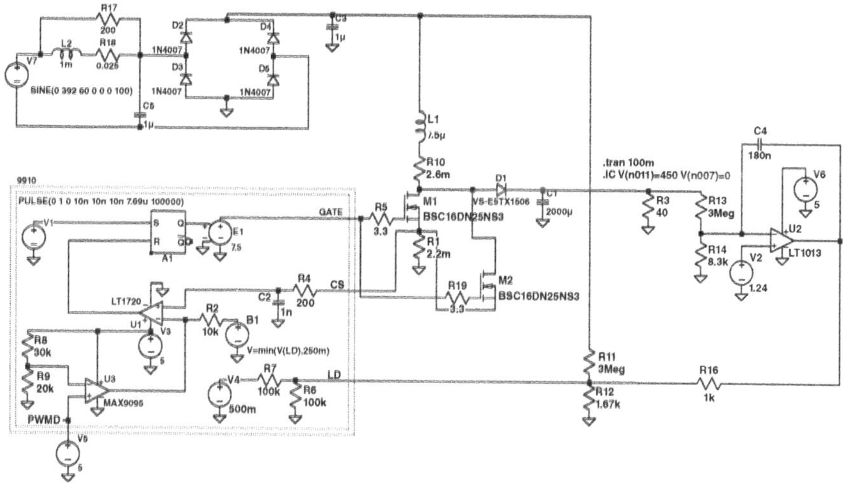

Figure 8.18 5 kW PFC using 9910.

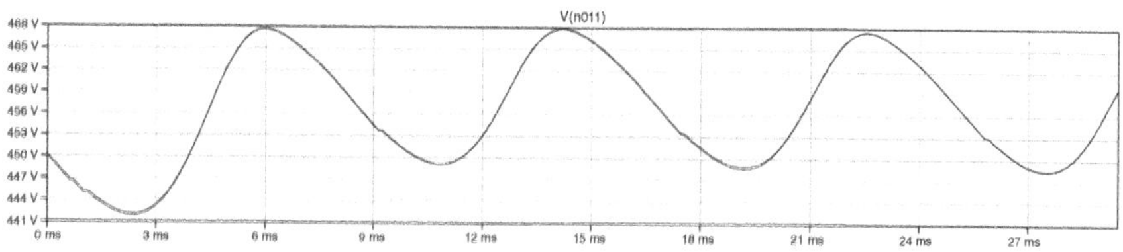

Figure 8.19 PFC output voltage with 9910 at 5 kW.

8.7.3 Electro-magnetic Interference

Since this is a constant-frequency rather than a constant-on-time converter, we expect the EMI to be worse. We add in a LISN to the model, with the result shown in Figure 8.20. Sure enough, the highest in-band noise level is at double the switching frequency, 260 kHz, of magnitude 51 dBμV, while the limit is 62 dBμV, leaving us with less margin, but still passing.

8.8 Notes for the Hobbyist

The most important thing for you to know when working on designs in this chapter is the material in Chapter 1. You should go read it right now.

The reality is that PFC is only required if you're going to be mass-producing your design, and even then, only if the power rating is 75 W or more. So, your personal projects don't require PFC, unless, of course, your project is to build one.

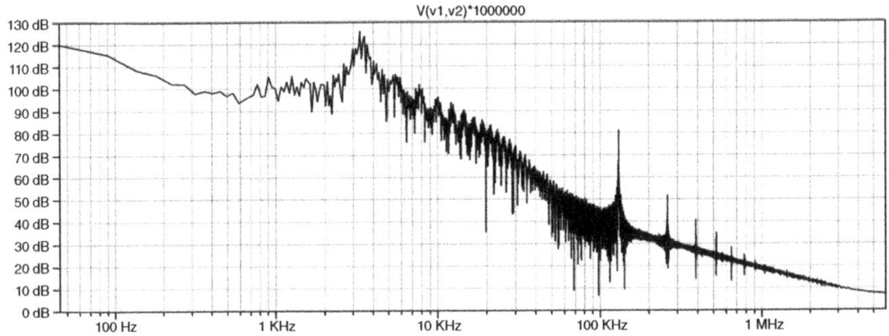

Figure 8.20 9910 PFC 5 kW EMI spectrum.

For many purposes, the passive PFC is good enough. Since it only involves four components and works well even above 24 W, this should be your go-to.

Finally, if you really want PFC, you can simply buy a board-mount, offline power supply from distribution. They always have PFC, and their outputs are isolated to boot. They're kind of pricey, but it may be worth it.

9

Isolated Converters

In all of the power supplies we've designed in this book up until now, there's been a single ground: the output power was referenced to the same point as the input power (or, in the case of offline converters, to the ground reference created by the diode bridge).

However, there is frequently the need to have the output be on a separate ground. This is almost universally true for offline converters for safety reasons. Although the first section of the power conversion may be a non-isolated power factor correction (PFC), a following stage is almost always required to convert from the primary referenced PFC output voltage to the desired output voltage referenced to a secondary, isolated ground. These are the types of supplies we will be designing in this chapter.

9.1 Isolated Power Configurations

Figure 9.1 shows a block diagram of the most common implementation of this concept. The AC is rectified by a diode bridge, and then a flyback is used to boost the voltage to a duty cycle (DC) above the peak input while correcting the power factor (PF). This output DC is then the input to the next stage, which, in its most common configuration, is shown as the primary of an isolated flyback transformer.

The flyback transformer consists of two windings of a transformer, both on a core of ferrite material, the core being shown as the two vertical lines. The two winding dots show the direction the wire is wound. With the placement shown, the transformer is a flyback; if one of the dots were moved to the other end (corresponding to winding the wire of that winding in the opposite direction), the transformer would be a forward.

Finally, the output of the secondary, on the right side of the transformer, is rectified to create the output voltage of the system. The secondary is referenced to a

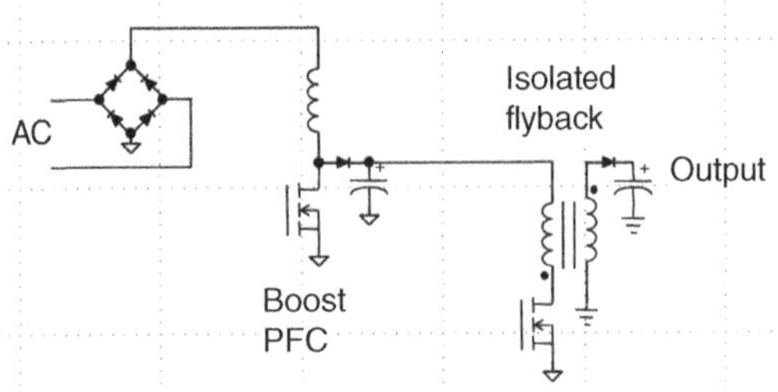

Figure 9.1 Concept of isolated converter.

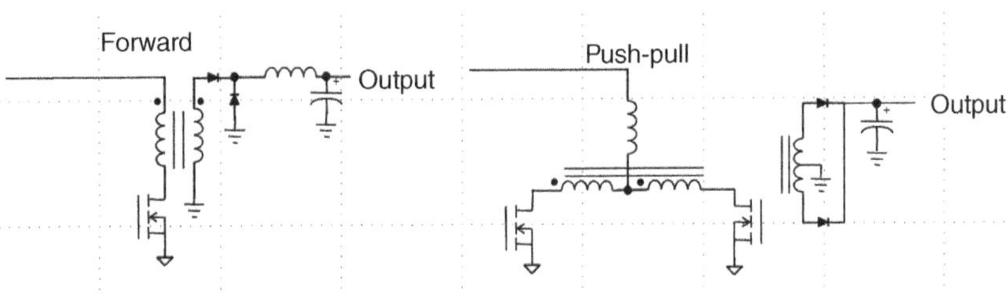

Figure 9.2 Other common isolated converters.

different ground than the primary, one that is created by the energy being transferred through the transformer. Ultimately, this rectified output and the isolated ground are delivered to the load.

Two other somewhat common configurations for isolated converters are shown in Figure 9.2. The forward converter shown on the left-hand side is different than the flyback in that the dots are on the same side of the transformer. What this means practically is that, whereas the flyback stores up energy in the primary when the transistor is on and subsequently releases the energy to the secondary when the transistor turns off, which limits the amount of energy transferred to the secondary per cycle, the forward transfers energy when the transistor is on. The extra inductor on the secondary is present because, otherwise, there would be no limit to how much energy would be transferred when the transistor was on, the primary voltage being applied directly (with the turns ratio) to the secondary. This then explains the extra rectifier on the secondary, to provide the inductor with current when the transistor is off. Thus, the flyback is simpler and cheaper than the forward, but in general will turn out to be able to transfer less total power than the forward and have greater component stresses.

The other configuration in Figure 9.2 is the push-pull converter shown on the right-hand side. This comes in two flavors, voltage- and current-fed; we show the latter because of its superior reliability.

In the current-fed push-pull, then, the input voltage is converted to a current by means of the input inductor. The two transistors then alternate as to which one of them is on; the two are never on at the same time. Suppose that the leftmost transistor is on. The current from the center tap of the primary of the transformer to ground through the transistor is the inductor current, IL. The voltage across this half of the primary is, say, V_{in}; since the other half is wound on the same core and has the same number of turns, the voltage on the far end where the other transistor is off must be $2 \times V_{in}$.

The split secondary, meanwhile, sees IL times the inverse of the turns ratio. (The voltage ratio is proportional to the turns ratio, and since energy is conserved, the current goes as the inverse.) When one transistor is on, one of the diodes conducts; when the other transistor is on, the other diode conducts.

The big advantage of all three of these types of converters over all the other types commonly used is that in all three of them the power transistor(s) are grounded. This makes it possible to drive them relative to the ground, and not require any sort of off-ground driver. The advantage of the push-pull over the flyback and the forward is that it has *two* transistors working with the primary current, so the power is split evenly between the two of them, whereas with the flyback and the forward all the power has to be handled by a single transistor. Of course, if you end up needing multiple transistors to handle the power anyway, this is not so much of an advantage. Since the flyback is the only one of the three not requiring an additional inductor, the examples in this chapter will all be flyback converters, except for a forward used to demonstrate synchronous rectification on the secondary side of an isolated converter.

9.2 The Transformer

All three of these isolated converters use a transformer. Although the transformers are all of different construction, what is the same about all of them is that they can all be designed with the number of turns on the secondary being different than the number of turns on the primary. The ratio of primary turns to secondary turns (in that order) is known as the turns ratio.

The turns ratio is of particular importance in the design of an isolated converter, as it sets the ratio of the input voltage and current to the output voltage and current. Power, of course, is conserved (at time periods greater than a single switching cycle), and so a turns ratio less than one produces a lower voltage and higher current than the input, whereas a turns ratio greater than one produces a higher voltage and lower current than the input. A turns ratio of one doesn't change the voltage or current but may still be used in cases where you want to just obtain isolation.

Starting with the forward converter, suppose that the turns ratio is N, so that for every one turn on the secondary, there are N turns on the primary. (Note again that N can be greater than, less than, or equal to one.)

When the transistor is turned on in the forward converter, the input voltage V is applied across the primary. Transformers work such that the voltage across the secondary is then V/N, because it's so many volts per turn. Of course, since power is conserved, the current is also transformed, but in the opposite direction. If the primary current is I, then the secondary current is IN.[1]

The end result of this is that the output voltage can be stepped up or down (or remain the same) compared to the input, so that the turns ratio is an additional separate knob that can be used to control the output voltage. This makes it possible to get fairly high output-to-input voltage ratios, which is a possible reason for using a transformer when isolation isn't needed. In practice, the turns ratio, N, is usually within the range of 0.1–10. (Meaning that the primary has one-tenth the number of turns that the secondary does, and, respectively, the primary has ten times the number of turns that the secondary does.)

The push–pull converter's transformer acts the same way. Here, the primary consists of two windings with a center tap, as does the secondary. Consequently, the ratio of input to output is again N.

A flyback transformer is different than in the other types of converters (indeed, this is why in Chapter 3 [Components] we suggested that it should have a different name). The flyback transformer acts on the primary side as an inductor while the transistor is on, but acts like a transformer from primary to secondary when the transistor is off.

For the forward and the push–pull, the transformer's output voltage is determined by the input voltage and the turns ratio. For the flyback, it's the opposite: the *input* voltage is determined by the *output* voltage and the turns ratio. And there's no concern about too much power being transferred, since that's limited by the energy stored in the primary inductance. This is the reason that no separate inductor is needed.

Transformer efficiency can be as high as 98–99% depending in most cases primarily on the output current. But be aware that the root mean square (RMS) current in the flyback is relatively higher than in the other types of isolation transformers, and thus its copper losses tend to be higher. This tends to limit the use of the flyback converter to relatively lower power than the forward or the push–pull. However, comments you see about the flyback being limited to 50 W or less

1 And thus power is conserved: the input is $V \times I$, and on the secondary it's $(V/N) \times (I \times N) = V \times I$.

aren't true; the power level amenable to a flyback depends on the input voltage. The author has successfully built isolated flyback converters with a single transistor at 600 W.

As a final note, we observe that it's not necessary for the primary and the secondary to be on separate grounds. That it is common to do so is because the transformer is used to accomplish safety isolation. However, the transformer is also used, as mentioned above, to accomplish large changes in input to output voltage. So, for example, our example in Chapter 8, converting 48 to 1 VDC could have been done with a converter including a transformer, which would have resulted in higher DC and on-time; in this case, the two grounds would have been connected together.

9.3 Control of an Isolated Converter

There are a number of unique issues with respect to controlling the voltage on an isolated supply, as the secondary ground is not usually connected to the primary ground.

9.3.1 Primary vs. Secondary-side Control

With the split of the ground system made possible by the addition of a transformer, a new question arises. Should the control of the power supply be located on the primary side or the secondary side?

An advantage of controlling the power supply on the primary side is that power is instantly available. As soon as the input turns on, it can be used to power the control circuitry, so that the supply is under control right away. If the power supply control is on the secondary, some power has to be transferred from the primary to the secondary before the control can activate. There could thus potentially be some period of time when the output voltage does something undesirable before the secondary is fully powered.

On the other hand, an advantage of controlling the power supply on the secondary side is that the "decision-making" part of the control, that is, the part that decides whether the switch should be on or off, is located on the same side that the output being controlled is on. Since the decision about the switch state is only an "on" or an "off"," the information that has to be sent to the primary side is digital; if the control is on the primary side, the information on the status of the output has to be transferred across the isolation barrier. Sending analog information across the isolation barrier is more subject to temperature, aging, and noise issues than sending digital information.

Over the years, a variety of controllers have been available that work on the primary side and others that work on the secondary side, with preferences among power supply engineers changing over the decades. In accordance with present-day preference, in this book we have our main control circuitry always on the primary side. Of course, non-main control circuitry, in particular that for secondary-side synchronous rectification, might still naturally be on the secondary side.

9.3.2 Isolated Feedback

One way or another, in an isolated supply, the information on the output has to be conveyed back to the input. In the case of primary side control, this information is analog, while for secondary-side control, this information is digital.

By far, the most common method of transferring information from secondary to primary is by using an optocoupler. And since primary side control is the most common, which requires analog feedback, the optocoupler most commonly has a BJT on its receiver side.

As shown in Figure 9.3, the most common scheme uses the 431 as an error amplifier. The output voltage of the converter on the secondary side is compared with the reference pin of the 431. If the voltage is too high, the 431 pulls more current through its cathode; if the voltage is too low, it pulls less.

The cathode current of the 431 is pulled through the opto-diode of an optocoupler, which produces light (internal to its package) proportional to the current. The light from the opto-diode is received by the base of the opto-transistor. And just as with discrete BJTs receiving current, this produces a higher current through the BJT from collector to emitter. In short, if the output is too high, there is more current through the opto-transistor, and this can be used as feedback (for example, by conversion to a voltage using a resistor) to the primary side control.

In this system, of which literally billions are made each year, the gain of the optocoupler is not too important, because it's in a feedback loop. As suggested in the chapter on components, choosing an optocoupler with a gain bin of (say) 50–200% is usually fine.

Optocouplers' gain, since it depends on an LED and a BJT, changes with temperature. This isn't too important, again, because the optocoupler is in a closed feedback loop. More significantly, however, the gain is strongly affected by the aging of the optically transparent material inside the package of the optocoupler. For this reason, it is recommended to always use 125 °C rated optocouplers.

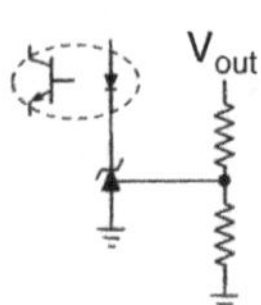

Figure 9.3 Isolation concept with 431 and optocoupler.

9.3.3 Bang–Bang vs. PWM

We have used both bang–bang and PWM control for converters in this book. The reasons for using one or the other remain the same when the converter is isolated. Now, however, the feedback for the bang–bang control is simplified compared with that for the PWM, since only a go/no-go signal has to be transferred across the isolation barrier, rather than an analog voltage. We give examples of both in this chapter.

9.3.4 Secondary-side Power

We've discussed in previous chapters how to generate operating voltage for the various ICs we use. Not infrequently, the same technique may be used to power the ICs on the secondary side of an isolated converter.

Indeed, it is often the case that the output voltage of the isolated converter is appropriate to run the secondary-side ICs directly. In such a case, the area of concern is whether the output voltage ripple and amplitude are of such values that they might possibly interfere with the proper operation of the ICs. In this case, in addition to the standard 100 nF bypass capacitor on the power pin of an IC, you can add a (say) 1 kΩ resistor between the output voltage and the power pin, which then forms an RC filter with the bypass capacitor of about 10 kHz. The integrated circuit (IC) usually draws very small current, so that the voltage drop on the resistor isn't significant.

If the output voltage is too low to operate the 431 and the opto-diode (say, less than 3 V), or if the voltage is so high that the ICs can't sustain it, something additional is required to provide power to the ICs. The most common choice is to add an extra winding on the transformer, and then rectify it (and possibly linear regulate it as well). Because of the difficulties of reproducibility of transformers mentioned in the chapter on Components, however, this is not the author's preferred method. When the output voltage is too high, it is straightforward to linear regulate down the voltage to an acceptable level using a resistor in series with a zener. This dissipates some power, though not a lot. In extreme cases, a small bang–bang converter could be implemented.

If the output voltage is too low, an LDO isn't an option. And it may not be possible to find ICs that work at a low enough voltage to build a switcher. What we do instead is indicated conceptually in Figure 9.4. We generate a square wave (here 100 kHz) and drive a small transformer with it. The primary is shown as being isolated from the secondary by a 10 MΩ resistor, which we use because LTspice can only have one ground. The load is shown as 1 kΩ, which at the 4 V produced is 4 mA, plenty to run, for example, the opto-diode and the 431.

We can go ahead now and stick in an actual 555 timer to make a real circuit to run the isolated IC power supply, as shown in Figure 9.5. We've set the timing capacitor to get 100 kHz), and have moved the isolated ground to the secondary

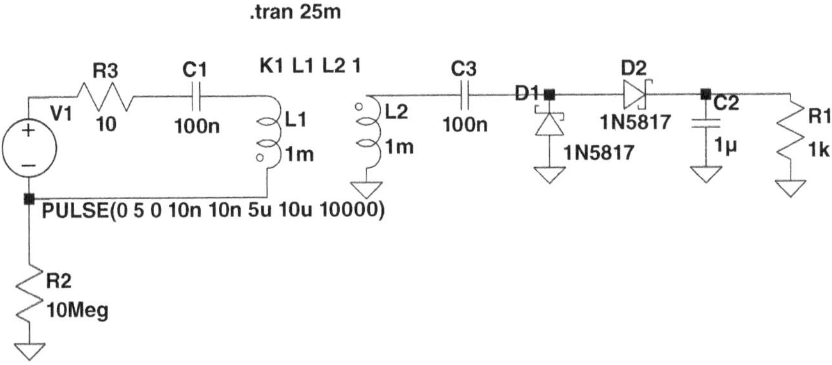

Figure 9.4 Secondary-side power supply concept.

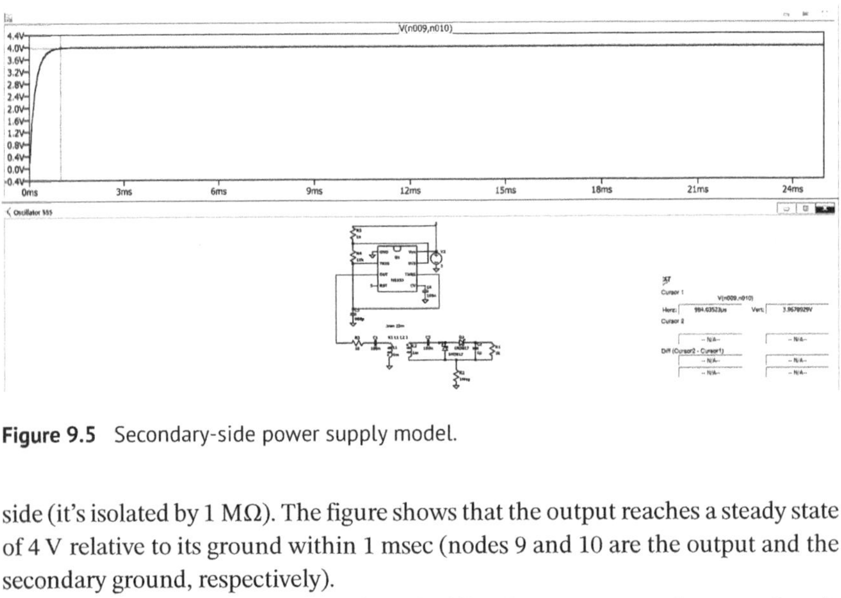

Figure 9.5 Secondary-side power supply model.

side (it's isolated by 1 MΩ). The figure shows that the output reaches a steady state of 4 V relative to its ground within 1 msec (nodes 9 and 10 are the output and the secondary ground, respectively).

The only other consideration when deciding how to power the secondary is the question of timing. You have to make sure that the output power doesn't go too high while waiting for the control power to come up. In general, this can be accomplished by slow-starting the main converter, giving the secondary-side control power circuitry time to come up.

9.4 Multi-output Converters

We've already suggested that having multiple outputs on a single transformer is a poor choice. The main reason is that it's difficult to get multiple vendors to produce transformers that are identical—there may be differences in the core material used, the winding or layering techniques may differ, and the insulation between layers may differ. The more windings there are, the better the chance that you'll have differences between your primary and your second source for the transformer.

Another issue with multi-output transformers is that the voltage on the various windings may vary. If one output is heavily loaded while a second is lightly loaded, the two output voltages may differ quite a bit from when the first is lightly loaded and the second fully loaded.

A possible way around this is to average the two outputs in the feedback. In this technique, the two outputs are added together (using a resistor from each as part of the resistor divider). This produces somewhat better performance on the second output, while making the performance on the first output somewhat worse.

Another possibility is to put an LDO at the output of the second winding. You have to make sure that the cross-loading doesn't take the input to the LDO below its headroom, and, of course, this might cause it to dissipate quite a bit more power than it otherwise would if the load current is high.

Yet another possibility is to put a bang–bang or PWM converter on the second output. But at some point you start to wonder whether it's easier to just have a second transformer. There really aren't any guidelines here; you have to evaluate each case individually, based on the currents being drawn on the different voltage lines and the output tolerances desired.

9.5 Cascaded Power Supply Stability

The input to an isolated power supply frequently comes from the output of a PFC converter. We've also noted in the previous section the possibility of following an isolated converter by another SMPS to improve output regulation. In any situation in which one power supply follows another, you have to consider the system stability.

As noted in the author's previous book [9], it's possible to have two SMPS, each of which is individually stable, but that, when connected in series, form an unstable *system*. While the reader is referred to the above-mentioned book for details, the bottom line is that such systems can generally be stabilized by having enough capacitance at the output of the first that goes to the input of the second. It's also possible to measure the gain margin of the system, but we won't attempt that here.

9.6 Example: PFC to 48 V at 2 A

In Chapter 8 (PFC Converters), we built a 100 W PFC SMPS producing a 200 V output. In the chapter on bang–bang controllers, we built a SMPS that took a 48 V input and produced 1 V at 100 A, which is also 100 W. The part that is missing is getting from the PFC front-end converter producing 200 V at its output to the third-stage converter with an input at 48 V, with isolation so that the 1 V output is safe (from the AC line). So in this section, we will design an isolated converter that takes 200 VDC in and produces 48 V at 2 A out.

9.6.1 Specifications Table

So let's write specifications, compatible with the output performance of the PFC and the input performance of the 48 to 1 V converter, Table 9.1

The load step is specified to match the load step on the 1 V output downstream.

9.6.2 Design

Since the 200 V is relatively large, we can make this converter an (isolated) flyback. Being a boost, this then needs to be a clocked system such as we've designed before to ensure the transistor isn't latched on. And in fact, we will use the 3845 type control IC running at 100 kHz in current-mode control. But now that there is a transformer involved, we need to re-do the equations for determining the power components.

The power delivered to the output in discontinuous conduction mode is

$$P = \frac{1}{2} L_{\text{pri}} I_{\text{pk}}^2 f \tag{9.1}$$

Table 9.1 PFC to 48 V at 2 A specifications.

Parameter	Specification	Measured
V_{in}, min	200 V	
V_{in}, max	220 V	
V_{out}, min	38 V	
V_{out}, max	52 V	
Minimum output current	400 mA	
Maximum output current	2 A	
V_{ripple}, max	1 V_{pp}	
80% step max. undershoot	1 V	
80% step max. overshoot	1 V	
Switching frequency	100 kHz	
Bandwidth	1 kHz	
Phase margin, min	30°	
Efficiency	90%	

where I_{pk} is the peak current on the primary, and L_{pri} is the primary inductance. If we assume that the (maximum) power is given and the frequency is known, we can solve this for L and I:

$$L_{pri}I_{pk}^2 = \frac{2P}{f} \tag{9.2}$$

We also know that the peak current is set by how long the MOSFET is on,

$$\frac{V_{in}}{L_{pri}} = \frac{I_{pk}}{t_{on}} = \frac{I_{pk}f}{DC} \tag{9.3}$$

so that

$$L_{pri}I_{pk} = \frac{V_{in}DC}{f} \tag{9.4}$$

We can divide the first equation by the second to find the required peak primary current:

$$I_{pk} = \frac{2P}{V_{in}DC} \tag{9.5}$$

which is as before, and from this we can now find the primary inductance,

$$L_{pri} = \frac{V_{in}^2 DC^2}{2fP} \tag{9.6}$$

as before.

Now, when we did the (non-isolated) flyback before, to ensure that we were in DCM we had to make sure that the reset time on the inductor was large enough, by calculating the Volt-second balance across the inductor. Now that we have a transformer, this is slightly more complicated: The on-time voltage across the primary of the transformer is still V_{in} × ton, but the off-time voltage is *the output voltage times the (inverse of the) turns ratio*, since this is the voltage that is reflected back to the primary side. Bear in mind that this reflected voltage is applied to the transistor *in addition to* the input voltage, and so the transistor has to be rated for this entire voltage. Again, to be clear, the reflected voltage is what is resetting the core; this voltage is in addition to the input voltage, which therefore doesn't matter for the reset.

In equations, then,

$$V_{in} \times t_{on} = V_{out} \times N \times t_{off} \tag{9.7}$$

where we have ignored the V_f of the output diode. We again solve this for the DC by adding $V_{out} \times N \times$ ton to both sides, yielding

$$DC = \frac{N \times V_{out}}{V_{in} + N \times V_{out}} \tag{9.8}$$

If $N = 1$, which is to say the primary and secondary have the same number of turns, this reduces to

$$\mathrm{DC} = \frac{V_{\mathrm{out}}}{V_{\mathrm{in}} + V_{\mathrm{out}}} \tag{9.9}$$

which is the same as for a non-isolated flyback.

For this design $V_{\mathrm{in}} = 200$ V and $V_{\mathrm{out}} = 48$ V, so

$$\mathrm{DC} = \frac{N \times 48\,\mathrm{V}}{200\,\mathrm{V} + N \times 48\,\mathrm{V}} \tag{9.10}$$

To keep the DC less than fifty percent, we must have $N \leq 4$; let's choose $N = 2$ (the primary has double the number of turns the secondary does), for which DC = 32%.

For this design, $P = 48$ V × 2 A = 96 W; frequency is 100 kHz; and the DC is 32%. Then

$$L_{\mathrm{pri}} \leq \frac{200\,\mathrm{V}^2 32\%^2}{2 \times 100\,\mathrm{kHz} \times 96\,\mathrm{W}} = 213\,\mu\mathrm{H} \tag{9.11}$$

with a peak current of

$$I_{\mathrm{pk}} \geq \frac{2 \times 96\,\mathrm{W}}{200\,\mathrm{V} \times 32\%} = 3.0\,\mathrm{A} \tag{9.12}$$

We select a primary inductance of 200 μH with a winding resistance of 1 mΩ.

The other power component is the output capacitor. Its value should be determined by the load step, which goes from 400 mA to 2 A and the requirement that this not exceed 1 V.

The bandwidth is supposed to be 1 kHz. The response time will thus be on the order of 1 msec, and looking at drawing the extra 1.6 A, the voltage drop will be something like

$$C \times V = C \times 1V = I \times t = 1.6\,A \times 1\,\mathrm{msec} \tag{9.13}$$

so that the capacitor should be something like 1.6 mF; we choose 1 mF and will evaluate its performance.

As for the transistor, it must withstand the 200 V input plus the reflected by N output voltage 2 × 48 V = 96 V output for a total of 296 V; so we choose a 400 V device. Current is low, so more or less any transistor will do. For the diode, the average current is low (96 W / 48 V = 2 A), but the peak is higher (2 × 3 A = 6 A). We can just pick a 10 A ultrafast recovery diode (this need not be a Schottky since the V_{f} is a small percentage of the output voltage).

We've selected the power components. As for the control, we'll put the error amplifier, in the form of a 431, on the secondary side, and so we need to send this analog signal across the isolation barrier. This can be done with an opto-isolator.

9.6.3 Modeling

We already built a boost converter model in the PWM chapter using the multi-sourced 3845, so we can use this as a starting point for this design. There are a number of significant differences:

1. We add in the transformer.
2. We add in an external error-amp and reference (in place of the 431), and then send the analog output back to the primary side with an optocoupler. The opto-transistor pulls current out of the COMP pin to set the DC, see below.
3. For the optocoupler we choose the LTspice model of the PC817A. Its datasheet shows it has 80–160% gain[2] at 5 mA, so we don't need much current on the opto-diode, which is desirable so that we can run directly from the 48 V output rather than having to generate a secondary housekeeping supply. In the event we use about 1 mA.
4. We need to generate the 12 V on the primary to run the 3845. Or, if this is part of the suggested complete power system, we could just use the one already in the PFC stage. We also need an isolated 12 V on the secondary.
5. We ultimately end up with an isolated secondary with an isolated ground. Since LTspice dislikes circuits that are not connected in any way, we add in a notional 10 MΩ resistor between the primary and secondary grounds. This will probably be there in your real design because of leakage, in addition to a few nanofarads of capacitance for common mode noise reduction. In the actual simulations, we set this resistor to zero to avoid the gigantic file sizes and run times we had in the previous example.

For the feedback, the reader should carefully note that the feedback pin of the 3845 isn't going to work: we need to find the error signal on the secondary. And, of course, we can't have a *second* error amp! Instead, we will ground the feedback pin of the 3845 and control the COMP pin directly to get the desired DC.[3] It turns out that the output of the 384x family COMP pin isn't really a voltage, but rather a current (this is common in IC design). So we will be pulling current from the COMP pin of the 3845 to adjust the DC. The maximum current the COMP pin can source is about 1 mA.

As for the polarity of the feedback, consider first the case where the output voltage is too high. In that case, the output of the error amp is low, and so pulls more current through the opto-diode. There is thus more opto-transistor current. Pulling this current from the COMP pin of the 3845, the COMP voltage is lowered, reducing the DC and therefore bringing the output voltage back down.

2 Most optocoupler have a gain range of two to one.

3 The reader should take **special note** of this, as it was a source of confusion back when the 384x family was released, and remains confused on the internet today.

Consider the opposite situation where the output voltage is too low. In that case, the output of the error amplifier is high, so the error amplifier pulls less current. The opto-diode gets less current, as does the opto-transistor, and so the COMP pin is allowed to go back up (since the feedback pin is grounded), resulting in higher DC, pushing the output voltage up.

The feedback polarity is thus correct, but there's one more thing to consider: What happens at startup? The secondary-side feedback runs from the output voltage, but at startup there *is* no output voltage, and thus there is no opto-diode current. So the opto-transistor has no current, and the error amplifier inside the 3845 is allowed to go to its maximum, producing maximum DC. Were there an occasion, we could have added what is basically a soft-start capacitor to the COMP pin. When VCC is first applied, the COMP pin will be at 0 V, so the 3845 starts switching at minimum DC and the output voltage starts coming up. Since the COMP sources only 1 mA, by selecting the capacitor appropriately, the secondary-side control system will be working before the output voltage reaches its set point.

We show our model in Figure 9.6. In addition to what we've already said, there are a couple of additional things to note:

- We've labeled the primary and secondary sides, with a dotted line representing the isolation barrier. This is just a label, it's not a component.
- Everything on the secondary side has been referenced to "SGND." To make measurements on that side, you should subtract V(SGND) from the waveforms you are looking at in LTspice. We've set the coupling impedance between the two sides to zero for ease of simulation.

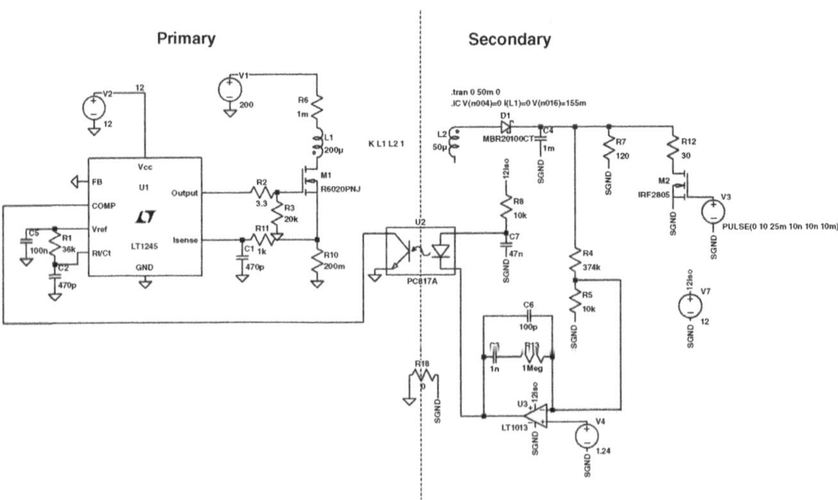

Figure 9.6 PFC to isolated 48 V at 2 A model.

- We haven't bothered to generate the state-space average model again; it's basically the same as in the chapter on PWMs, except you have to include the factor of N.
- And as a result, we've already built the compensation for the error amplifier.
- Note the additional 47 nF on the current source for the opto-diode, which further slows its change in current.

Running the simulation, we find that at startup, the output voltage overshoots to 50 V, and reaches steady-state of 47.6 V after about 16 msec. Peak-to-peak ripple voltage on the output once it achieves steady-state is about 6 mV at both full power and 20% power, suggesting that the output capacitor can be greatly reduced.

Load step response is shown in Figure 9.7. The load starts at 400 mA, transitions to 2 A at 25 msec and returns to 400 mA at 35 msec. Both undershoot and overshoot are about 170 mV. The output shows a bit of ringing at the transitions, suggesting a somewhat higher phase margin is desirable.

9.6.4 Transformer

What we haven't yet talked much about is the transformer in this model. There is no special model of a transformer in LTspice; you just put two separate inductors on the schematic and couple them.

As promised, the dots are shown on the schematic (you turn them on in the same window where you set the inductance value). We can check that the polarity (and the coupling) is correct by looking at the primary and secondary currents simultaneously; see Figure 9.8. Looking during the maximum load time, the primary inductor current rises with time, while the secondary diode current falls back

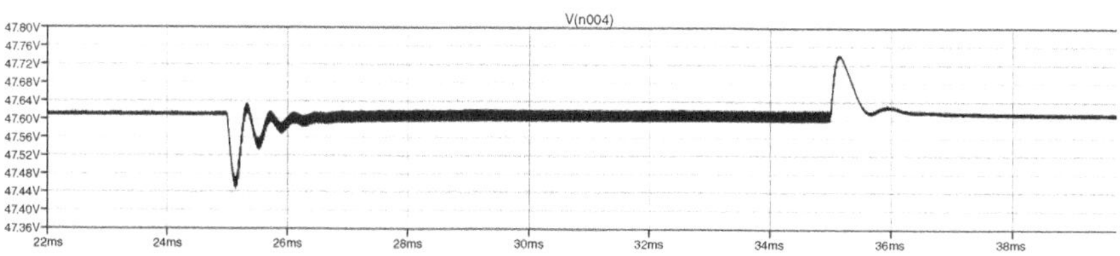

Figure 9.7 Ripple and load step of 200 to 48 V isolated SMPS.

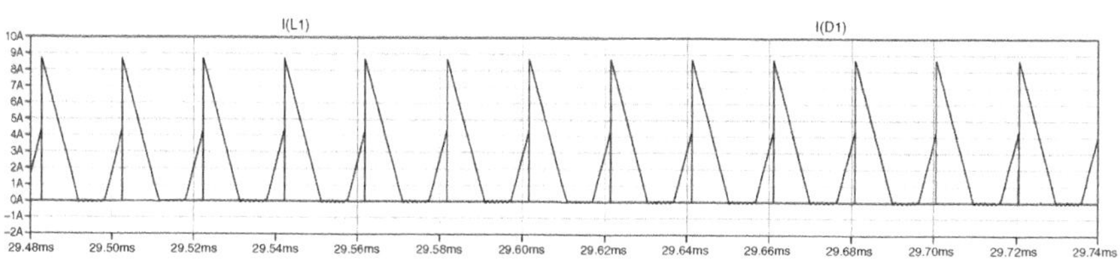

Figure 9.8 Primary and secondary currents in the transformer.

to zero with double the amplitude. Observe that the current goes to zero even at maximum load, so we are in DCM; it appears that the design could increase the primary inductance if desired.

The LTspice statement that causes the two separate inductors to form a transformer is “K L1 L2 1.” The last number in this statement is the coupling between the primary and the secondary. $K = 1$ means that the coupling is perfect, and there is no leakage inductance.

Depending on the details of the mechanical design of the transformer, the coupling will be imperfect, which is to say the coupling coefficient will be less than one. The reader may try changing it to, say, 0.98 or even 0.95 and observe the results on the transistor and diode. In practice, the transistor may, due to the leakage current, see rapid high-voltage spikes. Traditionally, these are suppressed using either a snubber or just a transzorb. Since all MOSFETs nowadays have avalanche ratings, it is fairly common to simply let the leakage energy go into the transistor by letting it avalanche. Of course, the transformer must be designed to minimize leakage to the extent possible, since this is wasted energy.

Since the design of the PFC to 48 V converter seems to have gone beyond what is necessary, the comparison with the specifications will be skipped.

9.7 Example: System Model

It is of interest to see what a *system* simulation does, that is, the performance of all three converters operating in series: the AC line to 200 VDC PFC, with the 200 to 48 VDC PWM we've just designed attached to it, and the 48 to 1 VDC converter we designed in the chapter on bang–bang control attached to that.

To make the top-level design clear, a system block diagram is shown in Figure 9.9. It might be objected upon looking at this, that perhaps we could have gone from the PFC 200 VDC output directly to an isolated 1 V at 100 A, not subjecting ourselves to the efficiency loss in the 48 VDC intermediate conversion.

Certainly the efficiency would be improved in going from three converters to two. However, the practical conversion would be difficult. Even with the maximum suggested turns ratio of ten on the transformer, the DC of such a converter would be something like 1 /20 V = 5%, which might be difficult to do efficiently at a reasonable switching frequency.

An alternative would be to have a synchronous, isolated half-bridge coming from the AC line down to, say, an intermediate voltage of, say, the geometric mean of the input and output voltages $\sqrt{1\,\text{V} \times 200\,\text{V}} = 14\,\text{V}$. Of course, a transformer would have to be designed, and the power stage would then have to be controlled suitably to accomplish PFC. The interested reader might consider how this could be done with a robust design.

While checking the stability of the whole system is necessary, it is beyond the limits of the subject matter of this book, as it is not different for robust designs than

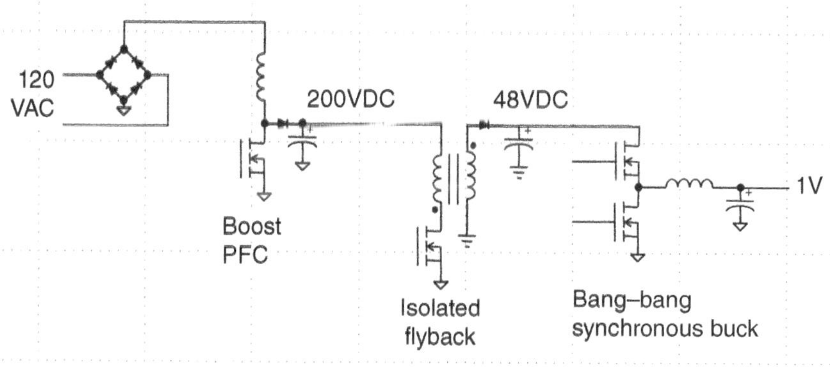

Figure 9.9 Block diagram of complete power system.

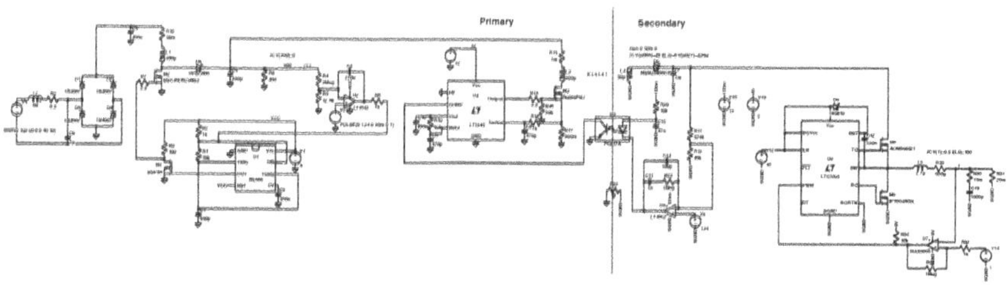

Figure 9.10 Three converter system from 120 VAC to 1 VDC at 100 A.

for others.[4] However, we can put all three converters together into a single LTspice schematic, shown in Figure 9.10, and bravely hope that LTspice and our computers are capable of running such a large, mixed (analog and digital) simulation.

Some notes on the combination of these three converters.

- All the grounds in the 48-to-1 V converter have been converted to SGND, since the converter is on the secondary side of the isolation barrier.
- We have, of course, cleaned up, so there is only a single .tran statement. We have retained multiple initial condition .ICs as per when they were designed; LTspice doesn't seem to mind. We've also, of course, produced names for the various nodes and changed the .IC names accordingly.
- We also changed the names for the K statement for the transformer.
- The PFC has retained its slow start.
- We haven't amalgamated the various internal supplies, though this could obviously be done (of course, primary and secondary supplies can't be combined).

4 Again, looking at the author's previous book [9] gives information on how to test system stability.

The result is an astoundingly slow simulation and a very, very large file: 1 msec of simulation used up 3 GB (sic) of memory. Indeed, this system is very stiff. A large part of this is due to the secondary-to-primary coupling: half of the circuit is referenced to "0" only through the 10 MΩ resistor. As a result, all of the secondary voltages are elevated by thousands of volts above "0," and subtracting large numbers from each other to get an accurate small number requires a large number of significant digits and is notoriously difficult in mathematics and for computer programs with limited precision.

To get around this, the author reduced the resistance between SGND and "0" to 0 ω. Now the circuit simulates quickly and only took 60 MB for that first millisecond. Further, the initial condition on the PFC converter was reset so that it was already running; that is, the output voltage was set to be initially 200 V. Similarly, the output voltage of the 48 V converter was also preset. As an intermediate step, setting the initial phase angle of the AC line to 40° provides enough input voltage for the system to get started.

What we got from all of this was that the three-converter system does indeed run. The PFC settles in to about 230 VDC in about 1 msec. The output of the isolation converter settles in to about 47.6 VDC after about 1.2 msec with some fairly well-damped ringing of small amplitude. The 1 V output is shown in Figure 9.11. It overshoots for about 100 µsec (which would be suppressed by adding a soft-start) and then regulates cleanly at 1.01 V with a peak-to-peak ripple voltage of 26 mV.

Starting the system from zero shows overshoots on some of the intermediate voltages, telling us that we need to sequence the power supplies; that is, the 200 VDC should be up before allowing the isolation converter to run, and its output, the isolated 48 V, should be up before the 1 V converter is allowed to start. This is left for the reader to design robustly.

For stability issues, a state-space average model of the entire system should be put together. This can be done by piecing together the state-space average models of the individual converters, and then using Ridley's criteria (as mentioned previously) to determine margin.

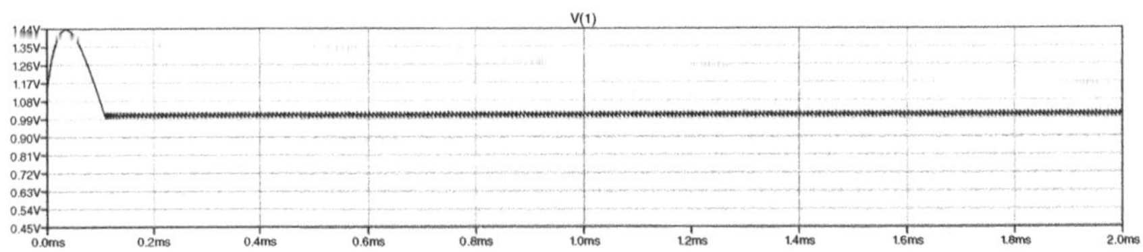

Figure 9.11 1 V output from 120 VAC input, complete system simulation.

9.8 Example: PFC to 1kV at 300mA

In Section 9.7, we used a 3845 constant frequency, current-mode controller, with a secondary-side error amplifier and an optocoupler transmitting analog information across the isolation barrier. Input was from the output of a 120 VAC PFC, while output was stepped down to 48 VDC at 100 W.

In this example, we're going to do almost everything differently. We're going to build a clocked boost bang–bang controller using a 555 and run it in voltage-mode. Input this time is from a universal input PFC so that it will be 375 VDC, while the output is going to be stepped up to 1 kV at 300 mA, which is 300 W. It's reasonable to suppose that a boost will work in this application, since the input voltage is doubled and the power goes as the square of the input voltage, while the power level has only tripled.

The secondary side will be doing the analog part of deciding whether the output is too high or too low (since that's so convenient on that side), but the optocoupler will only be transmitting the digital output of the comparator, on or off.

9.8.1 Design

We designed a clocked boost bang–bang converter in the chapter on bang–bang, shown again in Figure 9.12. It had a constant-frequency, constant DC 555 running a boost transistor and delivering a constant power to its output. When the output voltage went too high, as sensed by a comparator, the 555 was turned off until the voltage went back below the hysteresis. This makes a good starting point for this design.

The design for this chapter needs several adjustments.

1. Input is changed to 375 VDC, and output to 1000 V, with corresponding changes to the transistor, diode, and resistor divider.

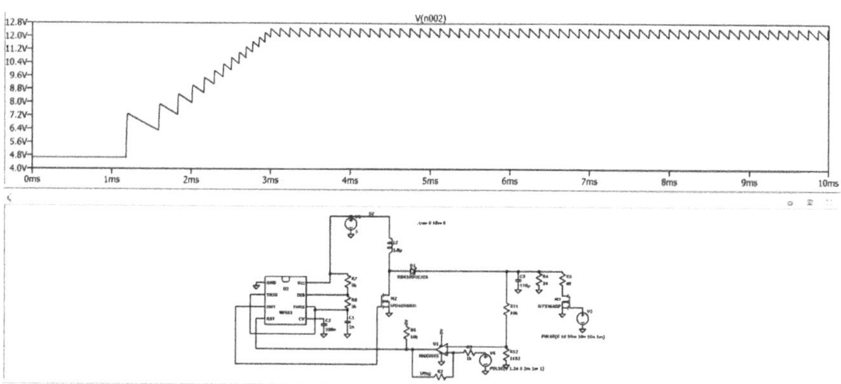

Figure 9.12 5 V to 12 V boost from chapter on bang–bangs.

2. Power level is changed to 300 W, so the inductance and output capacitor need to be changed.
3. The inductor needs to be replaced with a transformer.
4. The comparator needs to drive an optocoupler.

That's not so bad, right?

We start with the same set of calculations as the previous example. For this design $V_{in} = 375$ V and $V_{out} = 1000$ V, so

$$\text{DC} = \frac{N \times 1000\,\text{V}}{375\,\text{V} + N \times 1000\,\text{V}} \tag{9.14}$$

To keep the DC less than fifty percent, we must have $N \leq 0.375$; let's choose $N = 0.25$ (which is 1 : 4, an integer turns ratio often being the best choice), for which DC = 40%.

For this design, $P = 1$ kV × 300 mA = 300 W; frequency is 100 kHz; and the DC is 40%. Then

$$L_{pri} \leq \frac{375\,\text{V}^2 40\%^2}{2 \times 100\,\text{kHz} \times 300\,\text{W}} = 375\mu\text{H} \tag{9.15}$$

with a peak current of

$$I_{pk} \geq \frac{2 \times 300\,\text{W}}{375\,\text{V} \times 40\%} = 4.0\,\text{A} \tag{9.16}$$

if we want to be in DCM.

With V_{pp}=100 V (after all that's only ±5%), we have

$$\text{C} = \frac{2 \times 4.0\text{A} \times 40\%}{2 \times 100\,\text{kHz} \times 100\,\text{V}} = 160\,\text{nF} \tag{9.17}$$

We need the capacitor voltage to be at least 1200 V for margin. Generically, at very high voltages ceramic capacitors are disproportionately more expensive as their capacitance value increases, so that putting multiple small ceramics in parallel is cheapest. We can get two 100 nF 1500 V ceramics for US$1.29 each (1000 piece distributor pricing) with multiple sources.

If the ripple voltage was 10 V_{pp} instead of 100 V_{pp} the capacitance required would be ten times larger, 1.6 μF, and the cost of ceramic capacitors would be prohibitive. In this case, we could put two aluminum electrolytic capacitors in series (the maximum common voltage in which they are available is 630 V) and then add in balancing resistors to ensure that they shared the voltage appropriately.

Another choice would be a film capacitor. A 2 μF 1200 V metallized polypropylene is multi-source available at US$1.03, though it is through-hole and physically rather large (x 15 mm by 25 mm high). This is actually cheaper than the ceramics, but we'll stick with those and the 100 V_{pp}.

As for the transistor, it must withstand the 375 V input plus the reflected by N output voltage of 0.25 × 1000 V = 250 V output for a total of 625 V; so we choose a 1000 V device. Current is low, so more or less any $R_{DS,on}$ will do. For the diode, the average current is low (300 mA), but the peak is higher (0.25×4 A = 1 A). We can just pick a 1 A Schottky.

The optocoupler is a good choice for a digital signal, as you don't have to worry about gain variations. The only thing to verify is that the propagation delay (and the rise and fall time on the output) are fast enough that it doesn't add significant over- or undershoot.

9.8.2 Specifications Table

Time to write down a Specification in Table 9.2. Specifications are fairly loose, as this is not a very demanding application.

9.8.3 Modeling

We show a model in Figure 9.13. A number of selections and changes have been made based on the simulations:

- A 1000 V MOSFET isn't available in LTspice, so we settled for 650 V (recall that MOSFETs don't break in LTspice). This in turn has an $R_{DS,on}$ of 340 mΩ, which is quite high. We've consequently doubled up on the number of MOSFETs.
- The diode is 1200 V, which is cutting it very close; in actual usage you would probably use a 1500 V or 1600 V, for which there are many sources. The V_f doesn't matter too much, since it's a tiny percentage of the output voltage, the key is the reverse recovery time.

Table 9.2 PFC to 1 kV at 300 mA specifications.

Parameter	Specification	Measured
V_{in}, min	325 V	
V_{in}, max	375 V	
V_{out}, min	900 V	
V_{out}, max	1000 V	
Minimum output current	150 mA	
Maximum output current	300 mA	
V_{ripple}, max	100 V_{pp}	
50% step max. undershoot	20 V	
50% step max. overshoot	20 V	
Switching Frequency	100 kHz	
Efficiency	90%	

- To get 1000 V rating, we've put five resistors in series for the feedback voltage divider, each 1 MΩ and 200 V in a 1206 package. Six would actually be better.
- With primary inductance of 375 μH and a 4:1 turns ratio, since inductance goes as the square of the number of turns, the inductance ratio is 16 : 1.
- We've made the two grounds the same for simulation speed, of course, retaining the optocoupler and transformer. The reader can put all of the secondary-side grounds back to SGND and add a 10 MΩ resistor between it and GND.

The output voltage rises to 1 kV at the same speed as the reference does. A load step occurs between 5 and 7.5 msec. As shown in Figure 9.14, at half load the voltage is between 908 and 1093 V, while at full load it's between 903 and 1064 V, so it meets specs at both load conditions. There is no over- or undershoot as the load step turns on or off. It would be straightforward to increase the output capacitor to bring these numbers down if desired. A film capacitor has already been discussed for this purpose.

9.8.4 Transformer

A quick comment about the transformer needed for this design. Of course, for isolation, there will be some tape between the primary and the secondary windings. But if there are a significant number of turns (the number depends both on

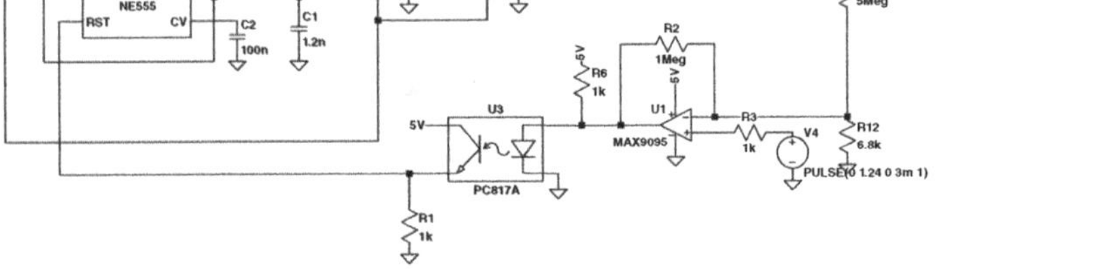

Figure 9.13 375 VDC to 1 kV at 300 W model.

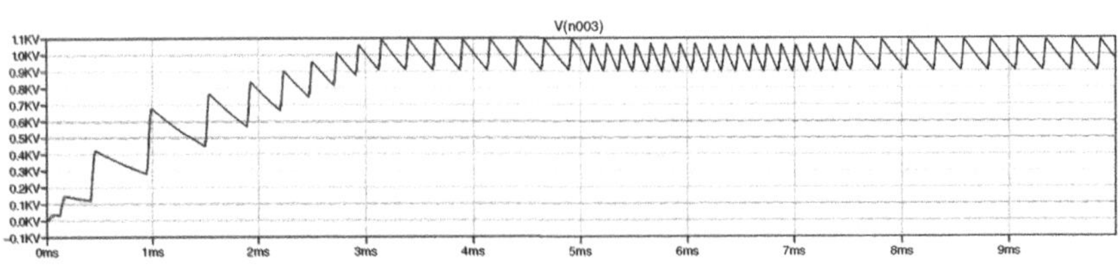

Figure 9.14 375 VDC to 1 kV at 300 W output voltage with load step.

the core material and the air gap), then the primary, which has quadruple the number, may require a layer of tape between layers of the winding.

To understand why, consider a winding that has two layers. One goes (say) from left to right on the bobbin, and then the next goes back from right to left, so that the winding terminations are both on the same side. That means the start of the winding, which is at 375 V, is right underneath the ending of the winding, which, when the transistor is on is at ground. 375 V may be more than the two layers of wire insulation can take. The solution is a layer of tape between the first half of the winding and the second. The secondary winding has one quarter the number of turns, but the voltage is higher and the wire is thinner (because the current is lower). So, if the primary requires two layers and a layer of tape between them, it could well be that the secondary does also.

9.8.5 Results

In Table 9.3, we show the results of the simulation. The biggest issue is the output ripple voltage, which makes the voltage dip down to 908 V. The hysteresis on the comparator is already quite small; the issue appears rather to be the propagation delay through the optocoupler, which looks to be something like 20 μsec. We, of course, didn't account for this when selecting the output capacitance. It looks like going to the metallized polypropylene 2 μF capacitor is probably necessary. A quick try shows that this change changes the minimum and maximum voltages to 950–972 V, respectively, thus meeting all the specifications with margin to spare.

Table 9.3 PFC to 1 kV specifications and results.

Parameter	Specification	Measured
V_{in}, min	325 V	325 V
V_{in}, max	375 V	375 V
V_{out}, min	900 V	908 V
V_{out}, max	1000 V	1093 V
Minimum output current	150 mA	150 mA
Maximum output current	300 mA	300 mA
V_{ripple}, max	100 V_{pp}	185 V_{pp}
50% step max. undershoot	20 V	0 V
50% step max. overshoot	20 V	0 V
Switching frequency	100 kHz	75 kHz
Efficiency	90%	92.0%

9.9 Example: Offline 900W Battery Charger

For our next example, we are going to build an offline converter to charge the battery of an electric scooter, suitable for home use. It runs from 120 VAC and is 900 W, and therefore must be both isolated and PF corrected, which we will do all in one stage.

The output is a 36 V, qty100A-hr battery, so that charging at C/4 (which will charge it in nominally four hours) will be a maximum charge current of qty25A, which will be a maximum of qty36V × qty25A = qty900W.[5]

This will require an I-V control system, producing a constant current until the battery reaches 36 V and then cutting back the current while holding the battery at 36 V. More detail on why this is being done is given below.

We will be doing the regulation with the 9910, as the power level is much too high for passive PFC. Using the 9910 to do PFC was already demonstrated using the author's invention in the PFC chapter. Using the 9910 to do *isolated* PFC has *not* to the author's knowledge ever been demonstrated before. The specifics of the I-V control for the battery have also not been shown with a 9910 before, though, of course, in other systems it is commonplace.

9.9.1 Specification Table

Since the output of this converter is a battery, a carefully written specification, Table 9.4, is necessary. The only thing requiring explanation is the output current.

Table 9.4 Specification for electric scooter battery charger.

Parameter	Specification	Performance
V_{in}, min	85 VAC	
V_{in}, max	132 VAC	
PF	>0.90	
V_{out}, min	25 V	
V_{out}, max	36.5 V	
$I_{out,avg}$, max	25 A	
Efficiency	90%	

5 The "C" is a very old and rather crude way of expressing charge rate of a battery. Nevertheless, it's standard among battery manufacturers and so you have to know what it means. "1C" refers to charging or discharging a battery at a current whose numerical value in amperes is equal to the numerical value of the ampacity (measured in Amp-hrs) of the battery. Seriously, this sort of means that it would take something like an hour to charge the battery at that 1C current ... even though manufacturers will explicitly tell you not to do this ....

Since the converter is PF corrected, the input voltage (and thus current) will go to zero at 120 Hz and peak at root two higher than the RMS. The intention here is that the current delivered, averaged over at least one-half of one line cycle, should be no greater than 25 A; the *peak* current during that period will be root two higher (averaged over a switching cycle; the current during a single switching cycle will be yet higher).

A specification usually also includes an *I–V* curve for proper charging of the battery. However, the way we have specified it makes the particular characteristics of the battery, in particular its equivalent series resistance (ESR), irrelevant. The *I–V* curve will be shaped by the circuitry depending on the characteristics of the battery and the maximum current and maximum voltage that the converter should provide.

This will be clearer if the electrical characterization of a battery is considered. A battery is *not* just an ideal voltage source. It also has some series resistance.[6,7]

So, when the battery is charging, its voltage can be thought of as its "no-charge" voltage plus the current times the ESR. As the battery approaches the fully charged state, the measured voltage will reach 36 V. To avoid going higher than this, the current will be reduced so that the voltage due to the ESR is reduced; eventually, the current is close to zero, and then the "no-charge" voltage is also close to that 36 V limit.

The specific *I–V* curve depends on the ESR of the battery, which varies from battery to battery, as well as with temperature and number of charge/discharge cycles of each individual battery. As an example, suppose our battery is at a no-charge voltage of 35.88 V and has 10 mΩ of ESR. When charging it at 25 A, that will cause an additional voltage at the terminals of 250 mV, so that the terminal voltage is 36.13 V. To get the voltage down to 36 V, the current has to be reduced. If we are now instead putting in 12 A, the additional voltage will be 100 mV, and the terminal voltage will be 36.00 V, which is what its maximum is supposed to be. As the battery approaches fully charged, the current continues to be decreased, and the terminal voltage approaches the "no-charge" voltage of 36 V. What we have described here works for almost any combination of these factors.

---

6 It's not really resistance, that's just a linear approximation. In fact, its *I–V* characteristic is better modeled by a hyperbolic tangent function.

7 A real battery also has some inductance. In typical batteries, though, this doesn't really become significant until frequencies above maybe 1 kHz. This will be, however, the reason we need to put capacitance at the output of the converter, since the battery will not be able to take the high-frequency current waveform the converter will generate. Measuring the inductance of a battery may require some special equipment. Need a guess? Free-air inductance of a wire is about 100 nH/m. So take the biggest dimension of your battery in meters and multiply by this number.

9.9.2 Design and Modeling

We've already designed a PFC circuit using the 9910 in the Chapter 8, so this will be our starting point, shown again in Figure 9.15.

The most important thing we need to do right away is to replace the inductor with a transformer. And the primary inductance needs to be decreased, since this original design was only 100 W, and now we're designing a converter at almost 10 times that power level.

We start by recalculating what is now the primary inductance of the transformer. If we wanted to remain in DCM, let's again suppose we would want a maximum DC of 40%. So at a switching frequency of 130 kHz, the maximum on-time is 40% / 130 kHz = 3.1 µsec.

We can put these numbers into the same equation as before, 900 W × $\sqrt{2}$ = 1275 W maximum, 168 V maximum, 40% DC maximum, 130 kHz:

$$900\ \text{W} = \frac{168\ \text{V}^2 \times 0.4^2}{2 \times L \times 130\ \text{kHz}} \tag{9.18}$$

and so the inductance should be less than 19.3 µH; since we actually want to be CCM for this high a power level, we instead take this value as a minimum. Selecting 25 µH, we find the peak inductor current to be

$$Ipk = \sqrt{\frac{2\text{P}}{Lf}} = \sqrt{\frac{2 \times 900\ \text{W}}{25\ \mu\text{H} \times 130\ \text{kHz}}} = 23.5\text{A} \tag{9.19}$$

Now we can consider the turns ratio. The output voltage is 36 V, so the reflected voltage on the primary will be 36 × N. The volt-second balance across the primary inductance (given that we are operating in CCM) is thus

$$169\, t_{\text{on}} = 36\ \text{N} \times t_{\text{off}} \tag{9.20}$$

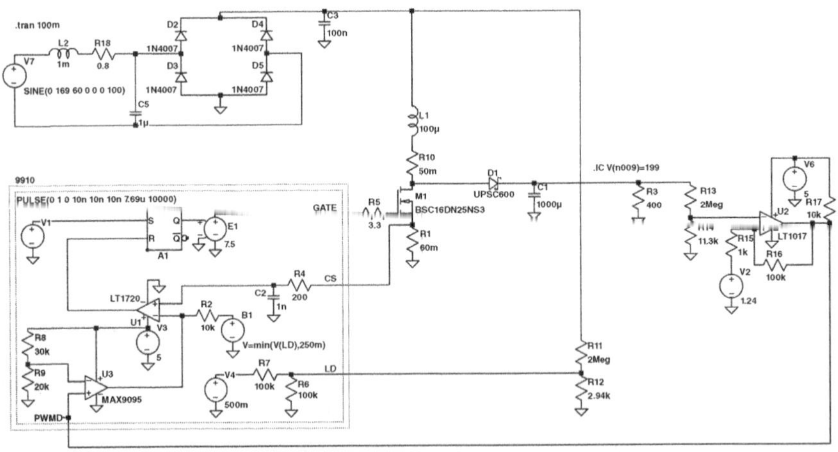

Figure 9.15 9910 PFC starting schematic.

so that DC will be

$$\frac{36\ \mathrm{N}}{169 + 36\ \mathrm{N}} \tag{9.21}$$

and so, to get a DC of 40% we need N to be 3.1, and we select the nearest integer, three.

The capacitor value cannot be determined in the same way as in the previous 9910 example, because it is in parallel with the battery. We will construct a battery model and then determine by simulation a desirable amount of capacitance.

Our starting model is shown in Figure 9.16; we have removed the feedback control as this is to be designed next. We've selected a heftier MOSFET and diode, and changed the current sense resistor for a 23.5 A maximum (250 mV/23.5 A = 10 mΩ). It will need to be further reduced when we actually simulate, due to limitations of the circuitry.

We've added in a unity coupling coefficient for the transformer for the moment. And finally, we've put in a battery model showing it having 10 mω ESR.[8]

9.9.3 Control Loop Modeling

Now we need to design the control system for this battery charger. You can follow along with this design we're showing in Figure 9.17.

As already noted, we want to have a maximum average battery current of 25 A, a maximum voltage of 36 V, and, of course, a smooth transition between the one and the other, so both the current and the voltage need to be measured.

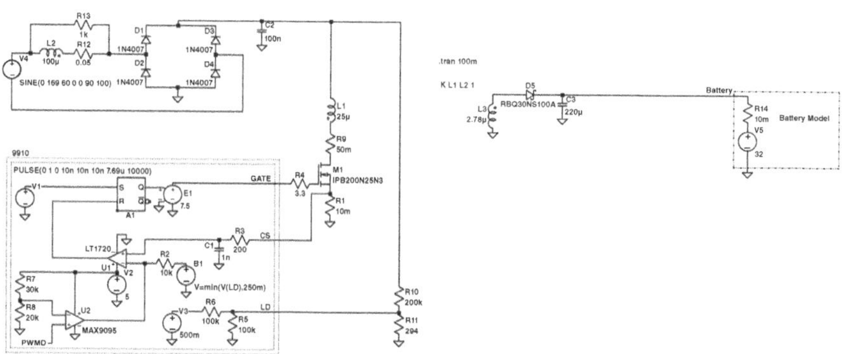

Figure 9.16 Starting model for 9910 battery charger.

8 We're running into yet another problem of a very stiff model. The battery charges in 4h, but the switching period is 10 μsec, so the stiffness is greater than one billion = 183 dB! In running the simulation, we'll just slowly sweep the voltage source to emulate a battery which is charging at a rate that can be simulated.

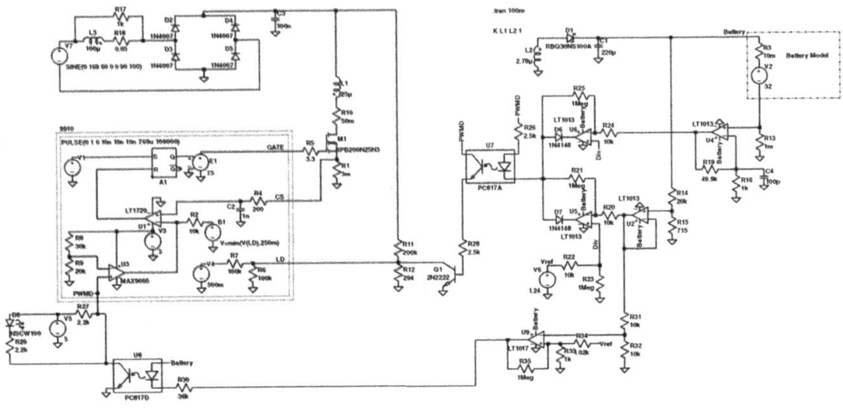

Figure 9.17 Model of 9910 battery charger.

Measuring the charge current is straightforward. If we wanted a 1 V signal at 25 A, that would be a power dissipation of 25 W! To keep it down to where we can use a 1 W resistor, we make 25 A dissipate (say) 600 mW, which would be a 1 mΩ resistor. This is a standard multi-source value. It is available at 1%, which would be 24.75–25.25 A, which is fine.

Measuring the output voltage with a resistor divider takes a little more thought. If we use 1% resistors, the tolerance could be as high as 36.36 V—which is higher than the effect of the ESR! So we need 0.1% resistors (as, it turns out, is common for battery systems).

Next, we're going to amplify the current signal to bring it to the 1.24 V from our reference (which also is going to need to be 0.1%, clearly). The (line-cycle average) current sense signal is a maximum of $25\ \text{A} \times 1\ \text{m}\Omega = 25\ \text{mV}$, so we need to amplify it by about 50. It's low impedance, so we don't have to worry about the input bias current of the amplifier, but we do have to watch that input offset voltage isn't too high. For example, 2% accuracy (which is probably all that is needed for the current) requires an input offset voltage of no more than 500 μV. Our standard LT1013 model is one-third of that. We can, fortunately, power the LM358 from the battery, since the IC runs up to 36 V, and to get some voltage margin on this, we could add a series dropping resistor and a zener.

What's more, the current sense signal is slightly filtered to get rid of switching noise that the capacitor in parallel with the battery doesn't catch. Recall that we're averaging over more than a single (half-)cycle, as the battery doesn't care how much current it gets in even a second or two. This could also be done with an RC filter, as long as the input current is very low. For the moment, we have a small capacitor on the current sense amplifier. It can't be done at the actual current sense because the sense resistor value is so small. The battery voltage itself is naturally filtered, as it isn't possible to change the voltage quickly except by changing the current.

The voltage sensor is again a little trickier. The resistor divider doesn't need amplification, but it *does* need impedance buffering in order not to load the divider and affect the measurement. The input offset voltage doesn't matter, but we have to be careful that the input bias current doesn't introduce an error. Using a 20 kΩ[9] and 715 ω divider (both 0.1%), the input bias current for 0.1% effect should be no more than (36 V × 0.1%)/20 kΩ = 1.8 μA. The LT1013 is better than that by orders of magnitude.

Having amplified and buffered the voltage and current signals to get 1.24 V at their maximum, we next need to determine how they compare with this reference voltage. For each of them, then, we build an instrumentation amplifier to find how much they differ from 1.24 V. The gain of each instrumentation amplifier is set to 100 using 0.1% resistors, the gain corresponding to an accuracy of 1%, which will dominate our accuracy sum. Since the inputs are buffered, the only thing we have to concern ourselves with is the magnitude of the resistor values; 1% of 1.24 V is 12.4 mV, and the resistors may see as much as 1 nA of input bias, so the resistor impedance should be less than 12.4 mV 1 nA = 12.4 MΩ. Of course, this is much too high a resistor value to be used in general, and so we choose 1 MΩ with a feed resistor of 10 kΩ.

At the end, you need to carefully add up all the tolerances and see if you've managed to hold the 36 V to 1%. Since this is not a function unique to robust design, we forego that exercise here.

Finally, the tricky part is left: we want whichever of the two signals implies a lower DC to actually control the DC. That is to say, we're going to diode or the two signals. Suppose, for example, that the voltage is close to 36 V; then the voltage sensing signal chain should have a lower voltage than the current sensing signal chain, so that the DC is cut back to limit the voltage (and, of course, this will also indirectly reduce the current).

In order to control the DC, we're going to pull current out of the divider from the AC line, which is controlling the line current to do PFC. Since the current being pulled is very slow, it won't vary over a line cycle, and so the PFC won't be affected.

What we do is to pull current from the divider using an NPN. Why not just use the NPN inside the optocoupler? It turns out that this opto-transistor is not especially optimized for low saturation collector voltage. When actually run in this circuit, it has an observed $V_{CE,sat}$ of 90 mV, which seems low until you remember that the signal range for the LD pin on the 9910 is only 250 mV. If we used this, then the maximum current into the battery could only be reduced to 9 A! Fortunately, our friend the 2222 goes (in the actual model) down to 20 mV, which would be a minimum charge current of 2 A.

9 At 36 V the 20 kΩ will be dissipating about 50 mW, so we need a 100 mW rated resistor.

The remaining issue with this feedback circuit (and it doesn't become apparent until you run the model) is that the 2222 has a gain of about 100. This multiplies the gain of the feedback loop, and so very small base currents are needed, which can run into issues with op-amp offset voltages and noise. To reduce the issue, we've reduced the impedance of the AC input voltage sense resistor divider by an order of magnitude to give a little more operating room for the NPN, while still dissipating only 200 mW in the top resistor. (The top resistor also needs to be two 0805s for voltage rating.)

The final control item is the circuitry at the very bottom left of the schematic. We just noted that, due to the very low maximum voltage of the LD pin, the minimum battery charging current is going to be 2 A. However, you can't go putting current into a battery once it's fully charged. The proper way to handle it (as is done in specialty battery charger ICs) is to make the battery charge current exactly zero. Then, when the battery drops down to a certain specified level, either due to a steady light load or eventually to battery self-discharge, you restart the charging.

To accomplish this, when the battery voltage reaches its maximum, we detect this with a comparator. The comparator activates a second optocoupler on the bottom left of the schematic, which turns off the 9910 using the PWMD pin. The comparator has an appropriate amount of hysteresis so that the 9910 doesn't turn back on until the battery voltage has dropped enough, something which may take hours or days.

You may observe that we've added in an LED in parallel with the PWMD pin. When the 9910 is charging, the light is off. When the PWMD pin is pulled low, the LED is turned on, indicating that the charging is complete. Of course, you could also add another LED that is active only when charging is happening; this is left for the reader.

9.9.4 Other Pieces

Before turning to measuring this system, there are a number of other features that you may want to add to this design.

Even at 90% efficiency, the power dissipation is going to be 100 W. Of course the unit has to be designed to take this much heat to ensure that individual components don't overheat. Still, if something unexpected happens, it's best to have an over-temperature circuit to shut down operation if something gets too hot. A convenient way to do this would be to add a thermistor to the PWMD pin. The thermistor could be mounted near a hot component (the transistor is a common choice), and would turn off the switching if it got too hot.

Hopefully, it goes without saying that, of course, this system needs an input line fuse. You might also want to include an on/off switch so that you can safely attach and remove the battery; soft-start would also be good.

Output overvoltage is already present. Short-circuit on the output is limited because the 9910 limits the maximum current the unit can produce; if your battery is shorted, you have other problems anyway.

Finally, not obvious at all is that you ought to have some sort of a battery recognizer, so that the user can't try to charge the wrong battery. At the most elementary level, you could have a custom output connector. Electronic ID is also a possibility, but it takes more thought about how to do that without a microcontroller while remaining robust.

9.9.5 Results

To start with, let's look at the current into the battery as a function of its voltage. What we expect is that at lower voltages, the current will be 25 A; as the voltage approaches 36 V, the current will start folding back until it reaches 2 A; and when the voltage is above 36 V, the current will go to exactly zero, as the switching is turned off. In the reverse direction, though it's less likely to be attached to a charger, as the battery voltage decreases, the switching will restart at 2 A, and then rise up till it's back to 25 A.

As we've already mentioned, it's of course impossible to simulate four hours of converter behavior when the switching period is 7 μsec. Instead, we are going to ramp our battery model from 25 to 37 V and back down over a period of 10 msec of simulation time. This is long enough to have many switching cycles at each voltage (in a manner of speaking), but short enough that the simulation time and the file size are not completely out of control.

In Figure 9.18, we can see that the current starts to decrease a little by the time the battery reaches 33 V terminal voltage. Normal charging shuts off at about 35.3 V and drops to about 700 mA. Above 37 V the current shuts off completely. As the battery voltage comes back down, at 37 V charging resumes at about 650 mA and at 34.8 V normal charging resumes, though, of course, this is not the normal operating condition.

Another thing we need to look at is the PF of the converter. We do this at a nominal 32 V output. Looking at Figure 9.19, which shows the input current and

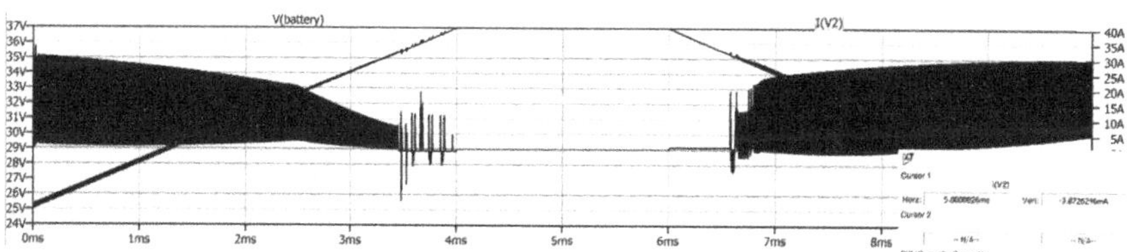

Figure 9.18 Battery current while sweeping the battery voltage.

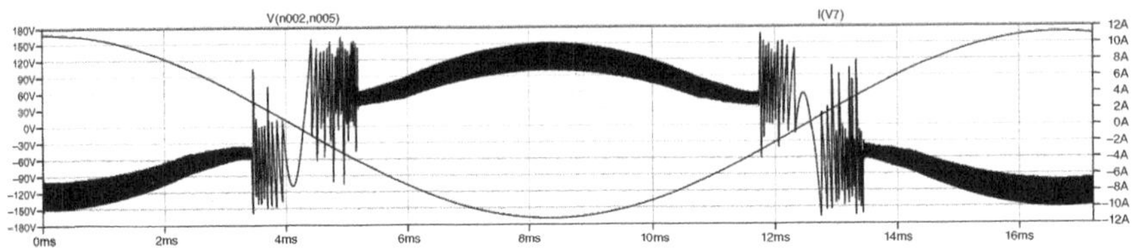

Figure 9.19 Input voltage and current of 9910 battery charger.

voltage, we can immediately see that the input current is sinusoidal and exactly phased with the input voltage. The noise at low voltages (below about 45 V) is due to the lack of dynamic range of the 9910 LD pin. But in fact, we measure PF = 0.935.

A few final comments. The input filter has been changed in this design from the 100 W model to limit the power dissipation. And the input capacitor has been removed to suppress a resonance with the input inductor. Whether the present filter is adequate to meet EMI is left to be checked by the reader.

The output capacitor is ceramic because of very high ripple current; aluminum electrolytics that can take it are quite bulky. For this design, we will choose quantity 10 of 22 μF, 50 V ceramics for a cost of US$5.10 (for all ten of them). Although ceramic capacitors don't "really" have any ESR, the designer needs to check in the lab that they don't overheat.

9.10 Isolated Synchronous Rectification

A not infrequent requirement in isolated power supplies is to have synchronous rectification on the secondary. With a low voltage output, the almost one volt of forward loss on a diode can greatly degrade the efficiency of the system and can generate very significant heat flux at high currents.[10]

Two approaches have historically been taken to secondary-side synchronous rectification. One has had the secondary-side MOSFET or MOSFETs driven from the primary side, through a small drive transformer or an optocoupler. Either can be made to work, but getting the timing right so that the synchronous rectifier is only on when it's supposed to be is a challenge. This method has generally fallen out of favor.

The other method is to control the synchronous rectifier by measuring the voltage coming from the secondary winding. When the voltage is positive, the synchronous rectifier turns on (possibly with some small delay); when the voltage is negative, the MOSFET is quickly turned off. While this seems straightforward,

10 On the other hand, synchronous rectification with a thyratron could lose 10 V or more! [1, 20].

there don't seem to be any second-source parts available for controlling such a design.

In this brief example, we construct a third route. We send the on/off signal to the secondary side using an RC and a diode, using a well-known method of level-shifting. The signal source is the gate drive to the primary-side switching transistor, and with appropriate RC values there is essentially no propagation delay. The capacitor itself, of course, needs to be of the high-voltage sort used for common-mode rejection across the isolation barrier; and in fact, we will use precisely this capacitor for both purposes.[11]

9.10.1 Design and Model

Our goal is just to show how to control the secondary-side synchronous rectifier MOSFET. So we won't be actually building a complete converter, but only modeling the portions necessary to the synchronous rectification: a primary running at a fixed DC and frequency, a secondary with a MOSFET rectifier, and the control circuitry for this device. We will specifically be demonstrating a forward converter. We won't need a specification table. The only criterion will be comparing the efficiency of a rectifier diode with the MOSFET synchronous rectifier.

Our starting point is a forward converter using a diode on the secondary side, which we do as a reference point. Looking at Figure 9.20, we are coming down from 48 V input. We have a MOSFET running at 40 kHz at 60% DC. The secondary

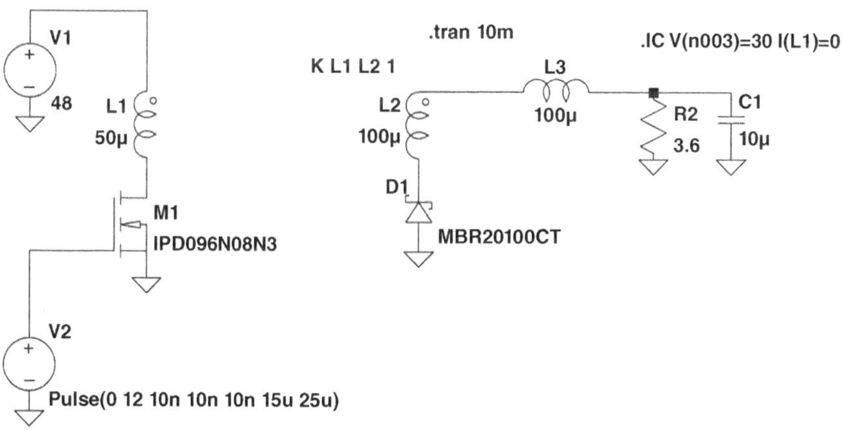

Figure 9.20 Forward converter with diode rectification.

11 Yes, intentionally transferring high-frequency signal across this will degrade common-mode noise performance. But this signal is small and high-frequency and common-mode noise rejection circuitry at the front end is always used anyway.

winding of the transformer is perfectly coupled to the primary, and the turns ratio is $\frac{\sqrt{2}}{2}$. Note the polarity dots on the transformer, which are opposite to those on a flyback. The secondary has a 100 μH inductor, a small capacitor,[12] and a static load.[13]

Measuring this system, the output voltage is about 21 V. Of key interest to us is that at the average current of 5.9 A, the power loss in the Schottky diode is 4.1 W, 3.0% of the total input power. At lower output voltage, the percentage would be higher, potentially much higher.

Now, we replace the diode with a MOSFET synchronous rectifier, as shown in Figure 9.21. There are quite a number of observations to be made on this model.

- Although this is intended to have an isolated secondary ground, we have just left it as primary ground to avoid the issues we had previously with LTspice.
- Observe that we have put the MOSFET in series with the secondary of the transformer, conveniently enabling us to use an n-channel. If we had tried to put it in series with the inductor, we would have had the usual problems with driving a floating gate.
- The isolation-crossing is done by an RCD circuit as is common; the capacitor is isolation-barrier crossing, the resistor provides a local ground reference, and the diode ensures that the signal doesn't go (much) below ground. The RC time constant is 47 μsec, slightly longer than the switching period. The output is shown in Figure 9.22. The flat-topped waveform is the switching transistor's

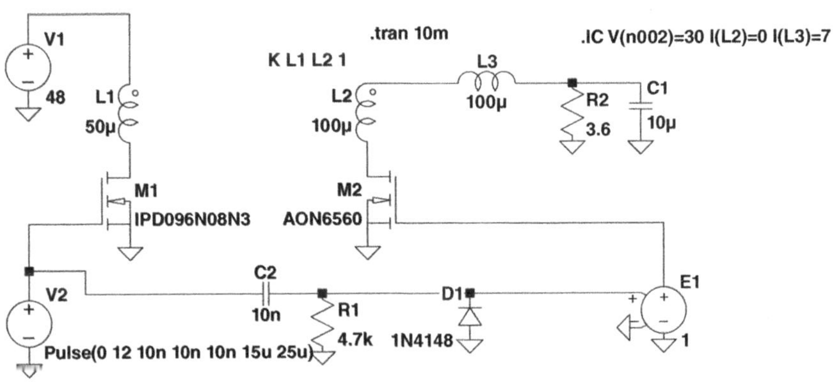

Figure 9.21 Forward converter with diode replaced by MOSFET synchronous rectification.

---

12 Ripple voltage is ±8%.

13 Don't be concerned with the unusual placement of the diode on the schematic. Since it is still in series with the secondary, it is exactly equivalent to the usual connection. It is done this way for convenience below in replacing the diode with a MOSFET.

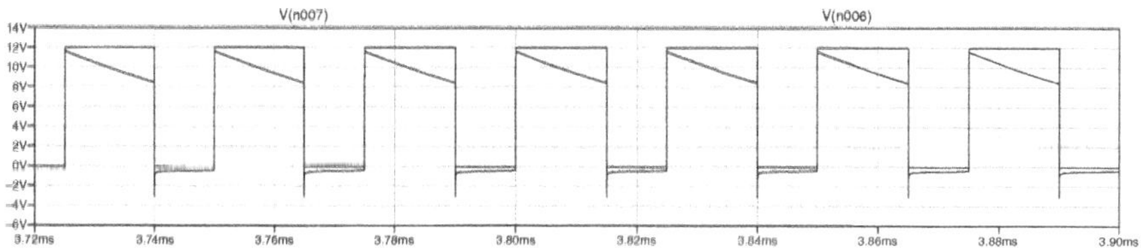

Figure 9.22 Switching transistor and synchronous rectifier gate drives.

gate, while the sloping-topped waveform is the synchronous rectifier's gate. Delay time is measured to be only a couple of nanoseconds.

- The synchronous rectifier's gate drive is just shown as a unity-gain amplifier; in reality something like a totem-pole might be used.

Measuring this revised system, the output voltage has increased to about 27 V. Of key interest to us is that, at the average current of 7.6 A, the power loss in the synchronous rectifier is 5.1 W, dropping to only 2.4% of the total input power.

As a final observation, observe that there is a limit to the improvement attainable. A diode will always have something like a forward voltage of 1 V regardless of its current; a MOSFET's voltage is linear in the current. Of course, no matter what you do, extremely high currents are difficult to rectify, or, for that matter, deal with in any aspect. Synchronous rectification is the usual choice for currents above about 5 A, but diodes can still be worth considering, just in case.

9.11 Notes for the Hobbyist

If you haven't yet read Chapter 1 (Safety), now would be a good time to do so. It's true that the secondary of an isolated power supply is isolated from the primary. You do, however, have to verify that it *really is* isolated. Having a transformer and an optocoupler isn't enough; there could be a connection between primary and secondary grounds unknown to you that would make the system non-isolated anyway. The only way to tell for sure is to put an ohmmeter between primary and secondary and make sure it reads open. And given Murphy's Law, you should do this every time you go to work on the converter; what if something has happened while you were away?

Even though the secondary is isolated, it still doesn't mean that it's safe. If it has a high output voltage and a significant power capability, it can still be hazardous. Follow the precautions in the Safety chapter even for isolated secondaries.

The biggest issue for constructing isolated supplies is going to be obtaining the transformer. Vendors have no interest in building anything custom for you. An alternative would be to build your own from wire, a bobbin, a core, tape ... there are a lot of things, and frankly a lot of specialized knowledge that goes into making a transformer, and so this isn't a good choice either.

The best plan is to get something similar to what you want that's available from a distributor. Some degree of variation in the turns ratio can be accommodated by changes in DC. Of course, it has to be at least *approximately* big enough or else the core will saturate. If you aren't sure, don't just pop it on at full load, try a minimum load first and keep your eye on the waveforms (and on the temperature of the transformer) to see if you're approaching saturation.

Naturally, if your transformer is way too big for the application, you probably don't have to worry about it saturating. But if it's *way* too big, before trying it, you should also check that it's designed for high-frequency operation; a 60 Hz transformer isn't going to work at 300 kHz! You should also check that it's the right *type* of transformer, not something like an auto-transformer; you can check this by verifying with a DMM that the primary and secondary are indeed isolated.

10

Special Topics

This chapter is intended to cover topics that are particular to certain power supplies, not all of them. Topics include implementing a current limit, a digital power good signal or one that lights an LED, power supply and system optimization, and others. All of these topics and more are addressed in this chapter while making the designs robust using multi-source parts.

10.1 Better Performance

10.1.1 High Speed

There's only so fast a SMPS power supply can be. A clocked SMPS is (even in theory) limited to something like 1/10 of its switching frequency, even if its error amplifier has enough bandwidth and slew rate. A bang–bang controller responds in a single cycle, but is still limited by the value of its inductance as to how quickly it can ramp up the current to maintain the voltage when the load suddenly increases. So what do you do if you need even higher speed, other than just increasing the switching frequency (which has its own limits, of course)?

On the other hand, an LDO isn't intrinsically limited by bandwidth. As long as the slew rate and GBW of its error amplifier are large enough, it can go as fast as you want. Of course, you can't use an LDO in place of a SMPS unless the currents are low, because the efficiency would be terrible. But this suggests a way to get very high speed in a power system: have a first stage which is some sort of SMPS (e.g., a PWM or a bang–bang), followed by a high-speed LDO. Since we know how to build very-low dropout LDO the efficiency hit of a second stage of conversion need not be particularly bad. Of course, the energy has to come from *somewhere* during a high-current pulse, so you need to include a big enough bank of capacitors (high-frequency, in practice ceramic in parallel with aluminum electrolytic) at the output of the SMPS. It won't be that bad—since we have large bandwidth, we have a high-speed response, and so the capacitance needs to support the voltage for only a short time, something like some microseconds, until the aluminum electrolytic is able to respond.

As a specific example, consider the 555 SMPS we designed in the chapter on PWMs. It had 12 V in and was supposed to have 5 V out at 2 A. Switching frequency was 100 kHz. Importantly for our example here, its bandwidth was specified at 1 kHz, so after a load step it could be expected to take about 1 msec to respond. But what if we needed it to respond in 1% of that time, 10 μsec? The bandwidth can't realistically do much more than double because of the resonance.

Of course, we could simply add enough capacitance that it meets specifications. How much would that be? We go (let's say) from 1 mA to 2 A, and it takes 1 msec to catch up, so the charge pulled is $Q = \Delta I * t = 1$ mC. The voltage shouldn't drop more than 1 V, so the capacitance needed is $C = Q/V = 1$ C/1 V = 1000 μF. This would be OK if it were an aluminum electrolytic, but they have too much ESL to respond that fast, and it's too much for ceramics of a reasonable size. What we'd like is to design an LDO that has 500 kHz bandwidth.

Of course, to run the LDO we need some overhead. Let's target an output from the SMPS of 5.25 V instead of 5.00 V, giving us 250 V of headroom. This will be a power dissipation of 500 mW.

Looking at our starting point LDO with these changes in Figure 10.1, we see that the increasing load step causes the output voltage to drop to 4.56 V. It takes about 5 μsec to begin to recover, and then its recovery to 1% (4.95 V) takes longer than 20 μsec. The peak of the unloading step is 5.22 V, so that's OK; it takes much longer to recover because the pass transistor can't sink current (neither could the SMPS as it wasn't synchronous).

The first thing we're going to do is do a better job with compensating the error amplifier. But the reality is that the best we can do still isn't good enough: what's

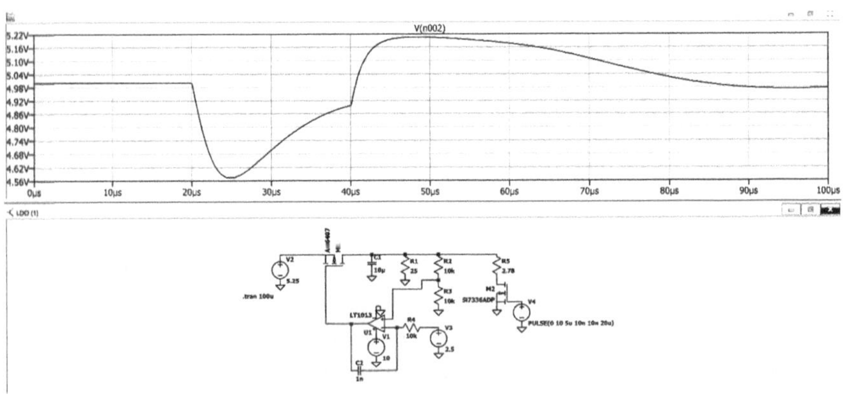

Figure 10.1 Starting schematic for high-speed LDO.

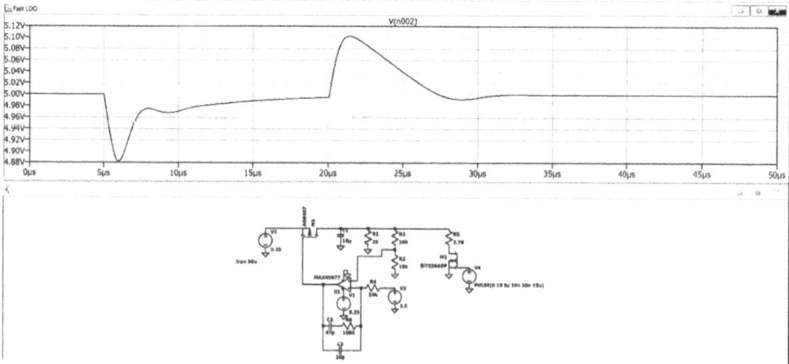

Figure 10.2 Schematic and response for high-speed LDO.

really slowing down the response is the op-amp. So we replace it with the high-speed MAX40077, and since that doesn't run on 10 V, we power it from the input.

Now with these changes shown in Figure 10.2, we see that the increasing load step causes the output voltage to drop to only 4.88 V. It takes about 1 μsec to begin to recover, and then its recovery to 1% (4.95 V) takes almost exactly 2 μsec from the start of the load step—10 times faster than before. The peak of the unloading step is now only 5.10 V, so that's OK; recovery to 5.05 mV takes only 4.32 μsec.

We've increased the performance of the power system by something like two orders of magnitude for a cost of about 5% in efficiency. And this wasn't optimized. There are certainly circumstances in which this would be a good trade-off.

10.1.2 High Precision

Having dealt with high-speed power in Section 10.1.1, we should now consider high precision. This often goes together with the high-speed requirement, and therefore we again consider a switcher followed by an LDO.

Getting *pretty* high precision in an LDO isn't that hard; it just costs slightly more. You use 0.1% resistors and a 0.1% voltage reference. You should make the impedances surrounding the error amplifier low, so that input bias current effects are low, and also use an op-amp with low input offset voltage to avoid having that affect the output.[1]

What if all that still isn't enough? You may have to trim the unit on the factory floor. Now this really does cost money, because it's often a manual operation. What

1 If you have to produce high precision over temperature, you, of course, also have to verify that the input offset voltage remains low over the temperature range; not all of them do. And then remember to check it over temperature for your second-source parts; this is an area in which second-source parts may differ.

you ought to do is have a resistor in the design that has the minimum value that is expected to be needed. You measure the actual output voltage[2] and then have a lookup table to add in series an additional resistor to the default resistor. Having the lookup table in place, along with a selection of resistors[3] already at hand greatly speeds up the trimming.

10.1.3 Spread-spectrum Switching

Our 5 kW PFC LED driver ended up with a fairly significant EMI filter. But there are a small number of control ICs out there that have a solution. They do spread-spectrum switching (SSS), which means that they intentionally periodically change the switching frequency of a PWM. The change occurs at an intermediate time scale, faster than the AC line but slower than the switching frequencies.

What SSS does is to spread out the noise spectrum. If before you had 60 dBμV of noise at a single frequency, you take half of the noise and stick it at a different frequency. The result is that the noise is now 6 dB lower at two frequencies. This clever technique was (to the author's memory) thought of by Raytheon back in the 1970s, but the author has been unable to find the original papers. Commercial parts that do SSS often have half a dozen or more frequencies.

Unsurprisingly, there are no multi-source parts available that do SSS. But it turns out to be easy! The basic idea is shown in Figure 10.3. Once in a while, perhaps every hundred switching cycles or so, you switch in a capacitor (for the 555 or the

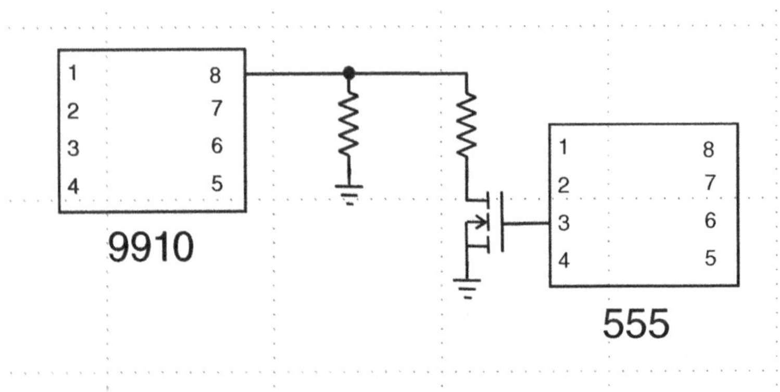

Figure 10.3 Spread-spectrum switching robust concept.

2 Be sure to use a calibrated meter with enough significant digits.
3 The trim resistor selection can also be produced beforehand by considering the spacing required to get the voltage within the right range. You then have all the trim resistors in labeled boxes.

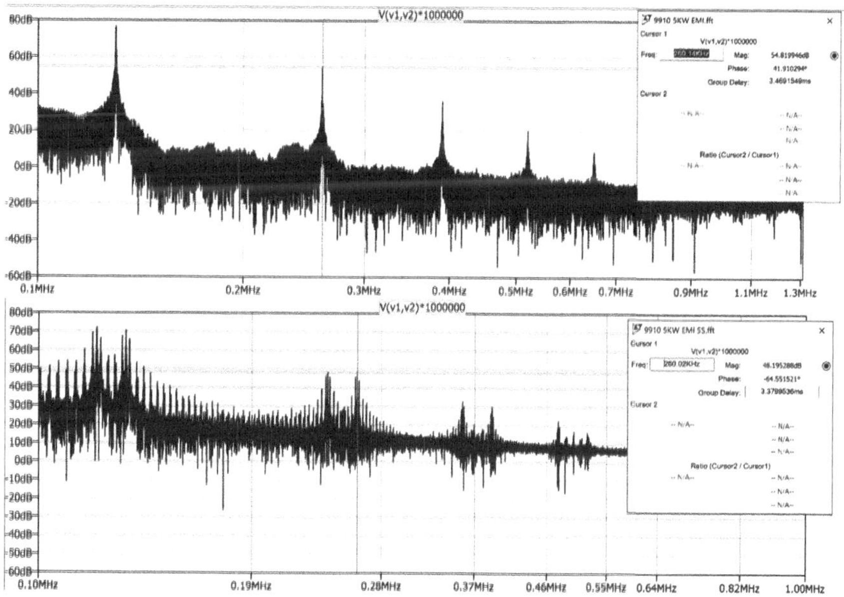

Figure 10.4 5 kW converter EMI. Top: No SSS. Bottom: With SSS.

3843) or a resistor (for the 9910) in parallel with the one that sets the frequency. And then, a hundred cycles later, you switch it back out. This gives you two different switching frequencies, which accomplishes the SSS. Make the frequencies different by maybe 10%, and make sure the lower frequency is still less than 150 kHz.

Without going into details about the construction (you can use a 555 and an n-channel MOSFET to switch the timing component in and out), the results are shown in Figure 10.4 (yes, more 5 GB files). On the top, the noise at the first in-band frequency has a sharp peak at 260 kHz of magnitude 54.8 dBμV. On the bottom, the noise has been split into two frequencies, peaking at 260 kHz and 238 kHz, with the largest magnitude being 48.5 dBμV.

This reduction of 6.3 dB means that your inductor—or, better yet, your line capacitor—can be reduced by a factor of two. For a capacitor that needs to be rated for line voltage, this can be a really significant reduction in both size and cost.

10.2 Current Limit

Some of the examples we've done so far have a problem that may be significant for applications: if the output draws more current than it's supposed to, the designs try to keep their output voltage regulated and produce more and more current until something breaks.

A related problem can occur during startup. If the input voltage comes up quickly (compared to the speed of the regulator), the current through the regulator will be very high, as the converter may try to produce its maximum current until the output voltage reaches its programmed value. Even if the input comes up slowly, the voltage may still overshoot, because it takes a certain amount of time, set by the compensation, before the error amplifier settles down to its steady state.

Now, of course, you can just blow a fuse somewhere upstream of the circuit if all you're worried about is an output short-circuit. And you're going to need a fuse in any case to protect upstream circuitry if there's a failure inside your power supply. But this doesn't help for start-up. And generally, power supplies ought to survive a shortage in their output without needing to change a fuse.

For both of these reasons, it's desirable to add in an output current limit. Conceptually, this isn't difficult. What we need to do is to sense the output current, and if it goes above a predetermined threshold, throttle back the control. The difficulty is in the implementation.

There are several possible ways to limit the current. When an overcurrent is sensed.

1. You can latch off the regulator until power is recycled. This is probably the safest thing to do.
2. You can turn the regulator off for a long time, then turn it back on until another overcurrent is detected. This limits the average output current and thus the regulator's average power dissipation.
3. You can have the regulator produce the maximum current continuously. However, if the output is a short, this will increase the stress on the power components.

In the following we give examples of designing current limits for the main types of converters we've designed in this book: buck, boost, and flyback, both isolated and not, using bang–bang, 555, 9910, and 3843 control. LDOs (as we will see) are the hardest to add a current limit to, and they will be treated last.

10.2.1 Bang–Bang Controlled Buck

The basic idea of current-limiting a buck converter is to pull-down the reference voltage and let it come back up slowly. This of course also provides a soft-start at initial turn-on.

As shown in Figure 10.5, we are adding current limit to the 5 V to 1.8 V at 2 A bang–bang converter, as being representative of buck converters, as well as bang-bang controls, in general.

The load passes through a 50 mΩ current sense resistor referenced to ground, and an added test short-circuit that's been added does so also (the MOSFET's

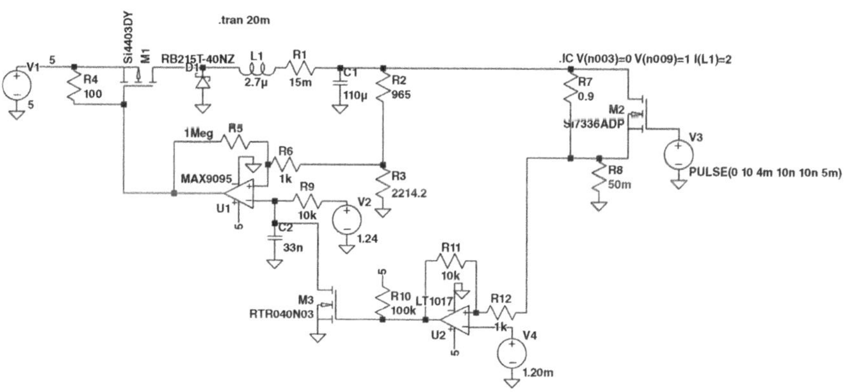

Figure 10.5 Current limit for a bang–bang controlled buck.

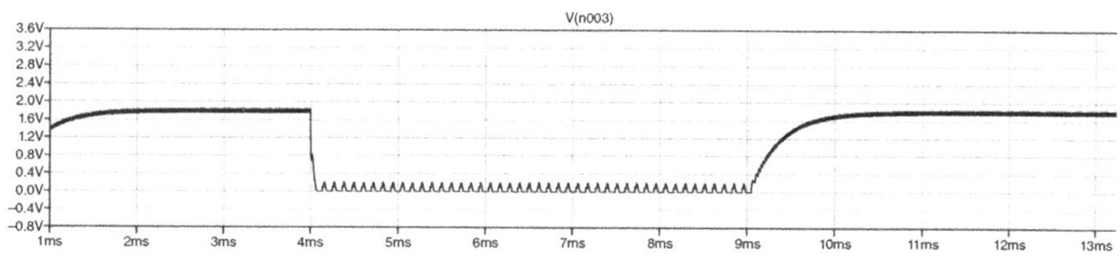

Figure 10.6 Output voltage of bang–bang controlled buck during short-circuit.

resistance is 2.4 mΩ). For simplicity, we compare the current sense signal with a 120 mV reference using a standard LT1017 LTspice model. This is weakly pulled up by a 100 kΩ resistor and inverted by an n-channel MOSFET. When the current is too high, the comparator output goes high, turning on the MOSFET, which discharges the reference voltage being applied to the bang–bang comparator. The reference voltage comes back up slowly according to an RC time constant.

The simulation shown in Figure 10.6 (run with RELTOL = 0.0001) shows the output voltage at its set point of 1.8 V.[4] The output short circuit happens at time 4 msec and the voltage drops close to zero. After 5 msec the short disappears, and the voltage returns to its set point in about 1 msec.

Current through the inductor (which includes the current through the short) is shown in Figure 10.7. Except for a brief surge of 18 A for 40 µsec, well within the MOSFET's capabilities, the average current during the short-circuit is only 1.12 A, less even than the normal current, and the MOSFET dissipates an average of 63 mW and the diode 384 mW, so the current limit completely protects the converter during an indefinite short.

4 An initial transient isn't displayed.

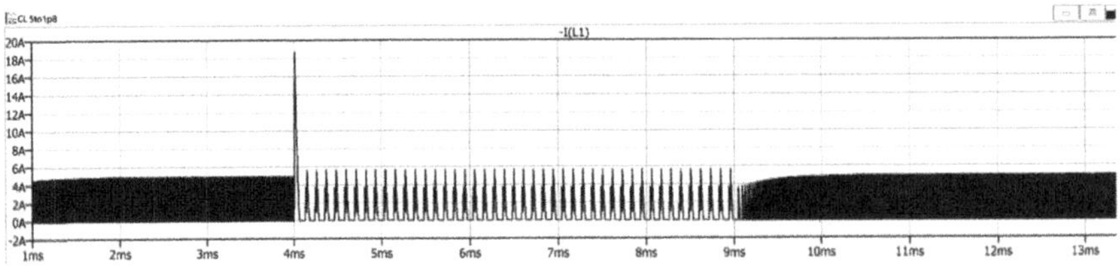

Figure 10.7 Output current of bang–bang controlled buck during short-circuit.

The reader should bear in mind that the two ICs are powered from the input; they have to remain on during the short, and so can't be powered from the output.

10.2.2 3843 Controlled Boost

In the last section, we showed how to design a current limit for a buck converter being bang–bang controlled. This time, we're going to design a current limit for a boost converter being controlled by the 3843, based on the model we built in the chapter on PWMs.

As is well known and pretty much obvious, there's no straightforward way to create an over-current protection on a boost converter, because during a short the output voltage will be less than the input voltage, and so current will flow through the inductor and the diode directly to the output, even though the MOSFET is off.

However, slightly more tricky things are possible, in particular shutting off the input power. We show a model in Figure 10.8. This is the boost converter we designed in the chapter on PWMs, converting 12–48 V at 2 A, using a 3843.

Here, we've added a p-channel MOSFET in series with the 12 V input power. When the output current goes too high (10% above maximum), the comparator goes low. We've again buffered it to make it open-collector, which inverts the signal. The NPN then turns off, and the gate of the p-channel switch rapidly rises, turning off the MOSFET. The current rapidly drops, but it takes a long time to discharge the capacitor through the resistor, and so the MOSFET remains off for a while. We've added in a fake 18 V source to compensate for the issues with the comparator; it presumably isn't needed in real hardware (or could be generated by another boost!).

The result is shown in Figures 10.9 and 10.10. After an initial transient (the circuit needs better compensation), the output settles to 48 mV. The output is shorted at 5 msec and remains so until 7 msec, after which it rises again to 48 V. While the current limit is active, the diode current is only 1 Arms, or only 282 A average, with peaks of about 6 A.

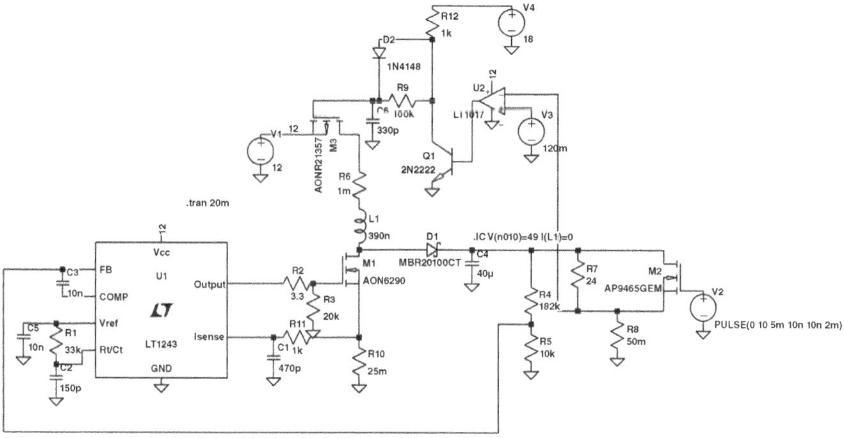

Figure 10.8 Current limit for a 3843 controlled boost.

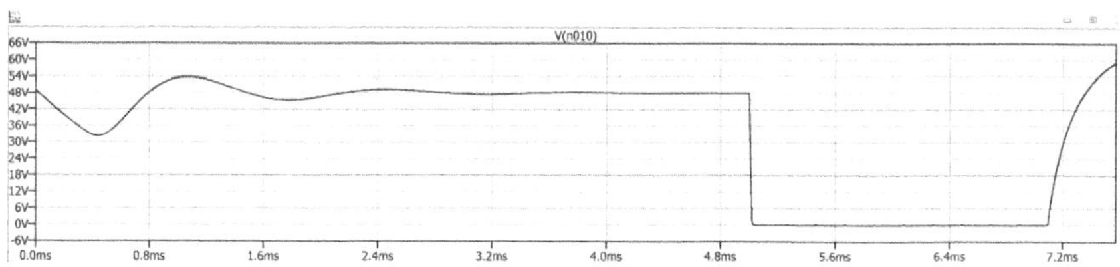

Figure 10.9 Output voltage of a 3843 controlled boost during short-circuit.

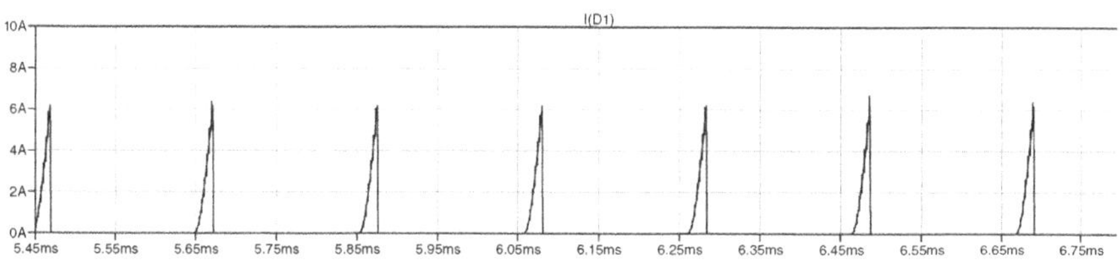

Figure 10.10 Output current of a 3843 controlled boost during short-circuit.

Almost all the power dissipation in the circuit during this time is in the p-channel switch, which dissipates a very high 37 W. This could be cut to 18 mW by putting two p-channel MOSFETs in parallel, while reducing the dissipation in each to only 9 W. During normal operation, power dissipation in the p-channel switch is only 1.6 W, affecting the efficiency less than 2%.

10.2.3 9910 Controlled PFC Boost

This time, we're going to design a current limit for an off-line PFC boost converter being controlled by the 9910, based on the model we built in the chapter on PFC. Input is 120 VAC, output is 200 VDC at 100 W.

By rights, turning off this converter ought to be easy, as the 9910 has a PWMD pin that acts as an on/off control. Unfortunately, however, just as with the previous 3843 circuit, this is a boost. Merely shutting off the switcher doesn't prevent the output from getting power from the input voltage.

This is particularly a problem in this case, as the input voltage is the rectified line. Doing as we did with the 3843 by putting a p-channel MOSFET in series with the input is even more difficult, since now the voltage goes up to 169 V and goes down to zero. How would you control the gate? Perhaps you could generate a floating voltage riding on top of the rectified AC and use an n-channel MOSFET, but this is going to require an isolation transformer and seems to be getting quite out of control for current limit.

Instead, what we're going to do is illustrated in Figure 10.11. Admittedly it's a bit more complicated than the other schemes we've shown so far, but it works very well. We've removed the EMI filter to speed up the simulation a bit. And as usual we've added in a very low resistance short on the output, in the form of a MOSFET, and are funneling all the current in the output through a current-sense resistor. Since it's 200 V at 100 W, the current is only 500 mA, and we can afford to generate 120 mV. Since we're going to a comparator, we don't need to amplify this signal.

What's new, however, is that we have an n-channel MOSFET in series with the current-sensing resistor (the resistor is in the source of the MOSFET). In normal operation, the MOSFET is on (the extra 120 mV in the source isn't enough to affect

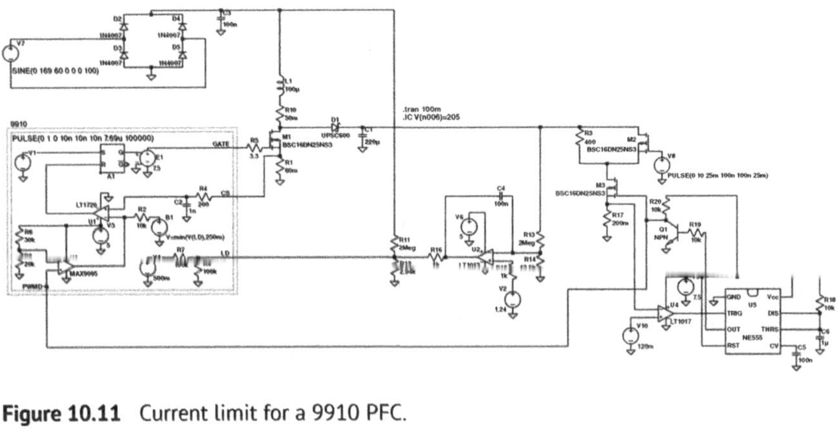

Figure 10.11 Current limit for a 9910 PFC.

the gate drive), so it just looks like a little bit of extra resistance in series with the current sense resistor, which in turn is a very small resistance itself.

When the output current goes too high, the comparator goes low. The comparator is powered from 7.5 V, which is available from the 9910. The comparator going low triggers a 555 one-shot, also powered from the 7.5 V. The output of the one-shot is inverted, and this then turns off the load MOSFET for the period of the one-shot, which is set for about 10 msec. The output current to the load during this period goes to zero, since there is no longer a path to ground for it.

However, this isn't enough. The 9910 has a minimum output duty cycle (not shown in the datasheet). This means that the boost will continue functioning, ratcheting up the output voltage. The delivered power has nowhere to go since the output is disconnected. Clearly, this would cause something to break, probably the MOSFET since its impedance is higher than that of the capacitor.

To fix this, we do a second thing. The same signal that controls the state of the load MOSFET is also used to control the PWMD pin of the 9910. In normal operation, the signal is high, and the MOSFET is on, and the 9910 is switching, so operation is normal. After an over-current event, the signal is low for 10 msec, during which time the MOSFET is off and so is the 9910. As shown in Figure 10.12, the voltage very gradually drifts down to approximately the line peak until the over-current condition is removed, and then the load is reconnected to the output and the PFC starts up again. The restart occurs within a maximum of 10 msec, depending on when, during the off-period, the short is removed.

In some ways, this is the best of the current limit schemes shown. Power to the output goes to zero and remains there as long as the short circuit is present. When the short is removed, normal operation resumes within 10 msec. The circuitry runs on internal power, the 7.5 V is generated by the 9910 directly from the rectified line voltage. Total IC count is two, a comparator and a 555, both of which are robustly available. The same scheme could be used for other circuits' current limits, so this is a really good choice.

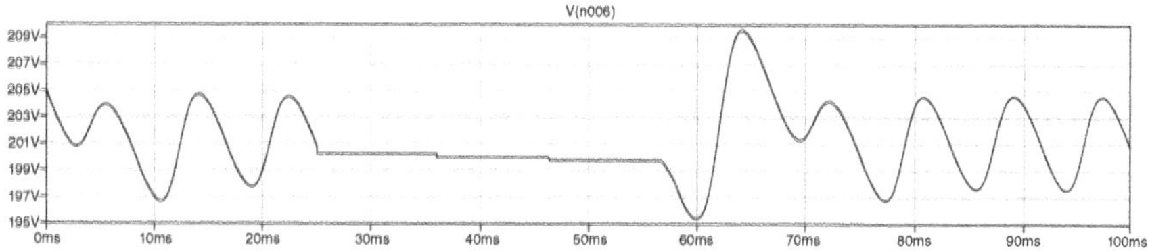

Figure 10.12 Output voltage during an output short for a 9910 PFC.

10.2.4 555 Controlled Isolated Flyback

For our last example of the current limit for a SMPS, we're going to examine an isolated flyback controlled by a 555. This is the example shown in the chapter on isolation, with the input being 375 V from a universal input PFC, and the output 1000 V at 300 mA for a power level of 300 W.

There is what seems at first sight to be a novel challenge here. The short circuit occurs on the secondary side, but we have to shut down the operation on the primary side. This presumably means sending a signal either through the already existing optocoupler or an additional one; we prefer the former since optocouplers add significantly to MTTF degradation.

So we need to send a digital signal through the optocoupler. This is not an issue *per se*, since this is a bang–bang control of the 555, and all we're sending across right now is go/no-go. But there is an issue. If the output is shorted, there's no power to run the optocoupler or the comparator.

What you have to do is to make it so that no feedback means the switching is turned off. This is a common requirement in isolated supplies so that the output voltage doesn't rise without limit if the secondary-side IC supply voltage or the optocoupler fails.

As shown in Figure 10.13, we've once again added in an output short (it's an 800 V, 350 mΩ MOSFET because LTspice doesn't have any with a high enough voltage rating). The output current passes through a 300 mΩ sense resistor and is compared to a 100 mV reference, corresponding to 330 mA, 10% higher than the maximum.

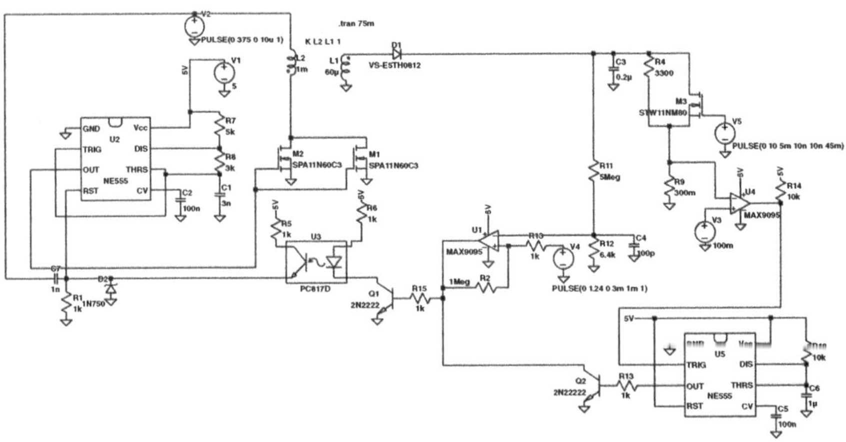

Figure 10.13 Current limit for an isolated bang–bang 555 converter.

Similarly to the previous design, when the over-current is detected by the comparator, it triggers a 555 one-shot. The output of the 555 is or'red together with the output of the bang–bang regulation comparator driving the optocoupler, either one serving to turn off the optodiode. The output of the optocoupler in normal operation pulls the RST pin of the 555 high so that the 555 is operational; when the optodiode is cut off due either to overvoltage or overcurrent, the resistor to ground on the RST pin turns the 555 off.

It should be observed that although the schematic doesn't show the secondary 5 V as being isolated, it can (and has to) be. When there is no main output power, there is presumably no secondary 5 V either, and so the RST pin is low and the converter is off. However, there are brief pulses of output, sufficient to momentarily run the optodiode, and this can restart the entire process. We've added in a startup capacitor from the input voltage (the PFC) to the RST pin to start up the converter the first time; a zener has been added to prevent the RST pin from going too high.

The resulting output is shown in Figure 10.14. The output voltage is present for about 1 msec due to the startup capacitor. Then the system starts regulating. At 5 msec an output short occurs. During this time there are very short (about 14 μsec) spikes once every 10 msec. During the shorted-output period, there is an average of 3.3 W drawn, 1% of normal power. After 45 msec the short ends, and the output voltage very quickly rises and starts regulating again.

This method, while again using a 555 to turn off power, does so by using the feedback path already present, so that an additional optocoupler isn't needed. It has the considerable advantage over some of the previous methods that there is no power dissipation due to the current limit while in normal operation, and minimal power while the short is present. And of course, it can start up and work even with no output voltage and recover once the short is removed.

10.2.5 LDOs

We finally turn to the current-limit for LDOs. As promised at the start of this section, this is harder than for SMPS. The reason is that all of those converters have

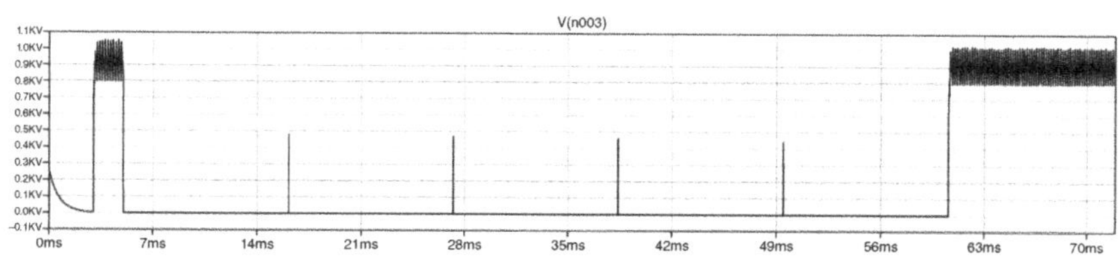

Figure 10.14 Output voltage during an output short for a 555 1 kV boost.

an inductor: When the switch turns on, the current through the switch doesn't rise instantaneously, but rather is limited by the inductance.

But for LDOs, there is no inductor. If the switch turns on, the current can go very high extremely quickly. An LDO running from 5 V with a 50 mΩ MOSFET could in theory pass 100 A if there's a short on the output.

What really happens is either that the source impedance is high enough that the 5 V source is in its turn pulled down, or else the MOSFET explodes. Either way, the result is something quite undesirable.

While we could do the same as we did for the boost converter and turn off the input power, instead, we're going to leave the output current at its current-limit value. We take two steps to make this happen. Just as with the other designs, we're going to sense and (usually) amplify the load current. We then compare the current with a limit voltage using a comparator. If the current is higher than the reference, we pull down the voltage reference for the feedback loop so that the converter starts over. The concept is shown in Figure 10.15.

The other important step is that we've replaced the more usual p-channel MOSFET with a PNP. This reduces the gain of the transistor, allowing more fine-tuned control of the output current.

To simplify the design to its basics, we have a 5 V input producing a 2.5 V at 1 A. We've set the reference directly to 2.5 V, so no resistor divider is used. Current sense is done with a 100 mΩ resistor, so that 1 mA produces 100 mV, and we've conveniently set the current limit level to 120 mV, corresponding to a 20% overcurrent of 1.2 A.

Finally, since there aren't any actual open-collector comparators in LTspice, we've built our own by running a 2222 NPN at the output. When the current is too high, the output of the comparator goes high, which turns on the 2222, pulling

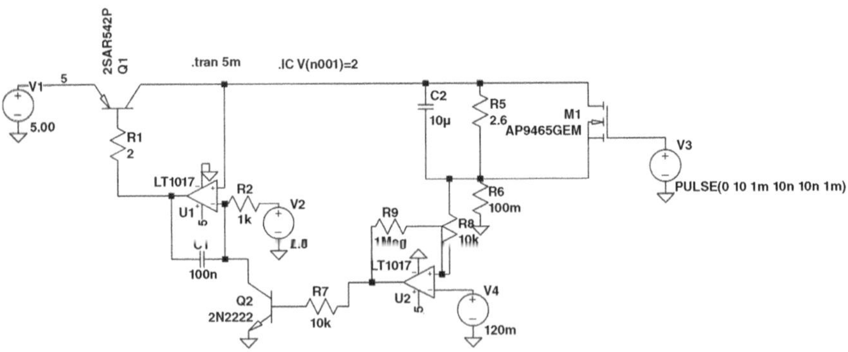

Figure 10.15 LDO current limit short-circuit schematic.

down the reference voltage to the error amplifier. It then takes a while for it to come back up, due to its low bandwidth.

For testing, we've added in a short across the load. The MOSFET has an $R_{DS,on}$ of 25 mΩ, which is a pretty good short and would draw 200 A from the 5 V if not otherwise limited.

Results are shown in Figures 10.16 and 10.17. (The small oscillations are due to inadequate compensation.) During the shorting event, average output current is 1.4 A, average output voltage is 168 mV and average power dissipated in the PNP is 6.7 W, compared with 2.5 W under 1 A load. Recovery after the short to 2.5 V takes about 500 μsec.

The only possible issue is the surge of current (and thus power) while the circuit responds to the onset of the short. It has a maximum of 7 A, which is within the rating of the PNP, and lasts for about 30 μsec at a power level of about 30 W. This is surely within the pulse-power limit of the transistor; a larger than normal transistor will always have to be picked for this type of application so that it can take both the increased power dissipation during the short (which may be indefinitely long) as well as the pulse of power while waiting for the short-circuit circuitry to turn on. The need for using the PNP rather than a p-channel MOSFET is clear from this performance, as the surge current and power would have been vastly larger with the latter.

As a final note, we may observe that the LDO could simply be latched off in the event of a short lasting too long. This could be done with a 555 timer running a flip-flop that (for example) removes power from the error amplifier.

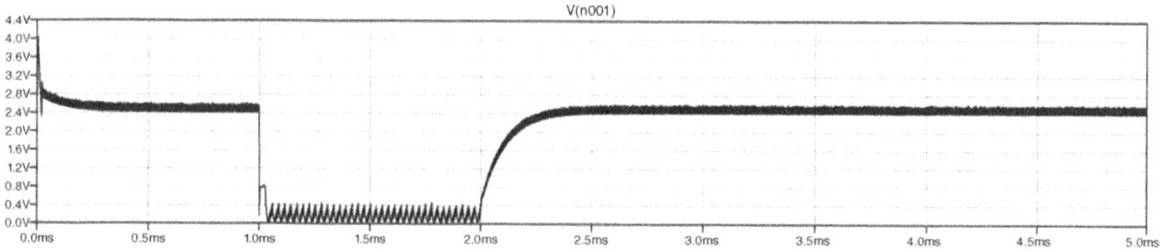

Figure 10.16 Output voltage of LDO under an output short circuit.

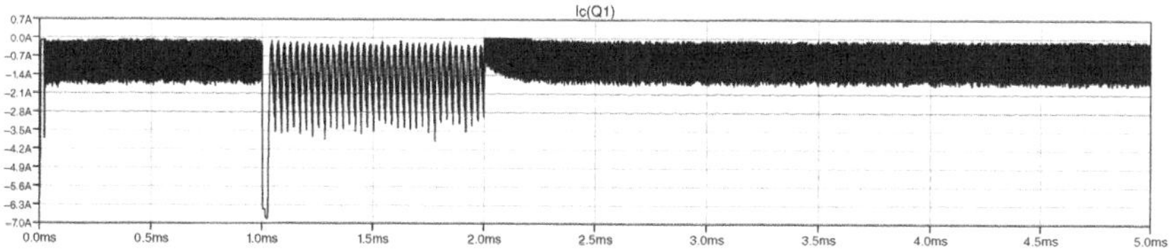

Figure 10.17 Output current of LDO under an output short circuit.

10.3 Negative Voltages

10.3.1 Example: 555 Negative Voltage 60 mW

Negative supplies aren't used very much anymore.[5] Once in a while you will find a split-supply requirement for an op-amp (though nowadays you can almost always find a single-rail device that will do the job). Just in case, we are here going to revive a negative switched capacitor circuit from the late 1970s [Jung], and use a 555 timer to generate an unregulated, but well-controlled, −12 V from +12 V by using a capacitor-diode arrangement. The output is rated to 5 mA, which is 60 mW, but will do even more with a bit of voltage droop. This is not a PWM, since we aren't modulating the pulse width. It's basically a switched-capacitor power supply.

As shown in Figure 10.18, we have a 555 oscillator running at 140 kHz and 50% duty cycle. Its output is capacitor coupled to a diode to ground, so that the signal is inverted. This signal is then rectified by a further diode and smoothed by a second capacitor. (In case you're wondering, neither capacitor sees more than the 12 V input, so 16 V capacitors would be fine.)

The output starts at zero and reaches regulation within about 1 msec. As shown in Table 10.1, the regulation performance is as good as many SMPS and the ripple is even better, and for a lot fewer components. The offset from an ideal −12 V is primarily due to the diodes' V_f, which is why we've chosen 1 A Schottkies for a 5 A load.

By the way, do you need an additional low-current −5 V? It's just a resistor and a zener away! Or if it's higher current, you can use a negative voltage as shown in the LDO chapter.

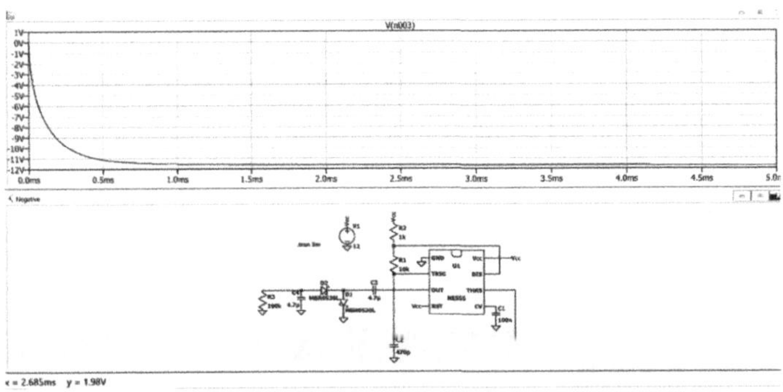

Figure 10.18 Negative voltage switched capacitor circuit.

5 Unless you're *starting* from a negative voltage, in which case a negative voltage higher than the input can be considered to be *positive* with respect to the input for the purposes of operating an SMPS.

Need exactly −12 V rather than dropping off with the current level? If you power the 555 from 15 V you then end up with about −15 V and can then either LDO or zener down to the −12 V accurately.

10.3.2 Tracking Supplies

Once in a while a power supply is needed that produces tracking positive and negative voltages. The idea here is that the absolute value of the two voltages should be very close, and not that the individual voltages need to be especially close to a particular voltage.

The previous section showed a capacitive switcher in which an input 12 V produced an output −12 V. However, the latter was −11.6 V at 1 mA load, which is an error of 3.3%, and when the load was increased to 10 mA the error increased to 6.7%. So this output doesn't track the sourcing 12 V very well at all.

Here we show a capacitor switching solution that has very tight matching. We build on the previous capacitive switcher by having two outputs rather than just one. As shown in Figure 10.19, we do the same capacitive coupling to generate both a positive and a negative voltage from the same source.

Table 10.1 Load regulation and output ripple voltage for −12 V switched capacitor converter.

Load current	V_{out}(V)	Ripple
120 μA	−11.7	280 μVpp
1 mA	−11.6	940 μVpp
5 mA	−11.4	3.8 mVpp
10 mA	−11.2	7.5 mVpp

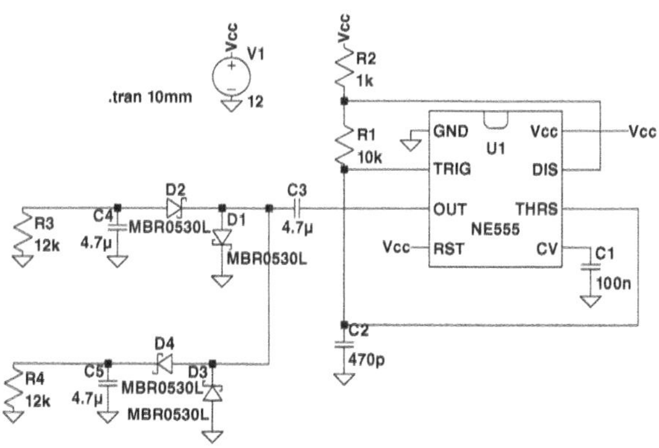

Figure 10.19 Tracking supplies.

Measurement shows that with both loads at 1 mA, the difference between the absolute values of the two outputs is 6.7 mV = 0.06%, and with both loads at 10 mA it is only 4.0 mV = 0.03%! Indeed, even in the completely imbalanced situation when one load is at 1 mA and the other at 10 mA the difference is still only 59 mV = 0.5%.

10.4 Lightning

In the chapter on Modeling, we built a model of an MOV (and in the Appendix we showed how to build your own model). In this section, we discuss building a lightning model so you can test your power supply's response, presumably with an MOV at the input.

The most severe type of lightning, and the type that you typically design for, is the 1.2 × 50 impulse. The first number refers to the rise time of the impulse in microseconds, while the second number refers to the fall time. The waveform definitions are shown in Figure 10.20.

The open-circuit voltage is typically 6 kV, with the short-circuit current of 3 A. Thus, an adequate model of the impulse is the voltage source with a 2 Ω source impedance.

To build a model, we use a PWL voltage source as shown in Figure 10.21, in series with the AC. The AC is phase shifted 90 degrees so that the impulse occurs at the line peak in the model.

The turn-on of the lightning in the PWL is delayed 2 μsec so we can see that the voltage at the output is starting at line peak, 169 V. It then rises to its peak in 1.2 μsec (rather than 10–90% in this amount of time, it won't make much difference). It then decays to half this value in a further 50 μsec.

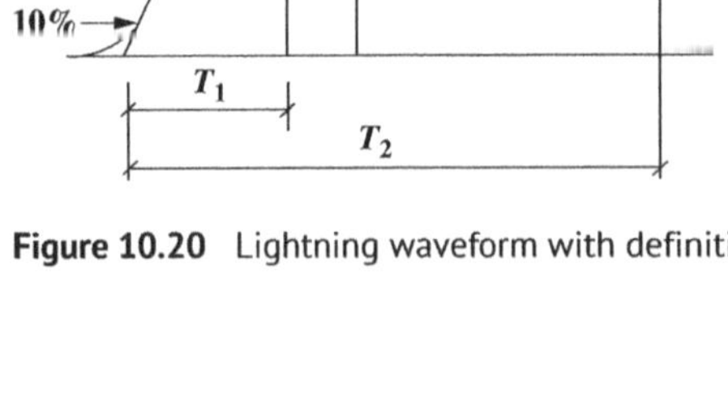

Figure 10.20 Lightning waveform with definitions.

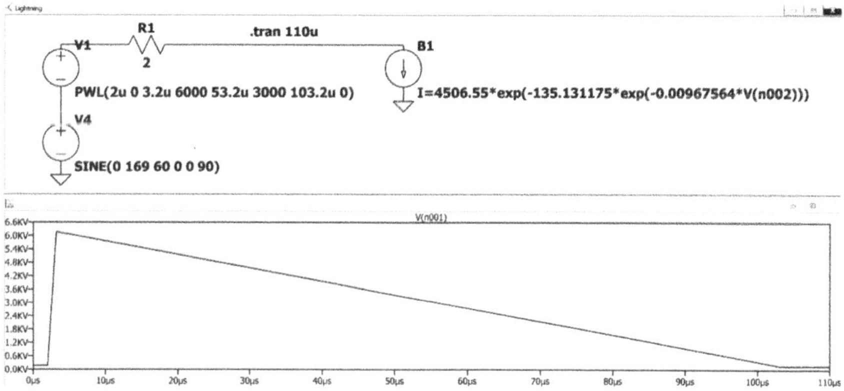

Figure 10.21 Lightning model and lightning voltage.

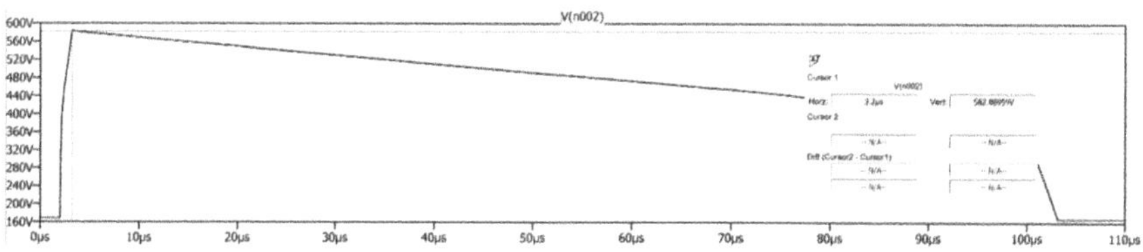

Figure 10.22 Simulated MOV voltage during lightning.

Since the specification says only that the full-width at half-max is 50 μsec, it's not clear what happens after that. Most authors assume it's an exponential decay, but the present author is inclined to think that if that were true it would be specified at $1/e$ of the peak voltage. In any case, it's easier to make it continue linearly from half the peak, which is what the PWL is set up to do. It's probably good enough. The resulting voltage is put in series with the AC line. In lab testing, the impulse is usually run 10 times at progressively different phases of the AC line; for modeling purposes, it is sufficient to turn on the impulse at the peak of the AC line and run it once (since we don't have data on aging of the MOV with impulse frequency). The resulting output voltage is shown in Figure 10.21.

We also have added in the model we constructed in the Appendix for a 10 mm disc MOV. When we run the lightning into this, the voltage (which would then be applied to the input of the power supply) is shown in Figure 10.22. It starts at line peak of 169 mV, and then peaks at 583 mV. After 100 μsec it returns to 169 mV. If none of the downstream components see an overvoltage, the design passes.

Finally, in Figure 10.23 we show the total energy deposited into the MOV during this lightning strike, which we measure to be 70 J. This can now be compared with

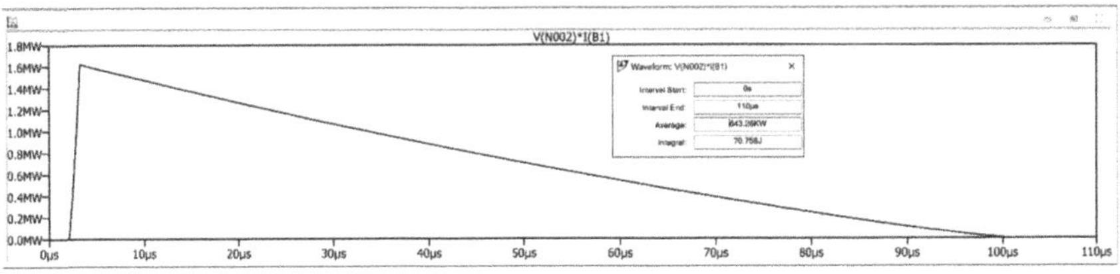

Figure 10.23 Energy deposited in MOV during lightning strike.

the datasheet, as total energy is one of the important listed parameters for selecting an MOV.

10.5 Communications

Communications with power supplies fall into two categories: commands sent to control the power supply, and reporting back from the power supply as to its status.

An example of the former would be an externally programmable power supply, in which some digital device tells the power supply what output voltage to produce. For example, this might be a VID code for telling a power supply what voltage to produce for powering a processor. It could also be as simple as a digital command to turn on or off.

An example of the power supply reporting back would be a power-good signal, indicating to some digital device when the output power of the power supply was in specification. This could be important if the power supply is commanded on externally, and then has some time required for a soft-start. Another signal going back might be an indication that the output short protection was triggered.

The first question about any of these methods is whether the signals need to be isolated. If not, then the signal level has to be determined: 0 V and 5 V? 3.3 V? TTL? However, many power supplies will be controlled from other parts of the system, and there may be (well-justified) concerns about signal degradation between the source and the receiver. For many cases, then, isolation between the power supply and the controlling device will be required.

A second question is how quickly the data communications are needed. Since we're assuming that the power supply doesn't contain a microcontroller, the speeds will generally be low. One millisecond is a reasonable guess for a soft-start time; VID codes change a few times per millisecond. So we're going to assume that the fastest communication needed is on the order of a few kHz. For isolated communications, the 6N137 optocoupler is plenty fast enough.

As a general rule, then, we don't want to send analog voltages in either direction. For setting a voltage, the digital signal can be decoded with a few robustly available logic gates and some small-signal MOSFETs controlling resistors. For transmitting data on output voltage or current, you don't need 12 bits of accuracy with an A/D. Despite what software people might tell you, all that is really needed is to send a digital signal that the voltage is good. The same is true for operating temperature; all that is needed is an over-temperature digital signal.

It should be noted here that the digital signals coming into the power supply will also need their own optocouplers to remove potential common mode noise. This can then be fed to a comparator with hysteresis to get rid of noise.

All of the data lines going in to or out of the board should have surge protection, probably in the form of a TVS diode.

Finally, if you just can't live without some software, the 8051 in 20-pin SMT is still available from multiple vendors. But you need to check pin compatibility between them, possibly allowing multiple footprints.

10.5.1 Power Good

It's often desired to have a signal from the power supply that its output power is in the specification range. Downstream circuits powered by the power supply can be held off until this power good signal is asserted. This can clearly be done with a comparator.

However, there are a few subtleties. Can the output overshoot during startup? Or does it ring after startup? Adding a capacitor to the comparator will hold the signal off for some time while initial transients disappear. And of course, all comparators should have at least some hysteresis to prevent chattering.

For board debugging, you can run an LED to signal status. Sometimes, too, the design engineer or the production engineer or the customer may want a visual signal that output power is present. A green LED driven by a comparator could be used for this. (It should be bright enough to be readily visible in indoor lighting). Or maybe you want a red light to show that the unit is plugged in, but the output voltage is not within specification. This could, for example, be a shorted output detector.

What if you want both, a red light for power not good and a green light for power good? One way to do this is shown in Figure 10.24. When the output voltage is low, the comparator output is also. By selecting the pull-up resistor as a (relatively) high value, the comparator output will be quite close to ground. This in turn causes the PNP to pull current through the red LED, while the base-emitter of the NPN is so low as to be off. (If the voltage remains too high, a base-to-ground resistor will fix this.)

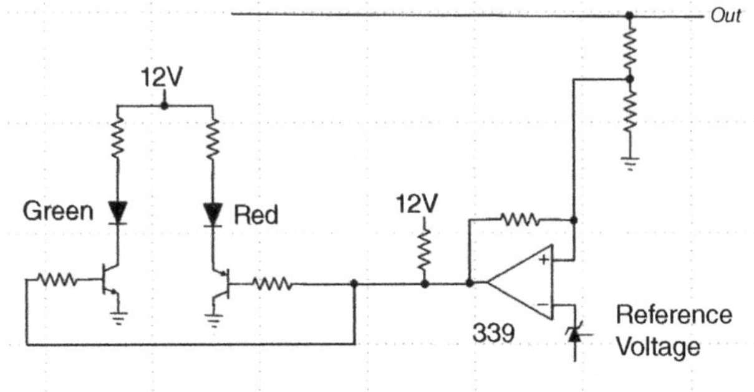

Figure 10.24 Power good/No good with red and green LEDs.

If, conversely, the output voltage is above its minimum specification, the comparator output will be pulled up towards the rail. This provides current into the NPN, driving the green LED, while the base of the PNP is sufficiently high that the PNP is off.

This all works well because the current through the LEDs is small (usually not more than 5 mA), and the gain of the BJTs is quite high, so that the comparator's pull-up resistor can be large.

10.5.2 Test Points

Sprinkle lots of test points on your PCB. They don't cost anything, and are far better than trying to poke at a component without slipping. Suggested spots are the input and output voltages, IC inputs and outputs, clock and ramp signals, V_{ref} etc. The only places where you may not want a test point are high-impedance nodes, where the capacitance of the oscilloscope probe can sometimes influence performance.

For nodes, you'll be spending lots of time on, you should place a hole for a pin that your oscilloscope probe can grab to. This is particularly valuable for ground. (Don't forget to pick ground on both sides if the supply is isolated; the oscilloscope ground has to be on the right side to make any measurements.)

10.6 Optimization

10.6.1 Power Supply Optimization

How do you optimize a design? Should it be the best possible efficiency?[6] The smallest footprint? Lowest cost? Highest reliability? If you simply ask this question

6 By the way, there's no such thing as "best possible efficiency." You can always make it more efficient by spending more money on the components and making it bigger.

upfront, the answer is likely to be "all of the above"; make it small, reliable, highly efficient, and the lowest cost. So what can actually be done?

As we've seen in numerous examples throughout this book, a robust power supply design already fulfills many of these requirements. Since, in comparison with more traditional power supplies, it uses only one or two small ICs, the board space required is smaller, the reliability is higher, and the cost is lower than in those other types of design. Since this book has shown how to design robust supplies operating at modern switching frequencies, the power components can also be at least not any larger than they would be in other types of design.

The one part where it is seemingly falling down is in efficiency. The general impression that one gets from the examples is that the efficiency of these designs is passable. For example, the bang–bang controlled examples (counting only those with output power greater than 1 W) averaged 87%.

The primary reason for this relatively low efficiency is the use of diodes for rectifiers, whose efficiency hit depends on the output voltage. Taking as an example the 5–1.8 V at 2 A bang–bang, the efficiency for the SMPS manages to reach 87% even without synchronous rectification. And this is where the output is fairly low, 1.8 V: the diode accounts for 428 mW out of 822 mW = 52% of total losses! We are, of course, using a fairly hefty Schottky, so there's not a lot of room for improvement unless you switch to synchronous rectification.

Once synchronous rectification is implemented, the efficiency of most of these converters is determined by the $R_{DS,on}$ of the MOSFETs. In the Examples, we've focused on fairly standard MOSFETs, and, of course, only on parts modeled in LTspice. Buying better MOSFETs will increase the efficiency at not much increase in cost.

10.6.2 System Optimization

In this book, we've designed a lot of power supplies and one power system (PFC followed by DC/DC followed by a third DC/DC), but we haven't discussed optimizing the design of a full power *system*, especially not when there are multiple outputs.

Consider a system with (to be concrete) 120 VAC input and four outputs (12 V at 10 A, 5 V at 2 A, 3.3 V at 1 A, and 1.8 V at 2 A), for a total power of just under 140 W, all referenced to the same ground.

Since the input is a line, the outputs have to be isolated from the input. But there are a number of possibilities for how to partition out the converter(s).

1. You could have a single transformer with five windings (primary plus four outputs). As suggested in the Components chapter, it can be challenging to have a multi-source transformer that will produce multiple outputs reliably. Since the currents are small, an isolated flyback could be made to work. Cross-regulation (how does the output regulation work when the main output is high current and the others low vs. when the main output is low current and the others high?) can be an issue also, even impossible if the non-main outputs need tight tolerance also.
2. If there were two main outputs (in terms of power), it might be possible to do the feedback from the sum of the two, which would reduce the cross-regulation issue. But that's not the case here.
3. We could have a single transformer to generate the 12 V and bring the others down using a switcher to get the 5 V and then LDOs for the other two outputs. This reduces the isolating transformer to only two windings, which is highly desirable. But now we need an additional switcher to get the 5 V. There will also be LDO losses of (5 V − 3.3 V) * 1 A = 1.7 W and (5 V − 1.8 V) * 2 A = 6.4 W, for a total loss of 8.1 W, which is only 5% of the total, but still ....
4. You could have two switchers, one from 12 to 5 V and a second from 12 to 3.3 V, and then an LDO from 3.3 to 1.8 V. Now you have two switchers, but only one LDO with a power loss of only (3.3 V − 1.8 V) * 2 A = 3 W.
5. Or you could have all three non-main voltages on their own switchers. This is the most efficient scheme using a two-winding transformer, but it involves a lot of switchers. Of course, they could all use the same reference.
6. This still isn't all the possibilities. You could have two isolation transformers, one for the main output and a second, much smaller one for all the lower-power secondaries, and again with all the various possibilities for LDOs or switchers for the other three outputs.

That's a lot of choices! How to choose? It's really going to come down to the tolerances of the various outputs, the power dissipation you can stand, and how much space you can afford for this power system.

The usual best choice (though of course not always) is to have a single isolation transformer delivering all of the power, as these transformers are the largest, most expensive, and longest lead-time component. The (usually single) output of this transformer is usually the highest power output. And then all of the other voltages are generated from this single output, either bucking or boosting as appropriate. If there are one or two other higher power outputs and some very low power ones, it often makes sense to generate those one or two with SMPS and then consider LDOs for the others.

10.7 Daughter Boards

A daughter board is typically a small PCB with some special function that plugs into a larger PCB via a connector. Although frequently used when the main PCB runs out of room, they can also be used to make an otherwise not-robust design at least somewhat more robust.

In particular, suppose that you have to have a microcontroller in your design. As we've observed in the chapter on Components, there really aren't any microprocessors or microcontrollers with second sources, and this is why we design everything without them.

If you put the microcontroller and its associated peripherals (memory, A/D, etc.) on a daughterboard, then when one of these components becomes unavailable, you can respin the daughterboard rather than the main PCB.[7] This is at least "semi-robust"; you only have to redesign and re-layout the smaller daughterboard rather than the main PCB. You could even do that in advance so that it's ready just in case.

10.8 Notes for the Hobbyist

While many of the items in this chapter won't be of interest to those building their own power supplies, the section on daughter boards potentially is. So is having test points, as you're likely to spend time debugging. And a current limit is always nice to have.

You might also want a power good signal with an LED; this is way more convenient than having to put a DMM on the output every time you want to know whether it's working or not.

The one recommendation the author wants to make is to avoid putting any of the power stage components on a daughter card. You shouldn't even put a single power component that you're not sure about (like a MOSFET or a diode that you want to test) on a daughter card. This is the highest current and highest voltage part of your design. Having these go through a connector, while not impossible, is a recipe for having problems. It will also introduce too much inductance in the power path, and it may not work solely because of that. If you really don't know which device package to use, lay the board out with multiple footprints,[8] and shrink the size once you've figured it out.

---

7 Of course, it's best to make sure that these components come from a different vendor than your first version of the daughterboard.

8 See the Components chapter.

Appendix: Modeling an MOV

10.9 Loading the Analysis Toolpak

We're going to be solving an equation with the Solver in Excel, and to do this you need to load the Analysis Toolpak[9]—it's not automatically loaded, presumably to keep the size of the executable down. Quoting from Microsoft:

1. Click the File tab, click Options, and then click the Add-Ins category.
2. In the Manage box, select Excel Add-ins and then click Go. If you're using Excel for Mac, in the file menu go to Tools > Excel Add-ins.
3. In the Add-Ins box, check the Analysis Toolpak check box, and then click OK.

Load the Analysis Toolpak before proceeding with the example in this Appendix.

10.10 Modeling an MOV

Because of the physics underlying the operation of a Metal Oxide Varistor, the author has demonstrated that (much to everyone's surprise) the appropriate model for an MOV is *the exponential of an exponential*! The form of the equation is

$$I = ae^{be^{cV}} \tag{10.1}$$

with a, b, and c constants. With three constants to be determined, it is straightforward to find the values of the parameters by selecting the data at three points from the MOV datasheet. The author recommends the data at the maximum voltage at 1 mA, the voltage at the maximum current, and the voltage at something closer to the top than the bottom currents, the first derived from the datasheet table, and the other two from inspection of the part's I–V graph. Naturally, there's some (unspecified by the manufacturer) variation from part to part, and the parameters change somewhat depending on how often the part is used,[10] but this selection is good enough to get you to approximately the right model.

As an example, consider the 10 mm disc from Bourns rated at 130 VAC continuous, the MOV10D201K. From the table in the datasheet, the maximum voltage at 1 mA is 225 V. Looking at the curve reproduced in Figure 10.25,[11] at 3000 A the voltage is approximately 600 V, and at 10 A the voltage is about 320 V

9 Yes, they spell it without the letter "c."
10 Typical parts are rated for 10 lightning strikes.
11 Be sure to count up to the right curve, the labels are shifted.

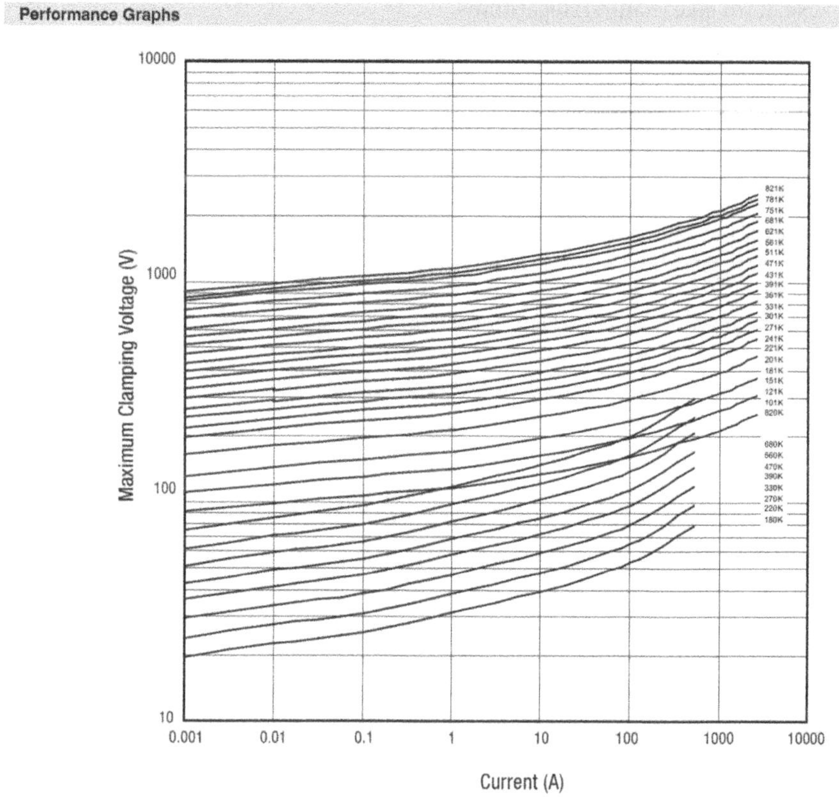

Figure 10.25 *I–V* curve of MOV.

This gives the three simultaneous equations in three variables

$$0.001A = ae^{be^{c225V}} \tag{10.2}$$

$$3000A = ae^{be^{c600V}} \tag{10.3}$$

$$10A = ae^{be^{c320V}} \tag{10.4}$$

This can be reduced to a system of two equations by dividing the second by the first, and the second by third to eliminate the variable a:

$$3 \cdot 10^6 = \frac{e^{be^{c600V}}}{e^{be^{c225V}}} \tag{10.5}$$

$$300 = \frac{e^{be^{c600V}}}{e^{be^{c320V}}} \tag{10.6}$$

Next, we take (natural) logarithms,

$$\ln(3 \cdot 10^6) = b[e^{c600V} - e^{c225V}] \tag{10.7}$$

$$\ln(300) = b[e^{c600V} - e^{c320V}] \tag{10.8}$$

And now we can reduce this to a single equation in a single unknown by dividing the two equations to eliminate the variable b

$$\frac{\ln(3 \cdot 10^6)}{\ln(300)} = \frac{e^{c600V} - e^{c225V}}{e^{c600V} - e^{c320V}} \tag{10.9}$$

This equation can be solved, for example, using Excel's Solver. We show the starting point of the spreadsheet in Figure 10.26. Build the spreadsheet as follows. In cell B1, put an initial guess; you can start with −0.01 (since c is multiplied by 600, it certainly has to be small). In cell C1, place the right-hand side of the equation, evaluated using the guess in cell B1 [=(EXP(B1 * 600) − EXP(B1 * 225))/(EXP(B1 * 600) − EXP(B1 * 320))]. And in cell D1, place the left-hand side of the equation minus one [=LN(3000000)/LN(300) − C1]: we want to minimize the difference between the calculated value and the actual so that this difference will be zero when we've found the solution.

If you haven't done so already, you now need to load in Solver, which is available in Excel but isn't automatically loaded.

Open Solver on the Data tab in Excel. As shown in Figure 10.27, the objective is going to be cell D1, and we want to find a value of zero by changing the variable cell B1 (containing the parameter c). Make sure you uncheck "Make Unconstrained Variables Non-Negative." Pushing the Solve button, a new box opens: Select "Keep Solver Solution" and push the button "OK" and the cells are altered as in Figure 10.28.

In cell B3 we have put the equation for parameter b using the solution in cell B1 for a [=LN(3000000)/(EXP(B1*600) − EXP(B1*225))], and then in cell B4 we have put the equation for parameter a using the solutions for b and c [=3000/EXP(B3*EXP(B1*600))]. Cells B4, B3, and B1 are now the values of the three parameters solving the equations, $c = -0.00967564$, $b = -135.131175$, and $a = 4506.55$, respectively.

	A	B	C	D
1	c	-0.01	2.68838	-0.0736

Figure 10.26 MOV equations spreadsheet starting point.

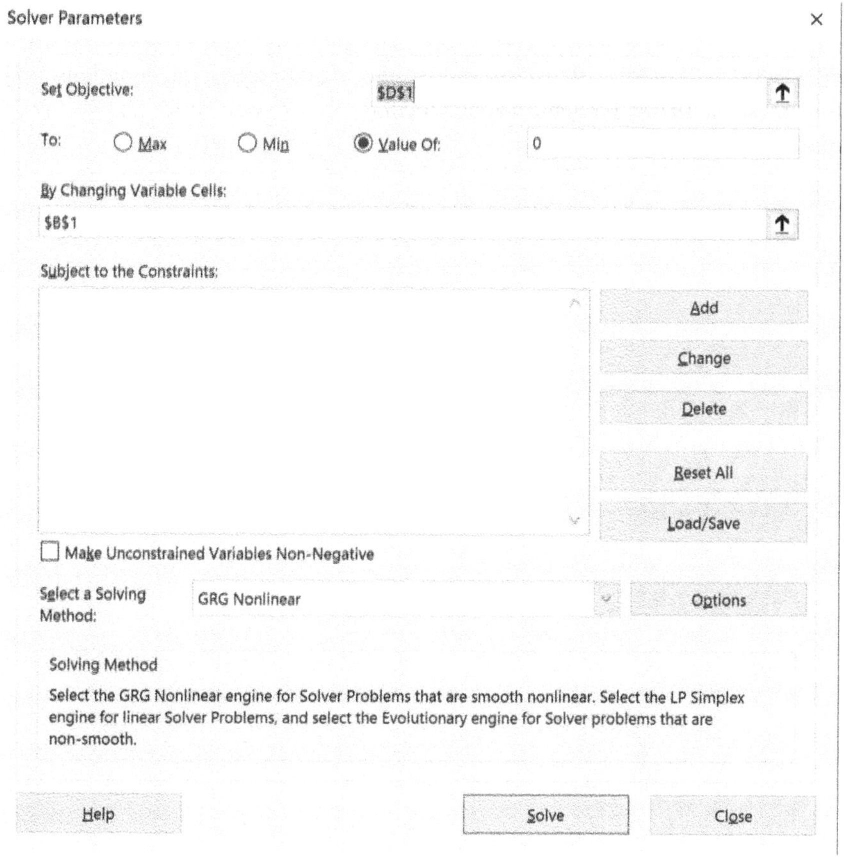

Figure 10.27 Solver window.

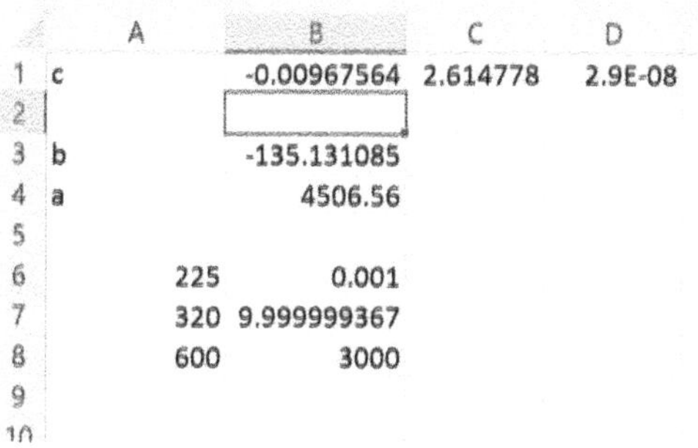

Figure 10.28 MOV model solved.

Finally, in cells A6:A8 we have the three voltage values with which we started, and in cells B6:B8 we have placed the original equations using our solution [=B$4 * EXP(B$3 * EXP(B$1 * A6))]. As shown, this exactly solves the equations, and thus these parameters provide a good model for this MOV.[12] Other MOVs may be modeled similarly.

12 Even MOVs of identical size and voltage can have quite different models, depending on their energy rating.

Bibliography

1 Shirriff K. Ken shirriff's blog: Glowing mercury thyratrons: Inside a 1940s teletype switching power supply, 2018. URL: https://www.righto.com/2018/09/glowing-mercury-thyratrons-inside-1940s.html.

2 Lenk R. and Lenk C. *Practical Lighting Design with LEDs*. John Wiley and the IEEE Press, 2011.

3 Linear Technology. *Linear Applications Handbook: A Guide to Linear Circuit Design*. Linear Technology, 1990.

4 Linear Technology. *Linear Applications Handbook Volume II: A Guide to Linear Circuit Design*. Linear Technology, 1993.

5 National Semiconductor Corporation. *Linear Applications Databook*. National Semiconductor Corporation, 1986.

6 National Semiconductor Corporation. *Linear Applications Handbook*. National Semiconductor Corporation, 1994.

7 Mims F. Experimenter's corner: The neon glow lamp, 1976. URL: https://www.nutsvolts.com/uploads/magazine_downloads/NeonLamp-Information.pdf.

8 U.S. Department of Defense. *MIL-HDBK-217F Notice 1, Military Handbook, Reliability Prediction of Electronic Equipment*. U.S. Department of Defense, 1991.

9 Lenk R. *Practical Design of Power Supplies*. John Wiley and the IEEE Press, 1998.

10 Babani B. *IC 555 Projects*. Bernard Babani (publishing) Limited, 1978.

11 Exar. *Timer Data Book*. Exar, 1981.

12 Jung W. *IC Timer Cookbook*. Sams, 1977.

13 Marston R. The 555: A versatile timer. *Electronics Now*, Sep. 1992.

14 Marston R. The 555: A versatile oscillator. *Electronics Now*, Oct. 1992.

15 Marston R. 555 oscillators. *Electronics Now*, Nov. 1992.

16 Marston R. The versatile 555. *Electronics Now*, Dec. 1992.

17 Anslow R. and Le Bras S. *How to Get the Best Results Using LTspice for EMC Simulation—Part 1*. Analog Devices, 2021.

18 Flugge-Lotz I. *Discontinuous Automatic Control*. Princeton university Press, 1953.

19 Venable H. The K factor: A new mathematical tool for stability analysis and synthesis. *Proceedings of PowerCon 10*, 1983.

20 Tung-Sol Electric. Thyratron (323B), 1960. URL: https://tubedata.edebris.com/sheets/127/3/323B.pdf.

21 Admiralty. *Handbook of Wireless Telegraphy*. His Majesty's Stationary Office, 1938.

22 Clarke P. and Scuderi R. *United States Patent #3,579,091, Switching Regulator with Random Noise Generator*. 1971 US Patent and Trademark Office,1971.

23 Department of the Air Force. *Radar Circuit Analysis*. Department of the Air Force, 1951.

24 General Electric. *Transient Suppression Voltage Manual*. General Electric, 1976.

25 Greenburg R., editor. *Power Transistor Handbook, First Edition*. Motorola, 1961.

26 Hegarty T. *An Overview of Conducted EMI Specifications for Power Supplies, SLYY136*. Texas Instruments, 2018.

27 Langford-Smith F. *Radiotron Designer's Handbook, Fourth Edition*. RCA, 1952.

28 Rosu C. Cispr 25 lisn types. *Journal Not Listed*, 2021. https://www.flexautomotive.net/EMCFLEXBLOG/post/2021/06/22/cispr-25-lisn-types.

29 Severns R. and Armijos J., editors. *MOSPOWER Applications*. Siliconix, 1984.

30 Standler R. *Protection of Electronic Circuits from Overvoltages*. Dover, 2002.

31 Terman F. *Radio Engineering, Second Edition*. McGraw-Hill, 1937.

32 Unitrode. *Applications Handbook*. Unitrode, 1997.

33 Maheswaran D. and Sreedevi V. T. A commercial low cost highly efficient UC3842 based high brightness LED (HBLED) lamp. *International Journal of Power Electronics and Drive System (IJPEDS)*, Vol. 9, No. 1, pp. 1–7, 2018.

34 Zimmer G. and Scott K. Spread spectrum frequency modulation reduces emi. *Analog Devices Technical Article*, 2015.

Index

Printed and bound by CPI Group (UK) Ltd, Croydon, CR0 4YY
17/09/2026
14964234-0003